The Wired Northwest

The History of Electric Power, 1870s–1970s

Paul W. Hirt

University Press of Kansas

Published by the University Press of Kansas (Lawrence, Kansas 66045), which was organized by the Kansas Board of Regents and is operated and funded by Emporia State University, Fort Hays State University, Kansas State University, Pittsburg State University, the University of Kansas, and Wichita State University

Library of Congress Cataloging-in-Publication Data

Hirt, Paul W., 1954–
The wired Northwest : the history of electric power, 1870s–1970s / Paul W. Hirt.
pages cm
Includes bibliographical references and index.
ISBN 978-0-7006-1873-6 (cloth : alkaline paper)
1. Electric power—Northwest, Pacific—History—19th century. 2. Electric power—Northwest, Pacific—History—20th century. 3. Electric power systems—Northwest, Pacific—History. 4. Electric power distribution—Northwest, Pacific—History. 5. Energy policy—Northwest, Pacific—History. 6. Electric power—Environmental aspects—Northwest, Pacific—History. 7. Northwest, Pacific—Environmental conditions. 8. Northwest, Pacific—Economic conditions—19th century. 9. Northwest, Pacific—Economic conditions—20th century. I. Title.
TK23.7.H57 2012
621.310979509′034—dc23
2012028594

British Library Cataloguing-in-Publication Data is available.

Printed in the United States of America

10 9 8 7 6 5 4 3 2 1

The paper used in this publication is recycled and contains 50 percent postconsumer waste. It is acid free and meets the minimum requirements of the American National Standard for Permanence of Paper for Printed Library Materials z39.48-1992.

Contents

Acknowledgments

I have benefited from the advice, assistance, and camaraderie of many people since I first conceived this project more than a decade ago. First, I want to thank the many archivists throughout the region that helped me locate and copy archival materials. Historians are absolutely dependent on archival collections and the professionals who staff them. In particular, I benefited from the extensive resources at the University of Washington's Pacific Northwest Collection. Staff there graciously allowed me to digitally photograph rather than photocopy archival materials. In two weeks I collected thousands of pages of primary source materials without having to purchase and lug around thousands of pages of photocopies. Likewise, I tremendously benefited from the rich collection of primary and published sources housed in the Bonneville Power Administration Library in Portland. Staff there also allowed me to photograph documents and granted me unrestrained access to their photocopier. It was undoubtedly the most open and accommodating archive I have ever used. I am especially grateful to Gene Tollefson, author of the ambitious coffee table book, *BPA and the Struggle for Power at Cost,* who was still at BPA when I was there doing research. He introduced me to knowledgeable library staff, pulled hidden gems from filing cabinets, and gifted me a copy of his book. I am also particularly grateful to Tina Kay at the BPA library who performed extraordinary service providing me high-resolution images for many of the BPA illustrations used in this book. BPA's openness and respect for researchers was remarkable and refreshing. I also want to thank the archivists at the Oregon Historical Society, Washington State Historical Society, and University of Oregon. Staff at the state government archives of Washington, Oregon, and Idaho assisted me in collecting the records of public utility commissions, conservation commissions, and other state agencies. In particular, I would like to thank Xan Allen at the Idaho State Archives, who personally located, photocopied, and mailed to me the first twenty years of Idaho Public Utility Commission reports. That was beyond the call of duty and much appreciated. Thanks, also, to my colleague Jeremy Mouat at the University of Calgary, who lent to me his valuable collection of BC Public Utility Commission annual reports, and to colleagues Laurie Mercier (WSU-Vancouver) and Katy Barber (Portland State University), who housed and entertained me during my trips to the Portland area archives.

In the days before the Internet, historians incurred personal debts to helpful staff at physical archives they travelled to. In the last decade, however, more and more archives have been digitized and made available on the Web. This makes archival research more anonymous than ever before, but more accessible than ever before. I took extensive advantage of digitized collections during the past five years, especially for British Columbia records. While I don't have anyone in particular to thank for the voluminous resources I collected from the Internet, I nevertheless want to acknowledge the organizations that digitize, house, manage, and store on servers for public access reams of historical records. I flinch at the thought of what it would be like if I had to get all my Internet-collected documents from the physical archives themselves. This study would have been significantly less ambitious or would have taken much longer to complete. The digital archives most helpful to me include those managed by the BPA, The Province of British Columbia (BC Stats), Environment Canada, Statistics Canada, the Center for Columbia River History, the Hathi Trust Digital Library, the Library of Congress American Memory Project, the US Energy Information Administration, and the Oregon Historical Society's Oregon History Project. I would also like to express my profound gratitude and admiration for the WorldCat, JSTOR, and Google Scholar search engines, which brought the world's scholarly publications literally to my fingertips.

Colleagues and friends have influenced my thinking about this project and helped improve the manuscript. I want to first thank the readers and reviewers of this manuscript Carl Abbott, Matthew Evenden, Harry H. Stein, John Yerger, and Greg Magee. It is a lot to ask someone to quickly read and comment on 600 pages of someone else's scholarly labors for little or no compensation. The final result is much improved thanks to these gracious friends and colleagues. My gratitude especially to Harry Stein, who went through the manuscript with a fine-toothed comb and provided me eight single-spaced pages of constructive suggestions. I met Harry in the BPA archives more than a decade ago and have enjoyed his friendship, document loans, advice, and support ever since. Thanks, Harry. And a hearty thanks to my editor at the University Press of Kansas, Ranjit Arab, for his unflinching encouragement on this final stretch of my multi-year marathon.

While not among the reviewers of the manuscript, several other friends and colleagues have shaped my understanding of the Northwest power system and encouraged my work. Most importantly, Eve Vogel, whom I met at a conference many years ago while she was working on her dissertation. We have since co-organized numerous conference panels and are co-authoring several papers. Her knowledge and enthusiasm have often rejuvenated my commitment to this project and helped keep my eye on the contemporary implications of the region's power history. Joshua Binus at BPA, who is another co-author, has similarly influenced my thinking about the energy transitions of the contemporary era. My colleagues in the Energy, Ethics, Society, and Policy working group at Arizona State University have been a constant source of motivation and inspiration, especially Clark Miller, Mike Pasqualetti, Joe Herkert, and Gary Dirks. My scholarly mentors Don Worster and Bill Robbins continue to provide intellectual inspiration for my work, while my colleague and friend Steve Pyne inspires my commitment to narrative and to getting this project off the shelf. As always, I am grateful for the intellectual and personal camaraderie of all my graduate students, past and present. They are the most important and enjoyable part of my academic life, and I thank them for their support and patience when I took longer than I should have to comment on their thesis and dissertation chapters. I want to extend special gratitude to the Felice-Bonanni family, my cousins in Italy, who hosted my wife and I for five days at their home in Segni and allowed us to stay at their vacation home on the Isle of Elba for a week while I finished writing the index for this book. I am twice blessed to have such a beautiful place to work and such generous cousins.

Finally, I want to thank my wife, Linda, for her constant support and encouragement throughout, especially this last year, while I was on sabbatical and crouched for months before my computer with the press deadline looming. We moved for the summer and fall of 2011 to our peaceful home in the beautiful Chiricahua Mountains of southeast Arizona, where I could write without distraction. Our small community of friends in Portal all graciously asked about my project and even more graciously accepted my obsessive focus on it. When friends or guests came over to the house, Linda would warn them ahead of time not to be offended that I would be chained to my desk and would probably ignore them. She and they took it all with good humor.

It has been a long and edifying journey, in good company. Thanks to all.

Introduction

Our lives in the developed parts of the world today are profoundly shaped by complex technological systems that we have created and reformulated over time to serve our needs and interests. The ones that work well become almost invisible to us, smoothly integrated into our daily lives. Electric power is one such system. The alarm clock wakes us, the water heater provides a comfortable shower, air-conditioning and heaters keep our homes cool in summer and warm in winter, kitchen appliances cook our meals and preserve our perishable food. Traffic lights guide our paths safely to work, shopping, or play. Phone lines, cell phone towers, computers, and the Internet facilitate much of our daily communication with others. For those who live in cities with rail-based mass transit, electricity runs the subways and light rail systems. Electricity powers our schools, offices, commercial centers, and factories. It illuminates our nights, our windowless interiors, our computer monitors, movie screens, and televisions. Amazingly, this crucial technology has been part of our lives only for the last 100 years or so, yet it has completely transformed modern life.

We hardly notice electricity's ubiquity or our dependence on it until it breaks down. And while a broken appliance or a burnt-out light bulb is annoying, far more troubling are interruptions to the larger power supply. Downed power lines and area-wide blackouts can cause millions of dollars in property damage, personal injury, and lost revenue. They are fodder for the news cycle, and they focus the attention of electric utilities and governing institutions whose job is to keep these things from happening. Our desire for and dependence on electricity have led us to design an electrical infrastructure and a set of rules governing it that ensure an acceptable degree of accessibility, reliability, affordability, safety, and fairness for users of this essential technology, as well as revenue and profit for the companies that generate and sell electricity. None of this came naturally or easily. This almost incomprehensibly complex socio-technological system that undergirds nearly every aspect of our lives was and remains a human invention that we have struggled over and constantly redesigned to meet the desires of both producers and consumers. Just like electricity itself, however, the human elements of this technological system become invisible to us when everything is working well. And it works so well so consistently that we rarely question the economic, political, social, and environmental foundations of the power supply: who provides it, how it is generated and distributed, how electric rates are set, and so on. Only periodic breakdowns in functionality reveal the human dimensions of this system. Electric power has not always been accessible, reliable, affordable, safe, or profitable. We made it that way, more or less, through creative effort, democratic action, and political struggle. As David Nye astutely states, "Electrification is a series of choices."[1] This book tells the story of how Americans and Canadians created and constantly reformed the electric power systems of the Pacific Northwest over the first hundred years since the invention of electrical lighting and power.

It is remarkable that so many of us can rely so thoroughly and consistently on the power supply in the United States and Canada that the system itself remains largely unnoticed in our daily lives. I have a summer home in a rural area of southeast Arizona served by a small rural electric cooperative. The co-op lacks capital to adequately upgrade and protect its facilities, so the power supply is vulnerable to periodic interruptions during the summer thunderstorm season. I notice the system's flaws when I'm using my computer or when I have meat or fish in the freezer and the power is off for more than a few hours. I have been to countries in the former Soviet Union where electricity is available only during certain hours of the day and countries in Latin America where fluctuating voltages make the lights dim and brighten as other users turn appliances on and off. Those times remind me of how human power systems are. We consumers, inventors, designers, investors, builders, regulators, scholars, investigative reporters, politicians, and public citizens are entirely responsible for the extent to which the evolving power system has or has not met our individual needs and social objectives. It is our creation, to our pride or disappointment. This human dimension is the focus of my analysis.

My purpose in this book is to explain how we developed and adapted this remarkable technological system over its first century to design success, stability, and satisfaction into the power supply; to provide ever more reliable profits; and to meet a variety of social objectives. I am particularly interested in the dynamics of problem-solving, governance, and the tense relationship between producers and consumers, between profit-seeking and the public interest. This relationship animates all technological systems that shape our lives, so the lessons drawn from this study of electric power apply to many other arenas of human endeavor. To make my task manageable, I focus on one bi-national region of North America: the Pacific Northwest, comprising Washington, Oregon, Idaho, and British Columbia. In most ways the story unfolds similarly on both sides of the international border, though occasionally at different times or tempos. In other ways, the evolution of the power systems north and south of the 49th parallel took different paths. Likewise, the paths taken in the three U.S. Northwest states were often similar but occasionally quite distinct, allowing a rich comparison of pathways and choices.

I chose the Pacific Northwest because I lived there for eleven years and came to admire the region's character and culture. While teaching western and Northwest history at Washington State University for more than a decade, I learned firsthand how integral the landscape of mountains and rivers is to the regional power system and how integral inexpensive and abundant hydropower is to the region's development. Electric power systems develop regional characteristics shaped by geography, the availability of natural resources, and culture. For example, coal-mining regions have many coal-fired power plants; regions with natural gas supplies or natural gas pipelines support many gas-fired power plants; regions with mountains and rivers, like the Pacific Northwest, have lots of hydropower dams; windy regions in recent decades have supported a great deal of wind turbine development; and sunny areas support solar power development. Regional culture shapes the power systems, too, influencing the presence or absence of nuclear power facilities; the location of dams, power lines, and wind turbines; and the mix of privately owned versus publicly owned electric utilities. Importantly, characteristics of the systems vary widely and are strongly influenced by historical context. Despite their deceptively apparent stability, electric power systems are in fact constantly evolving in response to social, technological, and environmental opportunities and constraints.

In the Pacific Northwest, regional conditions clearly shaped the evolution of the electric power system. Almost from the beginning, rivers and dams undergirded the Northwest power supply, although the presence of coal in western Washington contributed to the development of coal-fired power plants in that area through the early twentieth century. Also in the early decades of the industry, steam plants fired by wood waste were common in the heavily forested regions from the west slope of the Cascade Mountains to the Pacific Coast. But this power source grew less significant as hydropower increasingly dominated the electric supply during and after the 1920s. Not until the 1960s, after all the best dam sites were developed in the U.S. Northwest, did electric utilities there turn aggressively toward building what is generically called thermal-generated power: fuel-burning power plants. The presence of the U.S. government's Hanford Nuclear Reservation in central Washington undoubtedly stimulated the development of nuclear power in Washington and Oregon after World War II. Prodigiously watered British Columbia, with a smaller population, huge rivers, glacier-covered mountain ranges, and vast undeveloped territory in the twentieth century never turned much to thermal power. More than 90 percent of its electricity still derives from hydropower.

The whole region's heavy reliance on river-based hydropower put the electric system in direct competition with the salmon-fishing industry, generating a century of bitter conflict that still animates regional politics today. The presence of a vast fertile but arid region east of the Cascade Mountains and in southern Idaho married power development with irrigation development. Grand Coulee Dam provides the best example of this partnership. It is the largest single source of power in the region and supports the largest reclamation project in the region. By the mid–twentieth century, power development also had partnered with flood control. Libby Dam and Hungry Horse Reservoir in the United States, along with the three Columbia River Treaty dams in British Columbia, exemplify this partnership. In contrast, power development in the Northern Plains and Midwest is married to coal mining. Every source of electricity has its own allies and antagonists, which vary by region and shift over time, but the significance of those relationships is universal.

Just as the Northwest power system is characterized by the ubiquitous influence of hydropower, the regional character of the system is also influenced by the ubiquitous presence of public lands and public agencies. The large percentage of Crown-owned lands in British Columbia and the similar predominance of federal lands in the U.S. Northwest made both countries' national (and provincial) governments powerful players in the evolution of the regional electric system. This was especially important during the 1930s in the U.S. Northwest when New Deal programs initiated a precedent-setting and paradigm-shifting Federal Columbia River Power System with the construction of Bonneville and Grand Coulee dams.

All these regional characteristics created a unique regional power system. But at the same time, none of these characteristics by themselves are unique to the Northwest. Nearly all Canadian provinces have a predominantly hydropower-based electric system, as do many states in the United States. Power supply and irrigation are intimately entwined in California and the American Southwest. Power supply and flood control are entwined in the Rocky Mountain states, Missouri Basin, and the Appalachian region. The federal government plays a commanding role in the electric system of the Tennessee Valley and the Colorado River Basin. Some states and provinces have embraced nuclear power while others have rejected it. Some states and provinces have embraced nonprofit public power systems while others remain dominated by investor-owned utilities. Environmental problems and conflicts stemming from electric power development plague all regions, and the lessons drawn from problem-solving efforts in one area can easily inform problem-solving in other areas. Conflicts between hydropower and the salmon fishery in the

Northwest parallel conflicts between coal-fired power plants and opponents of air pollution and acid rain in the Midwest. Investments in salmon passage facilities at dams parallel investments in smoke-, particulate-, and sulfur dioxide–reduction technologies at coal-fired power plants. Problems and public demands led to both. Hydropower advocates faced an anti-dam movement maturing in the 1960s while advocates of nuclear power faced an anti-nuke movement maturing in the 1970s. Just as new large dams became socially and environmentally unacceptable in the United States after the 1970s and in British Columbia after the 1980s, coal-based power similarly became unacceptably dirty and nuclear power unacceptably risky and expensive. This has driven a new paradigm-shifting search for solving the future power supply challenge with much greater emphasis on conservation, efficiency, "demand-side" management, and clean, renewable sources of power. Thus, changing conditions, values, markets, and policies throughout North America are driving system evolution—as always—in regionally distinctive yet parallel and comparable ways. The process of change is what I seek to illuminate in this study, along with the lessons to be drawn from our intentional management of this complex socio-technological system over a century.

There is a rich and continuously flowing literature about the region's rivers and power systems, but no one has yet written an integrated history of electric power in both the U.S. Northwest and British Columbia. Existing studies of electric power concentrate on one or the other side of the border. Most studies concentrate on one institution, like the Bonneville Power Administration or the Federal Power Commission;[2] or one utility company, like Portland General Electric or BC Electric;[3] or one topic, like rural electrification or the struggle between the for-profit and nonprofit sectors of the industry;[4] or one river or dam or narrow time period.[5] I have tried to tell the story as it unfolded across three states and a Canadian province, multiple watersheds, and all significant utilities and institutions, over a century of time. I have also endeavored to place regional developments in their larger national contexts. This broad scope forced me to rely a great deal on secondary sources, making much of this study a work of synthesis, especially in regard to events in British Columbia.

Every author tells a particular kind of story. Some histories of electric power are triumphant stories of technological progress, the genius of inventors like Thomas Edison, and the acuity of business leaders and entrepreneurs like S. Z. Mitchell and Samuel Insull.[6] Alternatively, because many electric utilities are publicly owned not-for-profit establishments, admirers of these kinds of systems tell stories of triumphant progress in cooperative enterprise and advancement of the public interest over pecuniary interests.[7] Most of these partisans of public power are critical of investor-owned utilities, believing that for-profit enterprises cannot adequately or genuinely serve the public interest, even under government regulation. They often craft stories of decline in which the commercial power sector uses its wealth and political clout to suppress the public power movement, capture or intimidate regulatory agencies, and dupe the public into accepting corporate hegemony.[8] On the opposite pole are partisans of capitalism who believe that virtually all government regulation strangles business and that government-run economic enterprise is inefficient, politicized, susceptible to corruption, illegitimate, and un-American.[9] I approach these partisan attitudes with caution. Both business and government are human institutions equally capable of competence and ineptitude. Corporations and government agencies, along with nongovernment organizations (NGOs), are the institutions we invent to organize and accomplish large, complex tasks. They are really the only practical tools we have to manage our elaborate social needs. Both corporations and government agencies normally work day to day, accomplishing goals, adapting to changes and challenges, and occasionally performing brilliantly or failing spectacularly. Neither seems to be inherently superior to the other. Nor are they necessarily at odds with each other. In fact, most successful economic development in both the United States and Canada since the early nineteenth century has resulted from partnerships between the public and private sectors. There is good reason for this; therefore, I concentrate a great deal of analytical energy on the distinctive contributions of the public and private sectors to power system development and the alternately cooperative and conflicted relationship between business and government.

In North America, we have over the past 200 years very intentionally crafted a political economy that blends democratic politics with capitalist enterprise to facilitate private wealth accumulation while seeking social goals, public services, and a modicum of equity and justice. Corporations are adept at producing and selling profitable goods and services but useless in providing desired public goods and services that don't generate profits. Citizens demand many things that cannot be adequately provided by market transactions, so we invented taxes, public agencies, philanthropy, and NGOs to provide those: universal free K–12 education, police and fire protection, water and sewer services, public roads, parks and recreation areas, clean air, safe drinking water, soup kitchens, meals on wheels, medical research, and more. Electric power is one of many things in our modern world that hovers in the gray area between profitable commodity and nonprofit public goods, like hospital services, college degrees, and the Internet. Accordingly, much of the historical evolution of electric power systems engages this tension between electricity as a commodity versus electricity as a public service. The system we crafted rather adeptly blends these two approaches, involving a fraught but functional marriage between the private and public sectors.

As the record shows in the Pacific Northwest, both investor-owned and public-owned utilities have performed reasonably well most of the time, to the general satisfaction of their customers. Both have been susceptible to problems and failures, and neither can claim to be inherently superior. As a result, the public has accepted a mixed but integrated power system with a permanent place for both commercial and nonprofit utilities driven by both revenue goals and social goals. This in itself is a noteworthy accomplishment that reveals much about how Americans and Canadians use markets and governing institutions in partnership to achieve social goals. The resulting story is not triumphalist or declensionist, but rather one of mixed cooperation and conflict in the incremental elaboration of a progressively more complex and socially responsive technological system. The journey is the story, not the destination, because we are always on our way somewhere new—especially today, with the electric power system in significant transition since the 1990s.[10] We may not know exactly where we are headed, but understanding how we got here will make us better navigators.

Besides narratives of progress or decline, writers have adopted a variety of other lenses through which to assess the history of electric power in North America. The most common approach emphasizes politics and policy evolution. Governing institutions provide the arena where Americans and Canadians deliberately structure property rights, privileges, responsibilities, and incentives for the business sector, and individual rights, protections, and legal recourse for consumers to address injustices. It is also where elected representatives establish specific policy goals and create the public agencies that encourage or direct desired outcomes. Politics and policy are a crucial part of the history of electric power and a traditional analytical lens that has been applied well and thoroughly for over half a century.[11] Other scholars apply a social/cultural lens to their analysis, focusing on how electricity is understood by people, used in daily life, and integrated into popular culture. The master of this genre is David Nye, and many other scholars, especially since the 1980s, have adopted this particular approach, although not much of this kind of work has been applied in the Pacific Northwest.[12]

Still other scholars take an institutional or organizational approach focusing on how utilities or larger power systems are built over time, emphasizing origins, leadership, intent, infrastructure, technology, markets, and government relations.[13] Finally, there is the growing body of environmental histories that adopt many of the above analytical lenses but also pay more careful attention to the influence of the physical landscape, environmental impacts, and environmental protection policies.[14] All these approaches are valuable, so much so that I sought to integrate elements of all of them into this study. Each chapter emphasizes one or more of the political, social, cultural, institutional, and environmental approaches.

I have also compared developments between the United States and Canada whenever possible, although I readily admit British Columbia receives less attention than the U.S. states in this book. Scholarship comparing the United States and Canada is relatively rare compared to scholarship focused on one nation or one region within a single nation. While the number of studies of binational "borderlands" regions is increasing, they remain the exception to the rule. After attempting a comparative borderlands study in this book, I understand why. Despite the challenges, comparative studies illuminate more than the alternatives and mitigate all-too-common nationalist generalizations.

American John Findlay and Canadian Ken Coates edited a collection of essays published in 2002 and titled *Parallel Destinies*; it explores the historical significance of the U.S.-Canada border and interactions between the two nations in specific border regions. As the two editors explain in their preface, nation-state studies often assert, either explicitly or implicitly, that "each nation is different and that national background is of fundamental significance in helping explain who we are, what happened, and why it happened." But, they argue, these presumptions of national distinctiveness are usually untested and often exaggerated. They suggest instead that regional similarities in historical development on both sides of the international border are more common than nation-state studies presume and that a U.S.-Canadian borderland region may have more common characteristics shared across the international border than either area north or south of the border shares with their respective nation-states.[15] In other words, regions have distinctive characteristics that can influence development and identity more powerfully than political boundaries that often seem rather arbitrary to people living in borderland regions.

My study confirms this viewpoint. I found more commonalities than differences in the historical development of electric power systems in British Columbia and the U.S. Northwest. Areas on both sides of the international border in the Pacific Northwest often had more in common with each other than they had with their larger nations. But the record on this score is mixed and shifted significantly over time, making it difficult to identify clear trends or make consistent generalizations about the influence of region, nation, and the international boundary. This is not to deny the relevance of national, provincial, and state borders, which clearly shaped the geography of electric power systems and the policies that structured their institutional character. But over time, the ever expanding and increasingly integrated electric power system of the Pacific Northwest rendered political boundaries incrementally less significant. Electric power has become a regionally integrating force.

The international boundary in the Pacific Northwest is an imaginary line drawn on a political map in 1846 and does not follow any physical cues on the land itself. Similar environmental conditions (tall mountains, large rivers, and prodigious regional precipitation) along with shared technology, financial institutions, and macro-economic trends (booms, depressions, wars) affected system development in common ways on both sides of the border in the Pacific Northwest. Entrepreneurs in both countries were early adopters of new electric technologies. In fact, the Pacific Northwest was often on the cutting edge rather than the backwater. Still, some interesting distinctions are worth noting. City governments on both sides of the border quickly jumped in the game and offered franchises to electric companies to provide street lighting and electrified mass transit. But the relationship between city governments and electric companies was more fraught in the United States than in British Columbia, in part because B.C. electric companies often performed better and in part because Canadian customers and governments simply expected less and complained less about their investor-owned electric companies in the early decades of the industry. I present ample evidence for this in the narrative, but more research is needed to fully explain this difference. This distinction undoubtedly influenced the early adoption of state regulation of electrical services in the U.S. Northwest (1911–1916) versus the late adoption of provincial regulation in British Columbia (1938).

In the early decades of the industry, on both sides of the border small towns lacking electric service readily established municipally owned electric companies when privately owned electric companies would not provide service or when the service was unreliable or too expensive. In British Columbia, these municipal utilities usually bought their power wholesale from the investor-owned for-profit electric companies, and the two sectors (public and private) lived side by side without much conflict for the first six decades of the industry's development. Privately owned electric companies served all the large towns and all the major markets in British Columbia. This state of affairs remained largely unchanged until the province expropriated British Columbia's biggest electric utility in 1961, thereby converting the province to a predominantly public-owned power system. The situation was quite different in the United States. A great deal of conflict often accompanied the development of municipal electric systems, with private and public sectors treating each other as hostile competitors. Ideological and material battles between public and private power started early and never really ended in the U.S. Northwest, although overt antagonism between the two sides significantly diminished after the 1950s.

Both the United States and Canada experienced an epic boom in electric power development in the 1910s and 1920s, followed by a devastating bust in the 1930s. But their respective responses to the Great Depression differed significantly on either side of the international border. In the United States, the federal government stepped into the breach and launched major economic development programs, including two monumental dams on the Columbia River—Bonneville and Grand Coulee—to provide employment, stimulate the economy, and advance public power in the region. The Canadian federal government did not embrace public works or Keynesian economic policy as the Roosevelt administration did, and the provincial governments could not engage in significant deficit spending, so British Columbia limped through the Depression with virtually no new power system development and virtually no economic stimulus.[16]

World War II continued the contrast as federal war spending and military production programs supported by electric power from Bonneville and Grand Coulee dams remade the economy of the U.S. Northwest, causing an economic and population boom that far exceeded national averages across the United States. These two dams and regional war production totally transformed the Northwest economy, its cities, and its power system. While war mobilization finally lifted British Columbia's economy out of the 1930s Depression, war production in the province did not come close to matching war production in the U.S. Northwest, in part because the Canadian federal government's Wartime Industries Control Board refused to approve development of additional power supplies by the BC Electric Company serving greater Vancouver and Victoria. While south of the border, Federal Columbia River Power supported development of large aluminum-processing facilities in Washington and Oregon, north of the border a power-supply bottleneck exacerbated by the Canadian federal government inhibited this kind of development.

The trajectories of the Northwest states and British Columbia reconverged after the war with both areas enjoying a remarkable postwar economic boom that stimulated and in turn was fed by rapid expansion of hydropower and an integrated electric power grid. But interestingly, in the United States a conservative reaction against fifteen years of New Deal programs and aggressive federal economic intervention led to a slowdown in federal initiatives in the 1950s and a corresponding increase in private-sector power development over the next few decades. The opposite occurred in British Columbia. Decades of private-sector control of the power system gave way first to the creation of a provincial power authority in 1948 to support rural electrification (more than ten years after government-supported rural electrification was launched in the United States), and then in 1961 the provincial legislature took over most of the province's power supply in order to more aggressively pursue coordinated province-wide economic development similar to that pursued by federal and regional authorities in the United States during the New Deal. So we see similar initiatives and results but in different decades.

Overall, the record of similarities and differences between power system evolution north and south of the 49th parallel is mixed with no consistent distinctions, except for one. Canadians have always been wary of U.S. influence. Canadians resisted American ownership of their country's power industry and resisted selling power to or buying power from American utilities. Cross-border transfers of electrons rarely occurred before BC Electric joined the Northwest Power Pool in 1949, and even then power exchanges remained minimal. Still today, British Columbia generates and controls nearly all of its domestic power needs, neither importing nor exporting much electricity. U.S. investors and utilities, in contrast, have been more interested in open borders, seeing Canada as a potential source of supply and a potential market. In the Pacific Northwest, leaders of the electric power industry coveted British Columbia's resources and pressed for decades for joint development of the Columbia River. While American interests see the international border as an obstacle, Canadian interests see it more as a line of defense. Many of the authors in the book *Parallel Destinies* note this cultural difference, especially Ken Coates in the introduction ("Border Crossings") and Michael Fellman in his chapter, "Sleeping with the Elephant." Coates observes that while most Americans give little thought to the international boundary, Canadians have been preoccupied with it, "devoting enormous intellectual and political energy to peering southward from the boundary marker, fearful that the sleeping giant will awake and once again cast its expansionist eyes toward the subarctic." Fellman argues that, contrary to the cultural inclination of Canadians to focus on differences between themselves and Americans, both nations are substantially the same: "liberal, capitalist, democratic, federalist, and only reluctantly statist."[17]

The same mixed results are seen when comparing the U.S. Northwest states with the nation as a whole and when comparing British Columbia with the rest of Canada. In both cases the Northwest region sometimes reflects and sometimes bucks national trends. For example, besides being early adopters of electrical technology, utilities in the region often installed cutting-edge equipment: the largest turbines, the longest and highest voltage transmission lines, the highest-capacity transformers, the largest dams, and so forth. The region's utilities and engineers were not conservative. Rather consistently, the Northwest had higher-than-average electrification rates and higher-than-average per capita consumption, which abundant inexpensive hydropower enabled. It also generally had lower-than-average electric rates, enabled again by the fact that hydropower requires no fuel to burn and therefore provides cheap power generation after the initial investment is paid off. This economical foundation of hydropower made possible by the region's climate and geography led to early and rapid electrification, a manufacturing economy based more intensively on electric power than most other regions, and a vibrant culture of technological optimism that prevailed from the 1920s through the 1960s based upon the dreamed-of social transformations to be wrought by cheap, abundant electricity. But it also led to significant waste and inefficiency in the consumption of electricity. This condition came to haunt consumers in the 1970s and then served as a new frontier of technological innovation in the 1980s and afterward.

The Pacific Northwest followed national trends in the increasing scale of organization in the electric industry and the increasing concentration of ownership during the first half of the twentieth century. It followed national trends in the boom of the 1920s, the bust of the 1930s, and the recovery of the 1940s. But in regard to utility regulation and public ownership versus private ownership, both the U.S. Northwest and British Columbia bucked national trends. Compared to most other provinces, British Columbia came quite late to public regulation of private utilities and to provincial ownership. The Northwest states, in contrast, adopted regulation fairly early and held to public ownership more tenaciously than most other U.S. regions. It bucked the national decline in the number of municipal utilities in the first half of the twentieth century, substantially exceeded the national average in rural electrification from the 1920s onward, and in at least some parts of the region, embraced public power more enthusiastically in the 1930s and 1940s than the United States as a whole did. Here, it is important to note that the records of the three U.S. states that I examine differ from each other in many ways. Washington, Oregon, and Idaho cannot be treated as a homogenous unit, even though I generalize often about similar developments among the three states. Divergences between the three constantly complicate and enrich the story. Most strikingly, Washington embraced public power so vigorously after 1930 that by midcentury about half the state's customers received their electricity from publicly owned utilities. Oregon's public power sector also expanded during the 1930s and 1940s, but at a much slower rate and with less ultimate success than Washington's. Idaho came in a distant third, failing even to authorize public power districts as Washington and Oregon did in 1930. Idaho also was the last of the three states to establish a public utility commission to regulate electric services.

Amid all this diversity of paths and outcomes lies a common story of how we created, consumed, and constantly reformed this essential energy system at the foundation of modern life. It is a tale of intentional evolution, of how we adapted this technological system to better serve our desires, solve problems, and meet larger social goals. Often those efforts have been conflicted, inconsistent, inadequate, and misguided, but they nevertheless have led incrementally and progressively toward a more functional and responsive system. Each era of reform is built on the foundations of the previous era, which offers both the benefits of experience and the drawback of path-dependence. The complicated and contested evolution of the electric power system, if intelligently understood, offers valuable lessons about how each generation alters and adapts its socio-technological systems to solve problems and serve contemporary interests. History is instructive.

I start at the beginning, with the invention of the first electrical devices—spectacular lighting systems—and trace the path by which this technological novelty evolved into a profitable commodity and then into a major capitalist enterprise. Within two decades of the initial invention of the electric dynamo and the incandescent light bulb in the 1870s, electrical manufacturers and companies supplying electric power to customers had become one of the largest, most profitable industries in North America. At that point, this marvelous technology evolved into an institution far more complicated than a simple market commodity: it transitioned into a public service. Communities, states, provinces, and nations all provided or exercised oversight over a variety of public services and public utilities, such as city water supplies, sewer systems, public roads, mass transit, and waste and refuse disposal. Because these services all profoundly affect public safety, public welfare, commercial prosperity, and quality of life, citizens demanded they be effectively regulated or managed by democratically responsive governing institutions. During the first and second decades of the twentieth century, electric power became so integral and essential to modern life that virtually every city, state, and province in North America redefined it as a public service, designated electric power companies as public utilities, and either created agencies to regulate service, price, and profit or created publicly owned corporations to provide the service. At that point, electric power irreversibly became a complex socio-technological system the development of which was constantly negotiated among private-sector profit-seeking companies, their customers, and public-sector public-interest authorities.

My narrative weaves together stories from the consumer side, where electricity transformed mills, mines, households, commercial districts, urban transit, factories, and farms. I examine the ambitions and accomplishments of local and regional power companies that built the infrastructure to generate power and deliver it to customers. I trace the influence of national and international investment companies that financed and in some cases parasitized the operating companies. I examine the tensions between producers and consumers, the enthusiasm and disappointment of those buying and selling this new technology/commodity in a weakly regulated, erratic, and unreliable market. Along the way I highlight technological progress and its benefits, economic concentration and its mixed effects, wars, crises, and natural and human-caused disasters that all shaped system evolution. Finally, I dissect and evaluate the interventions of politicians and public agencies attempting to design a socially regulated electric power system that served as well as possible the combined needs and interests of investors; utility company managers; city, state, and national leaders; commercial interests; and residential customers all at once—a perhaps quixotic but nonetheless worthy endeavor. I demonstrate a modest but notable record of long-term progress in achieving this goal, despite compromises, backsliding, overreach, and intermittent failures.

I end my historical examination in the 1970s when a number of long-term historical trends reached their peak. The post–World War II generation brought the power system to a pinnacle of development designed to maximize production and consumption, which in turn brought unprecedented change to the region's economy and environment. But flaws in that development model undermined its continued advance. During the seventies energy crisis and the years after, the Northwest power system, painstakingly crafted over the previous century, underwent fundamental re-evaluation and reform. The structural changes to the electric power system occurring since the 1980s are comparable to the upheavals and transformations experienced during the Depression and World War II. As we are currently ensconced in a long era of substantial power system transformation, I hope that my careful examination of previous eras of struggle and change is instructive for those who seek understanding of our current challenges and especially for those who seek to shape the electric power system of the future. The system we inherited from the twentieth century shapes our options and choices today and will continue to shape our lives and energy regimes tomorrow as we move either blindly or intelligently toward a smarter grid and a sustainable, responsible energy system for the twenty-first century.

CHAPTER 1

The Brilliant Spectacle

Electricity was . . . the wonder of the age, the hallmark of progress.[1]

—David Nye

On May 15, 1879, while inhabitants of the city of Victoria on Vancouver Island were enjoying eighteen-hour stretches of summer daylight, a new kind of light appeared in their midst. The steamship HMS *Triumph* docked in Esquimalt Harbor, carrying an electric generator driven by a steam engine (called a dynamo at that time) attached to a bundle of wires and arc lights. After dark, the residents of Victoria would be treated to the first electric light shown anywhere in the Pacific Northwest. For the ballyhooed occasion, over 1,200 men, women, and children gathered at the harbor, many of them boarding four small steamships that positioned themselves in an arc facing HMS *Triumph* to witness the spectacle from the dark waters of the bay. When all was ready, the band on the *Triumph* commenced playing a waltz and a switch was thrown, sending electricity to the arc lamps on board. An attending observer reported, "The darkness, which had reigned supreme over Esquimalt Harbor was suddenly dispelled, and the harbor became as bright as day." A reporter for the *British Colonist* the next day exclaimed with evident amazement that the lights clearly illuminated one of the passenger steamers sitting 300 yards away, its decks packed with observers from town. Most remarkable, he noted, "the ladies' shawls and dresses were as clearly distinguishable as in daylight."[2]

For those first witnessing electric arc lights, the spectacle was beyond experience or words. Nothing on earth but a bolt of lightning could rival this kind of illumination at night. Two weeks later (May 28, 1879), another steamship, the *State of California*, outfitted with six arc lamps powered by the ship's steam engines, docked in Portland, Oregon, to give residents of the Willamette Valley their first glimpse of electric light. When darkness fell, a brilliant flood of white light illuminated the ship and the nearby docks. Through newspapers and word of mouth, most of Portland's 17,000 inhabitants soon learned about this new marvel of industrial technology. These two events marked the first time electric light appeared in the Pacific Northwest, and the technology would spread rapidly.[3]

A year later, in July of 1880, another new light appeared in the Northwest: the Edison incandescent lighting system—using vacuum-sealed filament light bulbs rather than arc lights. This softer glowing light would soon replace the brighter and more hazardous arc lights for indoor illumination. It arrived in Portland on a steamship owned by Henry Villard. At the time, Villard was owner of the Oregon Railroad and Navigation Company building his fortune in the transportation business. His ship contained a stand-alone dynamo for generating electricity, a passel of electrical wires, and Edison's newly invented incandescent light bulbs. About a month after the ship's arrival, lines were strung to shore and the Edison dynamo pumped electricity to light bulbs illuminating Ainsworth Dock and the Clarendon Hotel. The next day, the Portland *Oregonian* remarked, "The powerful rays lighted up the whole neighborhood to the brightness of day. Thousands visited the light and the vessel."[4]

Villard, a German émigré and railroad corporation executive who kept company with the nation's elite financiers, including J. P. Morgan, was an enthusiastic supporter of the Edison lighting system and a major financial backer of Edison's work.[5] In January 1880, just weeks after Edison ceremonially lit up his facilities at Menlo Park, New Jersey, with the new incandescent lighting system, Villard placed the first commercial order for this system and loaded it onto his new steamship, the SS *Columbia*, bound for Oregon that summer. Thus, while many easterners in the early 1880s viewed the Pacific Northwest as remote and marginal to the affairs of the American and Canadian nations (no transcontinental railroads had yet arrived in the region), Northwest business leaders were on the cutting edge of new commercial technology.[6]

As David Nye points out in *Electrifying America*, the first lighting systems, especially arc lights, were suitable mainly for staging spectacles.[7] People gathered from miles around to see this novel phenomenon of artificial light. For the entire history of humankind, illumination had been associated with only the sun, moon, and fire. Now there was a human-invented, human-controlled light source without flames, capable of being turned on and off by a key or a switch. Equally impressive—or unnerving—were the rumble and smoke of the steam dynamo that provided the electricity for the lights. To most people who still lived lives attuned predominantly to the daily rhythms of nature, the experience of snorting engines, spinning steel turbines, and artificial illumination was both frightening and impressive—a discomforting disruption of natural sounds and rhythms as well as a marvelous testimony to human inventive power.

Light and electricity have always held as much cultural significance as practical value. At its inception, people associated electric lighting not only with the ability to push back the darkness but with a larger sense of social progress and even cultural superiority. A 1931 history of the Washington Water Power Company imaginatively recalls the first light on the streets of Spokane Falls in terms reflecting both the romantic and the racist sentiments of that generation:

> That fall [1885] many a trapper, prospector and settler coming into the village for supplies, halted his cayuse and gazed in amazement at the strange lights—and wondered at the cause of such a celebration of illumination. Blanketed Indians regarded the blazing lights stoically and with their characteristic "ugh!" commented volubly on this achievement of the white man. In the dark pools below the Spokane Falls where for untold Indian generations swift spears hurled by marble-poised braves had impaled the writhing tyee (salmon), there now gleamed the faint reflection of these arc lamps—replacing the glow of campfires—presaging the time when powerful lights would bathe the entire falls in a flood of white light in which the mist from the tumbling waters would gleam with gauzy translucency.[8]

Such sentiments portrayed Indians as picturesque remnants of a bygone era, to be inevitably replaced by the benign advance of civilization. This popular story line served to obscure the purposeful and tragic dispossession of native peoples still ongoing at the time electric lights appeared in the Northwest. The Spokane Indian Reservation had just been established in 1877 (later confirmed by presidential proclamation in 1881). The Indians who lived and fished at the falls were unceremoniously removed from these valuable lands along the Spokane River, which were desired by European Americans, and concentrated on more remote, less desirable lands to the northwest of Spokane. The reservation would not receive electric lights or power for decades as the Spokane Indians continued to lose lands and resources to the new immigrants. In 1911, Washington Water Power built a dam at the Lower Falls of the Spokane River on reservation land, which blocked salmon and steelhead from migrating upriver. Then, in 1939, when Grand Coulee Dam was completed, it blocked all anadromous fish (e.g., salmon) from the entire Spokane River, and 2,500 acres of reservation lands were drowned beneath the reservoir. In 2005, the tribe was still fighting in the courts and Congress for reasonable compensation.[9]

As this example shows, economic and technological developments do not benefit all people equally. To the north, in British Columbia, similar politics of racial exclusion was at play. In the same decade that electricity initially spread its technological wings in the Northwest, the United States and Canada were embroiled coast to coast in industrial labor strife. In the North American West, this labor conflict included a violent, scapegoating campaign to disfranchise and deport Chinese workers who had been so crucial to construction of the transcontinental railroads. Once the Northern Pacific Railway and the Canadian Pacific Railway were completed in 1883 and 1885 respectively, many Chinese laborers found themselves out of work and unwanted. The 1886 legislative charter incorporating the Vancouver Electric Illuminating Company explicitly forbade the company from employing any Chinese laborers to build its electric system and even specified the fine to be levied for violations of this provision. The owners of the company were two Canadian Pacific Railway executives from the east, looking for new investments, and the government appended this racial exclusion policy at the behest of anti-Chinese agitators in British Columbia.[10]

Regardless of whether one was a have or a have-not in the Northwest, there were opportunities to observe the spectacle of electric illumination, and those spectacles were usually bound up with conceptions of social hierarchy. Many people first witnessed electric light at World's Fairs, especially the Columbian Exposition of 1893 held in Chicago, in which artificial illumination created a "Great White Way." The White Way mixed technological marvel with the notion of a racial hierarchy, as brightly illuminated cultural exhibits of the "white" race were aligned along the White Way, while exhibits of other, darker races were arrayed in descending order away from the European exhibits. Lighting also enhanced the monumental nature of World's Fair architecture, and as electric motors came into production, fairs added trolleys and other forms of electrical conveyances.[11] Similar fairs and expositions provided opportunities for the development of elaborate "spectacle" lighting systems in the Northwest, such as the Northwest Industrial Exposition in Spokane in 1891 and the Lewis and Clark Centennial Exposition in Portland in 1905.[12] The popularity of the Great White Way led to a burgeoning market for illuminating urban thoroughfares and central commercial districts. By the 1890s, Edison General Electric and other electrical manufacturing firms were briskly selling electric systems to cities to create their own "white ways" in downtown commercial districts.[13]

Lighting applications quickly shifted from novelty and show to more practical uses such as replacing the dangerous gaslights on downtown streets, illuminating the interiors of large buildings, and providing nighttime lighting at docks, mines, and mills. The longtime quest for functional nighttime and indoor illumination had led first to smoky, stinking torches and flickering candles, then to kerosene and coal oil, and eventually to natural gas. The last of these seemed quite a leap of progress over the former. As early as 1859 the city of Victoria issued a public franchise to a private company to erect lamps attached to gas lines for nighttime public street lighting.[14] The gas came to Victoria Harbour in shipping containers and was piped crudely along streets in the commercial district. Unfortunately, gaslight was often a flickering, disagreeable form of illumination that posed a constant fire danger. In comparison to electric light, gaslight fared poorly. By 1883, Victoria had contracted with a private company to use arc lamps to provide electric street lighting. The system was plagued with troubles, however, and the city eventually turned to incandescent streetlights in 1886. Again the Northwest distinguished itself on the cutting edge of innovation, as this was purportedly the first incandescent street lighting system in all of Canada.[15]

In Portland to the south, observers noted the superiority of electric light over gaslight, too. Upon viewing the first arc lights in 1879, a reporter for the *Oregonian* notes, "[It] is far more brilliant than that produced by gas but yet not painful to the eyes when steadily bent on it." Anticipating the bright future of electrical illumination and likely reflecting astute public relations by the owners of the arc lights, the journalist noted at the end of his article that electric light was "the most economical illuminator that can be used" and that a Mr. James Guild had been appointed "agent for the electric light in this city."[16] This casual observation foreshadowed how business entrepreneurs and city governments cooperated to simultaneously provide electric light services to townsfolk and profits to investors in the coming decades.

Beating out its urban rivals to the south, Seattle got the first "central station" incandescent public lighting system in the U.S. Northwest in 1886, the same year that Victoria's incandescent system came online. Installed by the Seattle Electric Lighting Company, this system had a central powerhouse connected by wires to street lamps and a variety of commercial and residential customers along the waterfront.[17] Entrepreneurs in other cities quickly followed suit, acquiring franchises from city councils to provide electric lighting to business districts, finding investors to cover the cost of equipment purchases, and negotiating deals with electrical manufacturing companies for delivery and setup of the generators, wires, and lighting. Catching the electric light fever, Victoria, Seattle, Portland, Tacoma, Astoria, Spokane, Vancouver, Hailey (Idaho), and other Northwest cities installed lighting systems between 1883 and 1887.[18]

The other main market for early lighting systems was big businesses, especially those that wished to increase their daily hours of operation. In September 1880, a dynamo with arc lights was erected in a large waterfront sawmill in Portland, with lines running to Ainsworth Dock. More Portland businesses ordered lighting systems in the next two years. In 1881, an Edison lighting system was installed at a silver mine in Hailey, Idaho. In 1882, on Puget Sound the Tacoma Mill Company installed lighting that permitted the mill to operate its machinery and work in the dockyards twenty-four hours a day. Also in 1882, Idaho's Philadelphia Mining and Smelting Company near Ketchum installed electric lights. A lumber mill in Astoria started burning waste wood from its operations to drive a dynamo supplying lights to its mill and the town of Astoria in 1885, setting a precedent in which wood waste from the ubiquitous lumber operations of the Northwest served as fuel.[19]

Even more significant for the history of electricity in the Northwest were the first examples of electric-driven motors employed in business and transportation. In 1888, two newspapers in Spokane and Vancouver electrified their presses. Vancouver's *News-Telegram* was the first newspaper in the dominion of Canada to electrify its printing press. In the same year, a hotel in Spokane Falls used electricity to run a circular saw to cut firewood for its wood-burning stove, setting a precedent that would be quickly adopted by lumber mills all over the region. In 1889, Seattle replaced several of its horse-drawn streetcars with an electric streetcar—one of many American cities to do so.[20] These examples foreshadowed the coming widespread electrification of manufacturing and transit that will be discussed in the next chapter.

In less than one decade a totally novel innovation, electric lighting, had penetrated most major commercial towns in the Northwest—a remarkably rapid diffusion of a new technology, especially in a region of the country still considered a frontier in the 1880s. As David Nye observes, "Spectacular lighting, begun as a device to attract crowds to World's Fairs, had become an essential part of commerce and the life of the street."[21]

This essential component of life and commerce was not, however, a democratic technology. More than other energy sources of the time—such as wood, coal, and kerosene—electricity required sophisticated mechanical components, managerial expertise, and a large measure of investment capital.[22] This is the familiar story of the industrial age. New machine-age technologies depended upon and in turn supported increasing levels of business organization and wealth concentration. Electricity and the corporation grew up together in America. Dynamos and electric lights were not something any blacksmith could manufacture in a workshed. Specialized factories manufactured the equipment needed for electrical generation and lighting, and those factories were owned by well-capitalized companies, which in turn were controlled by financiers on Wall Street and abroad. Edison's pioneering work on incandescent lighting, for example, was financed by Wall Street. Within a decade of his inventing the incandescent light bulb, Edison's Electric Light Company was taken over by Henry Villard, J. P. Morgan, and European investors in an 1888 merger with competitors to form the Edison General Electric Company. (Edison himself made $1.7 million on the deal and then bowed out; Morgan realized a $2.7 million profit on his initial investment in Edison Electric.) Shortly thereafter, another merger in 1892 created General Electric (G.E.), a household name today.

A chief corporate rival to G.E. in the 1880s was the Westinghouse Electric Corporation. These two companies manufactured equipment for lighting systems and sold them to customers across the nation. To encourage sales, they even supplied capital and equipment training to many customers. Struggles between these two giants over patent rights for generators and lights led to a veritable litigation war that extended until 1892, when the two corporations agreed to share patents, end the litigation, and concentrate on making and selling equipment. Controlling the most important patents, they kept out most other competitors and thoroughly dominated the electrical manufacturing industry for most of the twentieth century.[23]

Most electrical generators—"dynamos"—built in the 1880s were essentially steam engines, spinning a special turbine to create an electrical charge at a specific voltage. It required fuel (usually wood, coal, or fuel oil) and tended to be smoky, noisy, and dangerous. Air pollution control, noise abatement regulations, and occupational safety codes as yet hardly existed in America. Engine pressures might be improperly regulated, and gaskets could blow or the whole machine explode. And of course there was always the possibility of electrocution. The whole business was risky and demanded a fair amount of training and regular monitoring and maintenance of equipment. This led to the rise of specialized companies dedicated exclusively to installing, operating, and distributing electricity and lighting systems. In the early decades of the industry, these were mainly private companies in business for profit. Later, publicly owned utilities entered the market promoting "power at cost" and providing vigorous competition for the privately owned companies.

Another technology that existed in the 1880s for generating electricity did not employ the volatile steam engine, but until the turn of the century it was rarely used: hydropower. People had been using the energy in falling water to perform work for thousands of years. For over a century in New England and the southern United States, textile, flour, and cider manufacturers had built their mills along rivers, capturing the energy of moving water to turn spinning and grinding wheels. This mechanical water power is not the same as hydroelectric power, but the principle of using moving water to spin a mill wheel can be applied to spinning an electric turbine.

Marketing such a technology initially proved difficult. There were several problems: unlike a steam-powered dynamo, which could theoretically be located anywhere customers exist, the hydropowered generator could only be located on a river with a sufficient and reasonably steady volume of falling water. To capture falling water, facilities had to be placed at a waterfall or a dam. Most population centers did not have nearby waterfalls, and sites that did were often occupied already by mills tapping the falling water for mechanical power. The first hydropower generator at Spokane Falls, for example, was driven by water exiting a flour mill. The mill and the hydropower generator were owned by different companies, and the mill controlled the water and the water rights, so occasionally the mill owners would shut down the mill, cutting off the flow of water to the turbine and thus stopping electric service to the town.[24]

Electric companies could build their own dams if they could get control of a good site, but these were not always available near urban and manufacturing centers where the power could be marketed. Besides, dams were expensive to build, requiring engineering expertise, specialized equipment, and plenty of investment capital. In the 1880s the revenue potential of an electric power company was still uncertain, so capital could be difficult to acquire. Consequently, many of the businesspeople launching electric light and power services had to partner with investors associated with eastern electrical manufacturing companies and their financial sponsors, thereby giving up much local control over the system, its management, and its potential profits.

Securing funding for Spokane's first hydroelectric generator in 1887 exemplified this pattern. The Spokane Electric Light and Power Company executives could not raise enough capital from local sources, so they sent a representative, Henry Hoyt, to New York, where he met with the Edison Electric Light Company. Edison Electric and its financial partners provided the needed cash and equipment in exchange for a repayment agreement and a large chunk of stock in the company. Immediately after these agreements, Edison Electric's representative for the Northwest, Sidney Zollicoffer Mitchell, reorganized the Spokane company, renaming it Edison Electric Illuminating Company of Spokane Falls. Mitchell supervised the construction of the new hydropower facility in 1887 and 1888, along with the expanded electric distribution lines.[25]

Sidney Z. Mitchell had a hand in setting up most of the first electric light and power systems in the Northwest. He was an extraordinarily energetic and ambitious businessman who served as the Northwest representative for the New York–based Edison Electric Company from 1885 to 1902. He first encountered an electric light system while helping to install one on his ship during his service in the Navy in 1883. Giving up his commission in 1885, he sought out Edison in New York and spent much of the year taking various courses in electrical engineering at Edison's Menlo Park laboratories and helping set up lighting systems in New York. By September of that year, he had impressed the movers and shakers at Edison Electric enough to be offered Henry Villard's recently vacated position as Northwest representative. Villard was busy focusing on the completion of his Northern Pacific Railway. This put Mitchell in charge of sales and operations for the state of Oregon, the territories of Washington, Idaho, Montana, and Alaska, and the province of British Columbia. He was twenty-three years old. Mitchell immediately moved west with friend and business partner Frederick Sparling and set up a regional office in Portland. From that position during the next two years, he and Sparling sold and set up Seattle's first central station incandescent light system discussed earlier, as well as Tacoma's light system, Spokane's hydropower facility, and a dozen other systems from Port Townsend and Victoria in the west to Hailey and Boise in the east.[26]

In 1902, Mitchell returned to New York, and in 1905 he helped organize and became first president of Electric Bond and Share, a financial affiliate of General Electric and one of America's leading sources of investment capital, engineering expertise, and management advice for electric systems. Despite his return to New York, Mitchell remained deeply invested in the Northwest. Besides his oversight of all the Electric Bond and Share affiliates in the Northwest, he helped organize one of the largest electric power utilities in the region in 1910, Pacific Power & Light, serving as its board chairman from 1910 until 1933.[27]

Interestingly, while Mitchell's territory included British Columbia, his name does not appear in the histories written about electrical development in the province. Even though he sold many systems there for Edison Electric (later General Electric), he apparently did not participate in financing or managing any of them to the extent that he did in the United States. The primary entrepreneurs in British Columbia were Robert McMicking, who set up the first lighting system in Victoria; Harry Abbott and George Keefer, who set up the first lighting company in Vancouver; David Higgins, who launched the electric streetcar business in British Columbia; and Frank Barnard, who served in a role similar to Mitchell's in the 1890s by providing large infusions of capital from London to consolidate and expand electric power system development during the hard economic times of that decade (see next chapter).[28]

The increasing involvement of large eastern investment firms in electric system development paralleled the increasing use of falling water to drive electric generators. Electric light systems before 1887 were powered by steam dynamos burning fuel, but after that, electrical entrepreneurs increasingly tapped the great potential for water power in the Pacific Northwest. According to Thomas Hughes in *Networks of Power*, "Natural geography invariably figures prominently in accounts of hydroelectric projects."[29] This was especially true of the Pacific Northwest. With typical boosterism, a 1904 article in the *Pacific Monthly* magazine carrying the subtitle "An account of the marvelous motive power stored in the waterfalls and rivers of this section" expressed it best: "This whole region [the Pacific Northwest] is more favored in this particular than any like amount of territory in the United States, or in the western continent, and probably, than in the whole world."[30] Bruce Wolverton, the author of this article, followed this passage with a description of the mountain ranges, river valleys, and topographic diversity of the U.S. Pacific Northwest, which ranges in elevation from sea level to nearly 15,000 feet on Mount Rainier in the Cascade Mountains. In contrast, he noted, Illinois's topography ranged from 600 to 900 feet in elevation. Topography is important, because the more the rivers fall (the higher their gradients), the more total hydropower potential exists in them.

Equally important for electricity producers is the amount of water flowing in the rivers and the seasonal regularity of the flow. According to the United States Army Corps of Engineers, "the Columbia River pours more water into the Pacific Ocean than any other river in North or South America": just under 200 million acre-feet of water every year.[31] Annually, it carries the second largest flow of any river in the United States (the Mississippi River is the largest). The Columbia flows more than 1,200 miles from its headwaters at Columbia Lake in the Canadian Rockies of British Columbia to where the river meets the Pacific Ocean on the Oregon-Washington border, draining 258,000 square miles. Canada contains 15 percent of the Columbia River watershed but contributes 25 percent of the river's flow. This tremendous volume of water falling many thousands of vertical feet over its course makes the Columbia River the greatest hydroelectric power–generating river in the Northwest. In fact, the U.S. Bonneville Power Administration (BPA) claims that the Columbia supports "the largest hydroelectric system in the world."[32]

One of the reasons there is so much water in Northwest rivers is that at least three adjacent mountain cordilleras capture huge amounts of precipitation, both rain and snow, from storms sweeping in from the Pacific Ocean. In the United States, the Coast range (Pacific Cordillera), then the Cascades, then the Rockies each wring moisture from the wet Pacific air masses as they move west to east across the continent. British Columbia has an even more complex assortment of mountain ranges associated with these three cordilleras, plus the

Columbia mountain ranges in southeast British Columbia, the Cassiar mountain range in the northern interior, and mountain ranges on Vancouver Island and the Queen Charlotte Islands. Mountains cover more than 75 percent of the land of British Columbia. Some places in the Coastal and Cascade mountains receive over 100 inches (254 cm) of precipitation annually. Of course, the precipitation is seasonal, with nearly all of it falling between October and May—except in the coastal rainforests where it commonly rains year-round. Since the moisture comes mostly during the long Northwest winter, the majority of it falls as snow and collects in the mountains. As a result, Washington state has more glaciers than all other U.S. states combined. British Columbia, too, beats all its provincial rivals outside the Arctic in its abundance of glaciers. This means high river flows during the spring snowmelts. It also means difficult winter conditions in the mountains. Ever optimistic, Wolverton claimed in 1904 that "while the snows and ice fields of winter are numerous and extensive, yet the cold is not so intense as to interfere to any considerable extent with the operation of water power, even in the highest altitudes."[33] This is only partly true—and to the degree that it is true, it is attributable to the climate-moderating influence of the Pacific Ocean.

The second important environmental characteristic relevant to hydropower development is the consistency and reliability of a river's annual flow. If a river floods in the spring and is dry in the fall, power can be produced only part of the year. Characteristically for booster literature of the time, Wolverton exaggerated the region's qualities, suggesting that the year-round water flow of the Columbia River and its tributaries "tends to distribute itself more evenly than in almost any portion of the world." Contrary to this claim, the Columbia River in its natural state greatly fluctuates across seasons, with very high flows from April to June and relatively low flows from September to November. This fact led water power planners in the mid-twentieth century to advocate very large "storage" reservoirs in the upper Columbia River watershed to help decrease the variability of the river's flow (see chapters 10 and 12).

A technical limitation to electrical transmission posed the final obstacle to hydropower in the early years: Edison's lighting system used direct current (DC) at 110 volts, and its practical transmission limits were about half a mile. Higher voltages could be transmitted over cables for longer distances, but the risk of electrocution was correspondingly higher and the voltage had to be reduced to a more usable level at the point of consumption. Experiments with 11,000-volt alternating current (AC) transmission in Massachusetts in 1886 showed that higher voltages in AC allowed efficient transmission over a longer distance with less electricity loss through the line, but to Edison that seemed dangerous. He stuck with low-voltage DC systems until 1889, while George Westinghouse moved aggressively into AC systems. Buying a variety of leading electrical companies and patents, including new transformer technologies that allowed high voltages to be stepped down to 110 volts, Westinghouse came into the market in 1886 with a competing lighting system that within a decade had edged out Edison's more limited DC system. This famous competitive episode in history is referred to as "the Battle of the Currents." In the early 1890s, after corporate mergers created General Electric and the two rival companies began cooperating, AC and DC technologies combined into a "universal system."[34] With the advent of high-voltage AC transmission in the late 1880s, more remote water power sites could finally be exploited.

Once again, the Pacific Northwest was in the vanguard in applying electrical technologies. The facility installed in Spokane in 1887, mentioned earlier, was the Northwest's first hydropower generator, but it sent power only a short distance into the town of Spokane Falls. In 1889, the Willamette Falls Electric Company (a precursor to Portland General Electric) took the experiment one step further and set a nationwide precedent. It built a hydropower facility at Willamette Falls in Oregon City, fourteen miles from Portland, and built the nation's first high-voltage long-distance transmission line to send the power to rapidly growing Portland, which lacked a practical water power site of its own. Relying first on direct current (DC), the company switched the next year (1890) to the new more efficient Westinghouse alternating current (AC) technology, delivering power to Portland with very little line loss.[35]

This was just the beginning. In the next decade, the technology for long-distance transmission dramatically improved, making remote water-power sites increasingly valuable (see next chapter). With the Northwest's abundance of mountains and rivers, acquiring hydropower sites would be on nearly every electric company's wish list by the turn of the century. Exploiting these hydropower sites required building dams and penstocks, powerhouses and generators, distribution lines and substations. It required drilling equipment and explosives, surveyors and engineers, water rights and land purchases. In short, hydropower development involved more capital, expertise, labor, and political clout than most small, local businesses could muster on their own. J. P. Morgan, Henry Villard, Sidney Mitchell, Frank Barnard, and others were all poised to harvest those opportunities as they arose.

All these singular events—the first arc lights brought by the HMS *Triumph* and the SS *Columbia*, the first public incandescent lighting systems in Seattle and Victoria, the first electrified presses and sawmills, the first electric trolley, the first hydropower generator—marked the beginning of a new era, the dawn of the electrical age. Small steps such as these added up to a major transformation of life and livelihood within one generation. The marvel these innovations inspired in onlookers 120 years ago can scarcely be understood by a culture now inextricably embedded in electrical systems. Life before electricity was entirely different and was understood and experienced differently than the world with which we have become familiar. As David Nye argues, a succession of energy systems helped create a succession of distinct social worlds—from the world of muscle power supplemented by wind and water, dominant since ancient times, to the world of steam engines in the mid-nineteenth century, to the electrical age of the twentieth century. "Each energy system has been used to shape distinctive domestic patterns, work routines, urban structures, and agricultural methods, imparting particular rhythms and contours to the everyday round."[36]

In the world of muscle-powered labor, one's ambitions and expectations were limited to what could be accomplished by hand or with the aid of a draft animal. Production was local and small-scale. Change came slowly. A windmill, waterwheel, or horse-drawn millstone might do additional work for a farmer or other laborer, but these plodding mechanical devices worked on a scale and pace comparable to those of human muscle power. With the advent of steam-powered—and later gasoline-powered and electric—machines, the pace and scale of production dramatically accelerated. So pervasive was the understanding of animal power at that time that the ability of a machine to do work was expressed as horsepower. Likewise, people characterized the illumination given off by the new electric lights in terms of candlepower. Those were the most sensible metaphors available to a preindustrial culture.

As steam-driven machinery accelerated production, steam-driven conveyances accelerated the transportation of goods on rivers and rails. This, in turn, greatly stimulated market transactions, which supported the continued growth of cities, which placed increased demands on natural resources. The shift from an agriculturally dominated world of muscle power to an industrial world of factories and mechanical harvesters, steamboats and railroads, telegraphs and trolleys was based entirely on harnessing new sources of energy to run specialized machines designed to do work at an ever increasing pace (see Figure 1.1). All of this also led to new forms of concentrated wealth and political power.

The transition to a steam world led to changes in the city, farm, factory, and forest environments. Steam engines required fuel. At first, wood was the primary fuel, leading to extensive deforestation in the vicinity of factories, rail lines, and steamship routes. Eventually, coal came to play a dominant role as

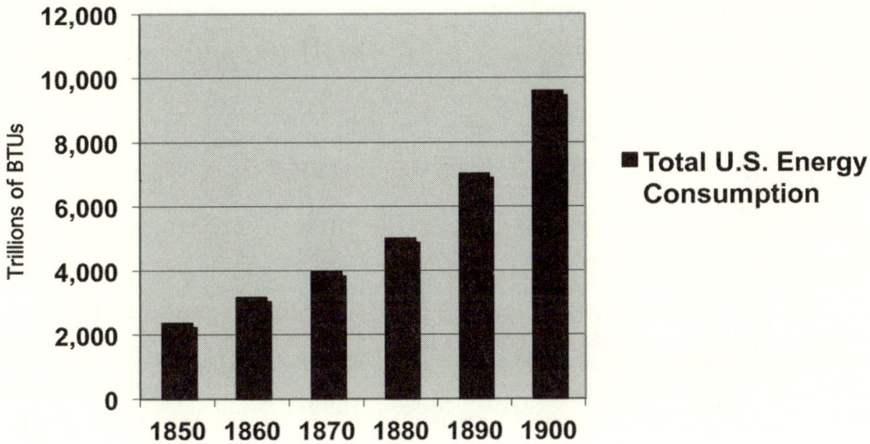

Figure 1.1. Total U.S. Energy Consumption, 1850–1900. Adapted from Energy Information Administration, *Annual Energy Review 2004*, 375. Graph by Paul Hirt.

the fuel of choice in the late nineteenth century. In fact, the rise of the coal industry directly correlates with the rise of steam-powered means of production and transportation. The shift to coal was stimulated by a number of factors. First, it was plentiful and relatively cheap. Fuelwood, on the other hand, was growing scarcer in many areas after decades of intensive cutting. Coal was a more efficient fuel than wood—it provided more heat energy per pound. Expanding urban populations and expanding industrial production needed fuel, and improved steamboat and railroad transportation allowed economical shipping of coal from coal-mining regions (see Figure 1.2).

The shift from muscle power to steam and the shift from wood to coal allowed greatly increased production, a growth in national wealth, improved transportation, and an expanding market. But the changes were not all desirable. The steam- and coal-driven world was busier, noisier, dirtier, more crowded, less healthy, and more impersonal. With industrialization, muscle power declined in value, and workers lost a great deal of control over their labor.[37] Self-sufficiency declined while consumption and waste increased. Environmental pollutants caused a myriad of health problems to workers and their families, who were often packed like sardines into cramped tenements and dank factory buildings. Crowd diseases like influenza and dysentery raged continuously; lung and skin diseases from air pollution (soot, particulates, and sulfur dioxide) were common; malnutrition plagued the urban working class; workplace injuries shortened lives and lessened quality of life; in sum, life could be "nasty, brutish, and short."[38]

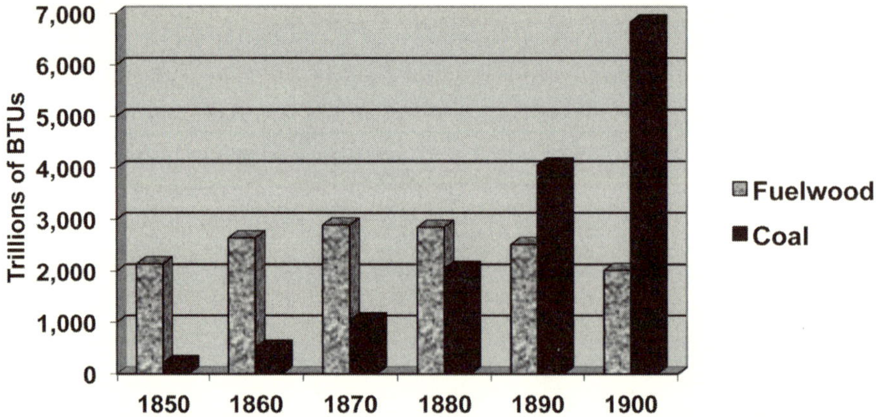

Figure 1.2. U.S. Energy Consumption: Fuelwood versus Coal, 1850–1900. Adapted from Energy Information Administration, *Annual Energy Review 2004*, 375. Graph by Paul Hirt.

Electricity, an extension of the steam revolution since dynamos were modified steam engines, offered a new source of power that visionaries hoped would mitigate many of the negative consequences of industrialization. Electric lights did not generate smoke, soot, and fire hazards in the home as candles and kerosene did. A dynamo could be housed in a building that was blocks away or miles away from the site of use, making electricity appear clean and noise-free to the end user. Electricity could move machines and motors with more precision and control than the belts and shafts of the steam-driven factory.[39] And, unlike a boiler for a steam engine, electricity could be completely turned off or on with the flick of a switch.

At the dawn of the electrical age, Americans were excited about the great changes accompanying industrialization but concerned about many of the negative consequences attending those changes. Electricity appeared to be a technology that could accentuate both the desirable and undesirable consequences of industrialization. Indeed, electric power did both and more.

CHAPTER 2

Power Incorporated

Cities, industry, and electric power expanded hand in hand in the late nineteenth century. Rapid urban growth created a lucrative market for organizations providing urban public services such as street lighting, water, natural gas, mass transportation, and telephone services. According to Christopher Armstrong and H. V. Nelles, "A whole new sector sprang into existence to supply these needs. Industries that had barely existed in the 1850s now absorbed huge quantities of capital, employed thousands of people, and serviced millions of customers. Entrepreneurs seized their opportunities."[1]

By the last decade of the nineteenth century, electric companies had emerged to dominate the field of municipal services, launching a new era in the production and consumption of electricity in the Northwest. Several important precedents were set in 1889, foreshadowing this new era: six of the Northwest's key cities (Victoria, Seattle, Tacoma, Portland, Spokane, and Boise) launched electric trolley systems for urban and interurban transportation that year. This not only altered the pattern of urbanization and city life but led to greatly expanded electrical use since trolleys consumed vastly more electricity than lighting systems. With rising demand for electricity, new utility companies—both privately owned and publicly owned—proliferated. One of the region's largest private utilities, Washington Water Power, incorporated in 1889. That same year, the Northwest's first municipally owned (nonprofit) electric system appeared in McMinnville, Oregon (with others to follow shortly). Competition between private and public utilities would lead to a major national debate over corporate power, government responsibilities, and citizen rights in the early to mid-twentieth century, but the conflicts remained muted in the Northwest until the second decade of the twentieth century.

The initial growth of electric systems in the late nineteenth century was often haphazard, but as profitability increased, the industry quickly became dominated by well-organized and well-financed private interests with extensive local political clout and strong ties to eastern electrical manufacturing companies and their associated financial syndicates. Investors acquired small isolated electric companies and merged them into large centralized corporations, often with investments in much more than electric light and power systems. In addition to owning the power plants and electric transmission lines used for lighting and for powering electric streetcars, the private companies that came to dominate the electric power industry often owned valuable city contracts and franchises; streetcar companies and their tracks and rights-of-way; extensive real estate holdings, including lands slated for urban development; water rights; dam sites; bridges; interurban rail lines; even water, gas, telephone, and telegraph facilities. Over time, many of these businesses would separate, be acquired by other companies or city governments, or be abandoned as they became unprofitable, but the trend toward big business and consolidated enterprises continued unabated.

This centralization and increasing capitalization of the electric power business started in the leading urban and industrial areas of the Northwest, where the best markets for electricity existed. The powerful and persistent corporations that came to dominate this industry in the Northwest in the twentieth century—Portland General Electric, Pacific Power and Light, Puget Sound Power and Light, BC Electric, West Kootenay Power and Light, Washington Water Power, and Idaho Power—all got their start in this manner in the late nineteenth century.

As the economy of the Northwest grew in the 1890s, so did electricity production and consumption. Even the devastating Panic of 1893 that bankrupted thousands of businesses caused just a brief slowdown in the otherwise steady growth of the electric power industry. In fact, the three-year depression following the Panic accelerated the consolidation of the electric industry as creditors took over smaller firms in financial trouble and consolidated them into larger, more efficient, less competitive, and more profitable regional electric corporations. By the dawn of the twentieth century, these corporations extended beyond the cities into the surrounding hinterlands, eliminating duplication and competition wherever possible, carving out virtual fiefdoms where they could securely provide this profitable new commodity with minimal competition.

A steady, remarkable increase in the use of electricity made all this growth and consolidation possible. Companies could hardly keep up with demand, especially as they demonstrated the benefits of incandescent lighting and electric motors, improved the reliability of electric service, and lowered rates. The constant need to build more power plants and transmission lines meant a recurring need for investment capital, and that usually meant taking on partners from Chicago, New York, Toronto, or London, who then exercised increasing control over the enterprises.

Building power plants also required either fuel or falling water to run the generators. The Northwest had plenty of wood to burn; indeed, many of the region's early dynamos burned sawmills' waste wood for fuel. Some low-grade coal was also available in the Cascade foothills near Tacoma, and more was importable on the transcontinental railroads completed in the 1880s and 1890s. Diesel, too, commonly fueled electric generators in the early decades of the industry. But in the well-watered Northwest, hydropower offered the most desirable and profitable source of electricity for those with access to adequate capital and engineering expertise. Companies scrambled to acquire the water rights and property necessary to control the best dam sites in order to secure a fuel-free and perpetual source of power. Most of the early hydropower plants were built at places like Spokane Falls in Washington (1888), Willamette Falls in Oregon (1889), Snoqualmie Falls in the Cascade Mountains east of Puget Sound (1898), and Swan Falls on the Snake River in Idaho (1901), where natural waterfalls provided gravity-assisted water pressure to spin a turbine. This led to local, regional, and national battles over who would control valuable hydropower sites (see chapter 4). The 1880s to the early 1900s thus marked an exciting period of precedent-setting developments and public debates that established trends for the remainder of the twentieth century.

ELECTRICITY AND URBAN MASS TRANSIT

Mass transportation in cities has long been a challenge. Before the twentieth century, most towns suffered with alternately dusty or muddy streets deeply rutted by horse-drawn carts. Early cities tended to locate along bodies of water that could be used for transportation. The advent of railroads offered an alternative to wagon wheels mired in ruts. Special cars rolled on iron tracks secured to wooden rails: a relatively mud-free, smooth ride, on which it was easy to move heavy loads. But noisy, polluting, coal-fired steam engines pulling railcars were inconvenient at best on a downtown thoroughfare. New Yorkers experimented with horse-drawn railcars as early as 1832. It worked reasonably well. By 1880, four hundred American cities had installed horse-drawn, rail-based streetcar systems. Electricity, once it was available, offered a much more attractive alternative to horses for moving streetcars. (The biggest pollution problem in New York City in the nineteenth century was horse manure!) Cable cars came first in the 1870s, some powered by steam engines and others by electricity. Then a superior technology, the electric-powered streetcar, or "trolley," made its appearance in the 1880s. In 1885, Baltimore installed a two-mile electric railway, the first American city to do so. Within a few years, electric streetcar systems had spread all over the country, with twenty urban electric trolley systems by 1888 and over a hundred three years later.[2]

Once again, the Pacific Northwest was in the vanguard of adopting new technology. As soon as efficient "traction" systems for powering electric trolleys could be mass-produced and shipped, northwesterners ordered them. (Traction referred to electrically driven transport.) As mentioned above, by 1889 cities in British Columbia, Washington, Oregon, and Idaho all had electric trolleys in place or under construction. In fact, Victoria's electric trolley was the third electrified city transit system in all of Canada.[3] This early and enthusiastic adoption of electrical technologies, along with the predominantly urban population of the far western states and provinces, resulted in the region being more electrified (per capita) than other regions of the United States and Canada, including the United States' industrialized Northeast and Midwest. For example, in the 1902, 1907, and 1912 electrical censuses published by the U.S. Bureau of the Census, the eleven western states as a whole consistently showed the most horsepower and kilowatt output per capita and the largest number of incandescent lights per capita. In fact, the American West had more than twice the national average of horsepower and incandescent lighting and more than three times the national average of kilowatt-hours per capita. Electric streetcars consumed a large proportion of that impressive electric load. Canada did not initiate an electrical census until 1917, but that year's census reveals a similar pattern. The Yukon and British Columbia had by far the highest per capita electrical horsepower in the dominion; British Columbia's was nearly three times the national average. The Canadian electrical census did not yet collect information on kilowatt-hours generated, per capita consumption of lighting, or electrical consumption by streetcars and railways.[4]

City boosters liked electric trolleys because they improved transportation and enhanced the modern image of the town. Businesses liked them because they stimulated commerce by tying business and residential areas together and concentrating traffic in downtown commercial districts. Town residents liked them because they eased shopping burdens, enhanced mobility, reduced the amount of horse manure in the streets, and even provided entertainment—people enjoyed riding the cars just for fun. To enhance ridership, trolley systems often built amusement parks on the edge of town where land was cheap to get paying customers out to the lesser used ends of the trolley lines.[5] Of course, electricity supply companies liked trolleys because they consumed so much power, they provided a daytime electric load demand to complement the nighttime demand from lighting, and the trolley companies' amusement parks provided weekend loads when the downtown business district was quiet.

Owners of electric generators looked constantly for opportunities to balance their loads, since the best profits were made by having as constant a demand as possible twenty-four hours a day, seven days a week. Fluctuating demands meant that expensive generators had to be idle or running at greatly reduced capacity during low-demand periods. Promoting trolleys was only the first of many strategies used by electric companies for balancing loads. Of course, the logic of this relationship quickly led to the merging of the two enterprises. Electric companies bought or launched their own trolley companies, and trolley companies bought or launched their own electric power companies.

Events in Portland around the turn of the century exemplify this close relationship between trolleys and power and light companies. During the 1880s the population of Portland more than doubled to over 40,000. Business boomed. People needed to get to work; goods required transport from producer to consumer. Use of electric trolleys rapidly expanded to fill that need. Ten transit companies formed in Portland from 1888 to 1891, each one receiving from the city government one or more franchises allowing it to lay track on city streets and provide public transit services. The franchises were legally binding contracts between the companies and the city, usually specifying the location of tracks, the level of service, passenger fares, and an annual payment to the city. The franchise protected the business from unwanted competition while securing for the city a public service that it did not have to build and manage itself. These franchises were supposed to ensure reliable service at a fair price while allowing for company profits. In the 1890s, however, profit-seeking exercised more control over these contractual arrangements than did public service. As Lewis Mumford later observed, capitalists of the late nineteenth century treated cities "not as a public institution, but a private commercial venture."[6]

The earliest electric trolley systems in Portland and elsewhere were speculative ventures, built piecemeal by their owners and urban authorities, often with little experience, and by numerous entrepreneurs hurrying to get in on the action. Rather than one coordinated public transit system, Portland got a proliferation of independent trolley lines owned by a dozen separate companies. Investors were generally after quick profits and often underestimated the costs of construction and maintenance. Shoddy construction and interrupted service plagued the early lines. Moreover, trolley companies built lines where they thought they could make the most profit, which was not always where city residents most needed them. The city council members who approved these franchises were sometimes owners or financial backers of the companies receiving the franchise or else political partisans who received rewards for voting as their benefactors wished. No independent board controlled transportation decisions for the city at this time. A kind of competitive free-for-all resulted, in which well-connected investors and pliant politicians determined the fate of Portland's public transit system. Some spectacular profits resulted but also many bankruptcies, delays, safety hazards, as well as inefficient development and inconsistent service.[7]

To solve some of these problems, Portland investors sought to consolidate the trolley lines so that they could be managed in a more coordinated and efficient manner, and, of course, so they might earn more profit. In 1891, Henry Failing and Henry Corbett, two of Portland's leading bankers, advanced a credit line to trolley company owner Charles Swigert to begin the consolidation. These were not run-of-the-mill businessmen. Corbett and Failing had owned highly successful businesses in Portland since the 1850s, and the two virtually ran the state Republican Party for decades. Both served in elected office: Failing was elected mayor of Portland twice; Corbett was a U.S. senator. Corbett even owned the *Oregonian* newspaper from 1872 to 1877, using it to champion his interests and flail his enemies. By the 1880s, Corbett and Failing owned the First National Bank of Portland, which financed many of the city's biggest projects. Swigert was a relative newcomer who arrived in Portland in the 1880s with family wealth and connections in San Francisco and in Bowling Green, Ohio. His entry into the electric trolley business in 1888 was immediate and ambitious. Within one year he owned several streetcar systems, an important bridge across the Willamette River, and the longest electrified trolley line on the West Coast. With the line of credit from Corbett and Failing's bank, Swigert merged five streetcar lines into the City and Suburban Railway Company, incorporated in 1891 and capitalized at $1 million. Failing became the new company's president and Swigert its general manager.[8]

These movers and shakers wanted a consolidated city government to facilitate franchise procurement for their sprawling enterprises and to promote coordinated regional development. After considerable lobbying, in 1891 the state legislature authorized the consolidation of east Portland, west Portland, Albina, and St. Johns. Many vested interests contributed to this political merger. Members of the state legislature who spearheaded the consolidation effort included Oregon Railway and Navigation attorney Joseph Simon, who was the Oregon state senate president that year; attorney P. L. Willis, who was Swigert's partner in the Electric Land Company and the state senator for the district that included St. John and Albina; and P. F. Morey, who was the president of Willamette Falls Electric Company and had just won a seat in the state legislature in order to promote city consolidation. With the greater urban area unified under one government, these men then successfully moved to unify electric power, transportation, and real estate development within the expanded city of Portland jurisdiction. The very next year, a well-connected and financially endowed group of investors formed the Portland Consolidated Street Railway Company, which combined four of the remaining street railways not yet under the control of the City and Suburban Railway Company. As Portland's historian Kimbark MacColl observes, "Economic consolidation was following political consolidation."[9]

The next big merger occurred in Portland's electric power industry in 1892, involving many of the same men. Morey (of Willamette Electric), Failing (of First National Bank), Swigert (of City and Suburban Railway Company), and others borrowed money from First National, General Electric, and Boston's Old Colony Trust Company to form Portland General Electric (PGE), which would dominate electric light and power service for the greater metropolitan region for the next century. PGE historian Craig Wollner observes that the new company "displayed a glittering array of the city's leading business lights who had been enticed into making major investments." Many of them were active in Portland politics and heavily invested in real estate. PGE had an initial capitalization of $4.25 million at a time when the average industrial worker in the United States earned a little over eight dollars a week.[10]

The stage was now set for the final consolidation. In 1906–1907, all the remaining Portland street railway companies combined together with PGE to form one giant monopoly called Portland Railway, Light and Power. It controlled nineteen distinct companies, twenty-eight rail lines, six power plants, and forty-three city franchises. It owned 431 electric trolley cars that moved sixteen million passengers in its first year of operation. Boston investors financed most of its $15 million capitalization and installed their own general manager, Benage Josselyn, to ensure the flow of profits to the eastern stockholders. He fought consistently with the city over franchises, fares, and service obligations during his six-year tenure.[11]

In response to these momentous changes, Portland's mayor and council sought to tighten up franchise requirements, raise service standards, and coordinate larger urban planning goals with trolley system development. After all, trolleys served people, and people needed a suite of city services such as water, sewer, roads, fire protection, and law enforcement. Coordinating at least some of these services with mass transit made sense. However, many city government leaders also invested in trolley and electric companies, which meant that the fox was guarding the henhouse. Portland government and Portland's business elite were so much the same at this time that the larger public interest often took a backseat to pecuniary interests. These men, however, believed that what was good for Portland National Bank or Portland Railway, Light and Power was good for Portland.

City leaders cared about the direction of streetcar development because trolleys profoundly influenced urban growth patterns, as all major transportation systems have done in urban history. Like the railroads of the mid-nineteenth century and the interstate highways of the mid-twentieth century, trolleys drew home-seekers and businesses along transit lines like ants to honey. Wherever the trolley lines extended, so went the future development of the town. Anyone who could afford to locate somewhere within convenient walking distance to a trolley stop did so. Thus, new home and business development tended to concentrate in strips along transportation corridors. Property values rose in the vicinity of trolley lines, allowing early investors with inside information the opportunity to make a killing buying low and selling high. Not surprisingly, many land speculators got involved in promoting trolleys, or at least investing large amounts of money in lobbying for extending trolley lines through their property. In Portland, these connections between real estate developers and trolley companies were ubiquitous. As historian Kenneth Jackson points out in his book *Crabgrass Frontier*, many owners of streetcar companies were more interested in land speculation than in "nickels in the fare box."[12]

Every city followed its own path in the development of electric power and electric trolleys, but the patterns revealed in Portland were common in urban areas throughout the region. In the Puget Sound area, numerous light, power, and traction companies formed after 1889 in Tacoma, Seattle, Bellingham, and other towns. There were at least twelve in Seattle alone by 1893, many of them launched by people more interested in real estate speculation than in running an electric power and transit company. Most of these companies were crippled or bankrupted by the Panic of 1893, which ushered in a severe three-year economic depression. This loss for some became an opportunity for others. Sidney "S. Z." Mitchell, the regional representative for Edison General Electric, saw that opportunity and brought on board the operational expertise and financial gravitas of the Stone & Webster Company of Boston, one of the nation's largest electrical engineering, management, and finance corporations. In 1898, owner Charles Stone came to Seattle at Mitchell's request and laid the plans that resulted two years later in the formation of the Seattle Electric Company. From 1900 to 1902 this eastern-financed corporation acquired virtually all the remaining electric light, power, and traction businesses in Seattle and their affiliated enterprises. Its president, handpicked by Mitchell and Stone, was Jacob Furth, a wealthy man who came to Seattle in 1882 from San Francisco to open a bank. His fortunes grew, and when Stone approached him, he was president of Seattle National Bank, one of the city's key financial institutions heavily invested in the ailing electric light and trolley companies. Stone & Webster continued to acquire electric and streetcar companies south to Tacoma and north to Bellingham from 1902 to 1904, consolidating them under Furth's management and connecting them with electrified interurban rail lines. In 1912 this sprawling collection of companies reincorporated as Puget Sound Traction, Light, and Power—the direct predecessor of the giant utility Puget Sound Power & Light that dominated electric service in western Washington for the rest of the twentieth century.[13]

Locally owned Washington Water Power (WWP), founded in 1889, started selling electricity to trolley companies in 1891. It averted a takeover in the preceding two years by acquiring or buying a controlling interest in the competing companies that S. Z. Mitchell had launched with Edison Electric Company financing. But the depression of 1893–1896 foiled its effort to remain independent. During the depression, three of Spokane's banks failed, eliminating WWP's primary source of capital. The company defaulted in 1895. One of its largest bondholders was a financier from New York named William White, who took over the bankrupt company and reorganized it in 1899 into a large, unified enterprise combining numerous formerly independent lighting, power, and streetcar companies. It virtually cornered the Spokane mass transit market by 1900, although it abandoned those investments a few decades later after streetcars became unprofitable. The newly reorganized WWP continued to absorb neighboring systems in the inland Northwest in the early twentieth century and eventually became part of the Electric Bond and Share holding company empire affiliated with General Electric.[14]

The situation was similar across the international border. In Victoria, a group of wealthy and influential men founded the National Electric Tramway and Lighting Company in 1889. Securing $250,000 in financing and a franchise from the city of Victoria, they built British Columbia's first electric streetcar line. Attendees at the ceremonial opening of the trolley line included the lieutenant governor and the premier of the province, members of the provincial cabinet and legislature, prominent judges, the mayor and city council of Victoria, and a wide array of the city's business leaders. Reporting on the pomp and circumstance the next day, the *British Colonist* proclaimed, "Yesterday will be looked upon as an era in the History of Victoria. The starting of street cars shows that the City has arrived at an important stage in the path of progress." This same editorial explained the material significance of the streetcar to citizens of Victoria and hinted at the link between streetcars and land development:

> The establishment of the street car service will at once very considerably increase the area of the city, and will make available many lovely spots which were considered a few months ago too distant from the centre of business to be valuable for the sites of residences, and it will bring the country near to the poor man, and the poor man's children.[15]

Receipts from streetcar service in the first full year of operation totaled $78,000.[16] Clearly this was an attractive arena for investors with big dreams. Just as clearly, while the "poor man" benefited from having a mass transit system in Victoria, a portion of his and his family's trolley fares further enriched the financial fortunes of a privileged few. So long as service remained good and fares affordable, these class tensions remained muted. But complaints about windfall profits extracted from workers' meager salaries were never far from the surface. Those complaints grew in proportion to the growing wealth and power of electric streetcar corporations.

Hot on the heels of Victoria, investors in Vancouver and nearby New Westminster launched three electric streetcar companies in 1890, all incorporated by acts of the provincial legislature, with a combined capitalization of $1 million. In New Westminster, the two newly incorporated companies had overlapping directors, and most of the company directors had either made fortunes in the Fraser River gold rush or had the requisite political connections to help secure a city franchise. The two New Westminster companies merged the following year.[17] Unfortunately for these businesses, Canada slid into a nationwide depression starting in 1891, two years before a similar fate hit the United States. Economic growth, population growth, and the number of riders failed to increase as expected, and the companies' overbuilt trolley lines hemorrhaged cash. The streetcar companies in Vancouver and New Westminster struggled along for a while and then went bankrupt, only to be resurrected by investors with deeper pockets in England, who bought the assets of the trolley and lighting companies at bargain prices. The struggling Vancouver Electric Railway and Light Company, for example, offered to sell all its assets to the city of Vancouver in 1893 for the assessed fair market value of $477,000. The city declined. After bankruptcy and service cuts, the company again offered to sell to the city in 1894 for $380,000. Voters rejected the offer. Finally, later that year, London investors stepped in with a newly incorporated company and bought the Vancouver Electric Railway and Light holdings for a mere $290,000. The same investors, represented locally by businessman Frank Barnard, bought the similarly bankrupt New Westminster and Vancouver Tramway Company in 1894. Barnard became general manager of the consolidated businesses.[18]

Two years later, London financier Robert Horne-Payne and several partners, including Barnard, formed Consolidated Railway and Light to accomplish the ultimate merger: purchasing all of the electric light, power, and railway companies of Vancouver, New Westminster, and Victoria and combining them into one centralized regional monopoly—a merger completed and legitimated by an act of the provincial legislature in 1896. Victoria still owned its street-lighting system, which it had taken over from McMicking's lighting company in 1885, but all the other light, power, and railway systems were acquired by Horne-Payne, Barnard, and their partners, who then formed the British Columbia Electric Railway Company (BCER) in 1897. Registered in London, England, BCER controlled local assets worth $2.5 million. Barnard assumed the position of managing director, but only for about a year, as the real source of capital and control was Robert Horne-Payne and his board of directors in London. The London directors did not want an independent-minded company manager in Vancouver, so they replaced Barnard in 1898 with the somewhat more deferential Johannes Buntzen. The company thrived, even under absentee ownership, and became the dominant electric utility of the province until 1961.[19] In this study I usually refer to the company as BC Electric, which became its common moniker after the decline of the electric streetcar and railway business.

While the electric streetcar industry in British Columbia did not start with as many separate trolley companies as American cities did, the end result was the same: local entrepreneurs built local businesses that eventually attracted outside capital to finance expansion and consolidations creating regional electric power and transportation monopolies that were endorsed and rather weakly regulated by local and provincial governments.

Historians of the electric utilities portray the events of the 1890s in quite different terms depending on their sympathies. Histories of the major for-profit utilities, like PGE, WWP, Puget Sound Power & Light, and BC Electric habitually portray company owners and managers in a positive, altruistic light. The authors of a history of Puget Sound Power & Light call the acquisitions by S. Z. Mitchell and Stone & Webster a "rescue operation of major proportions." Maiden portrays English investor Horne-Payne, who helped Barnard build his electric empire during the 1890s depression, as a "genius" in finance who was "immensely stimulated" by the opportunities in British Columbia and determined "to throw all the force of his character into the development of Canada." Wollner's history of PGE refers to the financial consolidators as "visionaries" engaged in "a heroic gamble . . . requiring courage, commitment, and creativity. Men steeped in the characteristic enthusiasm and values of the middle class of the late nineteenth century, they led a city into the twentieth."[20] Perhaps these business moguls embraced certain middle-class values, but it would no doubt surprise the typical small-business owner of the time to see the oligarchs portrayed as their peers. Other historians whose sympathies lie more with the cities and consumers served by these commercial enterprises pay more attention to the troubling concentrations of wealth and political power controlled by these financial movers and shakers and to the shortcomings in services that resulted from concerted efforts to wring more profits from these enterprises.[21]

Besides inner city trolley systems, many companies built electric interurban rail lines connecting two or more cities and powered by electric lines. Regular railroads using steam engines avoided building short runs because starting and stopping a steam engine consumed a great deal of time and energy—and therefore money. For that reason, railroad companies usually built only one train station in each town and maintained long distances between stations. Electric railways had no such limitation. Electric trains could start and stop at will and as often as desired with relatively little loss of energy or time. Consequently, interurban electric railcars began filling in the service gaps left by the main steam-powered railroads built decades earlier. By 1904, a thirty-eight-mile electric interurban system called the Oregon Water Power line connected Portland with half a dozen communities up the Willamette Valley.[22] By the second decade of the twentieth century, lumber businesses outside Portland were the biggest customers of the interurban system and used the electric railway to send logs to mills and markets.[23]

Interurban lines also connected Spokane with surrounding towns in the "Inland Empire" and with Lake Coeur d'Alene and the mining districts in north Idaho.[24] An interurban line connected New Westminster with Vancouver, as mentioned above. Electric interurbans flourished in the Puget Sound region, connecting the many growing port towns there. All of these connections had the expected effect of stimulating commerce and communication between cities and between urban and surrounding rural areas. Wallis Nash, in a 1905 Oregon booster publication, clearly expresses the hopeful connection among water power, electric rail lines, and economic development: after assuring his readers that Oregon is "fertile [and] full of resources," he explains, "there is no more favorable field for the electric road than in Oregon, and especially western Oregon. Water power is running by in almost every river bed and creek bottom. The main arteries of communication already exist, for which the electric lines will supply feeders and develop traffic."[25]

Running electric lines for these railways also provided an opportunity to extend other forms of electrical service to people in the vicinity. The trolleys required high-voltage current, which was excellent for long-distance transmission of electricity, so rail companies and power companies often cooperated or even combined into a single company to make these trolley wires both power the trains and serve as feeder lines to bring electricity to nearby homes and businesses. The rail company owning the Willamette Valley interurban, for example, built and owned its electric lines but bought power from PGE. PGE in turn contracted with the rail company to use the trolley wires to transmit electricity to communities served by the interurban.[26] Municipalities served by both companies participated in approving the locations of the trolley routes and electric lines and awarded service franchises to these and other companies. Eventually, the Portland Railway, Light and Power conglomerate merged all the power and traction companies into one. Still later, after the electric streetcar business declined, the electric power business separated into Portland Electric Power Company (PEPCO) in 1924.

As many of the examples above reveal, the rise of electric traction in mass transit forged such a close relationship between electric power companies and railway companies that the two often merged. From the 1890s until at least the 1930s, it was difficult to distinguish between the two industries. This trend extended nationwide. The U.S. Census of Electrical Industries noted 169 electric railway companies in 1912 that owned and earned significant revenue from electric light and power companies. Between 1902 and 1912, income to these companies from selling power to customers outside the railway business jumped fivefold, from $6.2 million to $31 million.[27] With the decline of electric trolleys during and after the 1920s, however, this marriage became strained; eventually the two industries separated as the increasingly profitable electric power business divested itself of the financially ailing streetcar business.[28]

SOLVING PROBLEMS

Because a variety of entrepreneurs and governments worked in a sometimes haphazard or competitive manner to develop electric systems in these early years, a plethora of social, economic, and infrastructural problems arose.

When the electricity business was wide open and unregulated, competition could be fierce, multiple companies often fought each other for franchises from cities, cities often took kickbacks to award franchises, multiple utilities sometimes strung electric wires on the same streets and competed for the same customers. Waste and inefficiency ran rampant. Moreover, cutthroat competition often led companies to cut corners, leading to unreliable service, shoddy construction, and dangerous facilities. As electric lighting and trolleys became an increasingly significant component of urban life, public complaints and demands for better service increased along with calls for tighter municipal regulation and even for municipal ownership of the trolleys.[29]

To improve service and get more rational control over all the interconnected components (electrical generation, lighting, wiring, and trolleys), both city governments and the electric power and traction companies attempted to establish more systematic and comprehensive organization of these services. There were several means available for improving organization: individual businesses could consolidate into unified regional enterprises capable of planning and administering integrated systems over large areas; local governments could develop urban growth and public services plans for their cities and then award exclusive franchises to companies that agreed to follow the plans; or local governments could get into the business of owning and managing their own electric generation, lighting, and transit systems and bypass the private sector. Each of these options was pursued simultaneously all across the country, including the Northwest. No single solution ever entirely dominated. From the 1880s right up to the present, America's electric supply industry has been a complicated mix of private enterprise, public regulation, and public ownership.

Each kind of ownership posed particular opportunities and pitfalls. Private businesses could be well organized and relatively effective in achieving specific objectives in pursuit of profits. They tended to make quicker decisions, adapt to change more rapidly, and incorporate new innovations more readily than government agencies. Of course, this was only true of businesses with skilled and honest managers and employees, and only if the business was successfully earning profits that supported maintenance and improvements. On the other hand, unregulated private enterprises primarily pursued their own self-interest, which could often be in conflict with the public interest. For example, unless forced by government regulators to do otherwise, profit-seeking electric companies provided electric service only to areas with enough customers of sufficient financial means to generate an acceptable profit. As a result, most of the working class, the poor, and people in rural areas never received electrical service until governments intervened. Private business was also highly vulnerable to market fluctuations, which often led to unreliable service or even no service in the event of a bankruptcy. Finally, private companies 100 years ago were largely immune to public oversight and public pressure. They could charge whatever the market would bear and serve only those they wanted to serve.

Government cooperation with the private sector through the awarding of municipal franchises offered a different set of opportunities and problems. In exchange for exclusive service agreements (essentially a legal monopoly), a city government through contract stipulations could direct where utility and transit lines would run—and thus who received services. Larger urban design and public welfare objectives could be pursued in this way. Companies generally liked the arrangement because it eliminated outside competitors and essentially guaranteed them steady customers in a protected environment, even if they chafed against the franchise stipulations and trolley fare and electric rate controls. Companies often lost some potential for higher profits but also escaped many business risks in the process. City governments generally liked the arrangement because they could achieve some of their municipal infrastructure and service goals without having to involve themselves in the difficulties of raising capital, constructing, and managing electric utility and transit systems.

On the other hand, this kind of public-private partnership could be problematic. While private enterprises needed to earn profits, city council members needed to earn votes. Voters' desires and shareholders' desires could easily be incompatible. Sometimes the partnership worked well, sometimes the partners fought each other continuously.[30] The potential for corruption posed another problem. With lucrative franchises to dole out, city leaders were prime targets for bribery. Gilded Age politics were rife with corruption at the city, state, and federal levels, and franchises in particular bred kickbacks and vote-buying on a large scale.[31] Cities sometimes issued franchises with no restrictions, no time limit, and no payments for privileges. Over time, however, reformers inserted more public interest clauses, term limits for franchises, service specifications, options for the city to purchase the system, and annual payments. But even these kinds of franchise stipulations could be difficult to enforce. As one analyst in a 1912 study explained,

> The administration and enforcement of franchises, in American cities, has been left to the companies themselves. Save in the few states that have recently adopted public service commissions, no public official or body has had ample power to see that franchises are administered in the interests of the public and the consumer.[32]

Figure 2.1. "Franchise Hog." Source: *Oregonian* (1906). Reprinted from E. Kimbark MacColl with Harry H. Stein, *Merchants, Money, and Power: The Portland Establishment, 1843–1913* (Portland, OR: The Georgian Press, 1988), 426.

Railroads and land speculation laid the foundation for this kind of corruption in the mid-nineteenth century until it became so rampant that it led to forty years of concerted nationwide efforts to clean up government and business.[33]

A utility or transit system wholly owned and managed by the city avoided the potential conflicts involved in the public-private partnership. There was one owner with the single goal of providing a valuable service to the community. Corruption was still a danger, however, especially during the construction phases. Publicly owned systems usually provided service at a lower

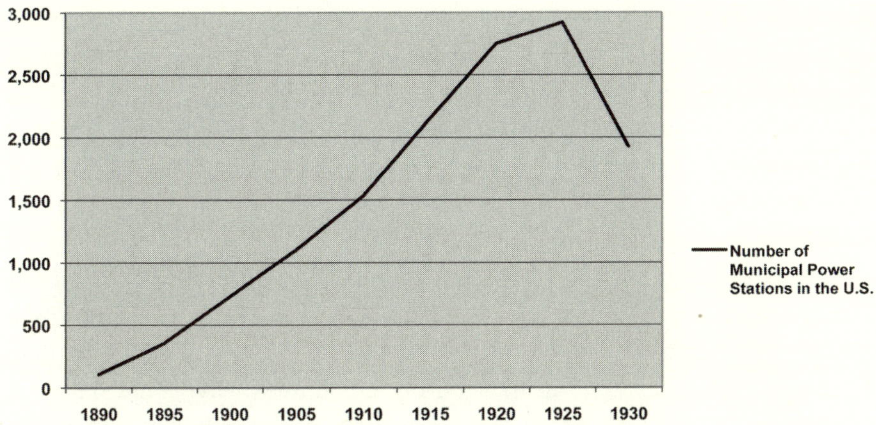

Figure 2.2. Municipal Power Stations in the United States. Data source: David Schap, *Municipal Ownership in the Electric Utility Industry—A Centennial View* (New York: Praeger, 1986), 9–11, 21–28. Graph by Paul Hirt.

price since they did not need to earn profits, they could get financing at lower interest rates, and they could draw on the tax base for expansion or repairs when needed. If the city government was honest and efficient, people usually liked these municipal utilities because service was reliable and cheap. Most customers believed they were getting power or transportation at cost rather than being gouged for profits by big businesses. However, if city government was dishonest or inept, or if the city lacked the financial resources needed for expansion and maintenance, these systems could be poorly designed, unreliable, and in disrepair; in short, they could be just as bad as the worst privately owned systems. Finally, while democracy is good politics, it is not always the most effective tool for running a complicated technological system. Politicians worrying about the next election did not always provide adequate support for city-owned utilities or appoint the ablest administrators. The regular turnover of elected officials sometimes stymied the long-term planning, technical expertise, and consistent follow-through necessary to keep a public utility running smoothly.

All three of these institutional arrangements appeared in one place or another across both nations in the late nineteenth and early twentieth centuries. From the 1890s to the present, however, private utilities under government regulation have dominated the electric utility industry in the United States, with publicly owned systems providing significant competition in the Northwest and the Southeast.[34] Municipally owned systems proliferated from the 1890s to the early 1920s, but big private utilities serving much larger markets began buying up city systems in a sweep of consolidations after the turn of the century, so that after 1923 the total number of municipal systems went into a sharp decline all over the country—except the Northwest.[35]

A variety of factors explain why cities established or did not establish municipal electric systems. Much of it was a matter of timing, circumstance, capability, and sentiments in the community itself. The two most common motivations for creating a municipal system were intractable problems with a privately owned supplier of electricity (poor service, unreasonable rates, disrepair, refusal to expand services, franchise violations, etc.) and the complete absence of an electrical system and no investors willing to install one. Both situations were common, but that did not guarantee a city would set up its own electric power utility. Until at least 1912, U.S. courts had ruled that cities could not own public utilities unless state law explicitly authorized it. This stymied many hopeful efforts. Even those cities granted this authority by the state often were subjected to onerous requirements that inhibited municipal ownership. (Often, these requirements were put in place at the behest of the private power interests.) Even if a city surmounted all these obstacles, financing posed another hurdle. Voters sometimes balked at approving the bonds (indebtedness) necessary to construct an electric power system or to buy out a privately owned system to improve it. Finally, the private utilities viewed municipal systems in their market areas as dangerous competition that threatened their expansion and profits. Consequently, they aggressively fought "municipalization." Despite all these difficulties and complications, many municipal systems were launched and many persisted. At their peak in 1923, there were 3,083 municipally owned power stations in the United States.[36]

Municipal ownership flourished in the U.S. Northwest. Between 1889 and 1905, at least nine towns in Washington and Oregon established municipal electric systems, including McMinnville, Milton, and Forest Grove in Oregon, and Ellensburg, Waterville, Port Angeles, Tacoma, Centralia, and Seattle in Washington.[37] As mentioned earlier, Victoria retained its own street lighting system, too, despite all the private consolidations in the 1890s and early 1900s. Nelson, British Columbia, launched a municipal lighting system in 1898 and within a couple of years found itself at war with the private West Kootenay Power and Light Company, which was expanding its market into Nelson. Nelson also acquired its electric streetcar system in 1909 after the private company running it went bankrupt.[38] Tacoma bought its water and power system from a private company in 1893, one of the first cities in the nation to do so, although the privately owned Puget Sound Power and Light served parts of Tacoma, too.[39] Seattle got into the municipal ownership game in 1902 when voters approved a $500,000 bond to construct a hydropower plant at Cedar Lake, which delivered its first electricity to Seattle in 1905. The city council belatedly created Seattle City Light in 1910—a city-owned electric company that continues operation to this day.[40]

The story of Tacoma's launch of a municipal system is typical. A private company owned by the Northern Pacific Railway had received a franchise from the city in 1884 to supply Tacoma with water and electricity. Tacoma was the Puget Sound terminus of the Northern Pacific Railway. Absentee ownership of the water and power system by a St. Paul–based railway corporation led to poor service and high prices, as well as recurrent health problems (e.g., typhoid) associated with polluted water delivered by the company. In 1893, voters approved a bond to purchase the system for $1.75 million. After the purchase, Tacoma inspectors discovered that the water delivery system was decrepit and the water source polluted, as well as that a dam had failed and the electric system was outdated and in disrepair. The city sued the company but was able to recover only a small amount of its exorbitant investment. Despite these problems, the new city-owned utility reduced electric rates by 25 percent that year, another 25 percent the following year (1894), and a whopping 75 percent in 1903. Tacoma quickly built a water and power system that was the envy of the region. It remains a publicly owned, state-of-the-art system today.[41]

In Idaho, municipal electric systems got a later but equally enthusiastic start. The town of Idaho Falls established a municipal system in 1901; then a cascade of additional municipal systems in southern Idaho followed the 1910 decision of the U.S. Bureau of Reclamation (BoR) to sell low-cost power from its Minidoka Dam on the Snake River to farm towns in the vicinity of the dam. Minidoka was the Northwest's first federal hydroelectric dam. The towns of Burley, Declo, Heyburn, and Minidoka all established municipal systems buying surplus power from it. The connection between Minidoka Dam and the establishment of these Idaho municipal utilities was not mere coincidence. A 1906 act of Congress that expanded the authority of the Bureau of Reclamation had established the principle that hydropower from federal dams should be sold first and at preferential rates to publicly owned utilities.[42]

As the examples above show, municipal electric systems were sometimes created from scratch and sometimes acquired from private owners. In the United States as a whole, between 1890 and 1900 about one in seven municipal electric systems got started by acquiring a pre-existing private company. Between 1901 and 1920 that ratio doubled; almost one of every three new municipal electric systems got started by acquiring private companies. This stimulated much of the heated rhetoric coming from the private utility industry about the threat of municipal ownership and creeping socialism. But the reverse was also true: private electric companies often bought municipal electric systems. Between 1886 and 1900 only 21 municipal systems changed to private ownership, but in the next two decades, 329 municipal systems—about one of every nine municipal systems—switched to private ownership. These numbers obviously favored municipal ownership in these early years, but after 1923 the ratios reversed, resulting in a steady decline in the total number of U.S. municipal electric systems (see chapter 4).

The proliferation of municipally owned electric systems in the late nineteenth and early twentieth centuries masks the real degree to which private businesses dominated the industry. Municipal power plants were generally small, and nearly 90 percent of them were in towns with populations under 5,000. Private for-profit companies generated the vast majority of electricity in the United States, in all regions of the country. In 1912, for example, 30 percent of all electrical stations were owned by municipalities, yet they produced less than 5 percent of all electrical power.[43] Moreover, private power ownership concentrated increasingly in the hands of large corporations as opposed to local independent businesses. The U.S. Census of Electrical Industries specifically noted this trend between 1902 and 1912, observing that large corporations controlled three quarters of the electricity produced by commercial establishments in 1912. Even more remarkable, these incorporated enterprises "reported 98.4 per cent of the total income, 98.1 per cent of the horse power . . . and 99.1 per cent of the kilowatt-hour output shown for all [privately owned] stations."[44] In Canada, for the closest comparable year of 1917, the difference was not quite as stark. Just over a third of central electric stations (36 percent) were municipal-owned; they collected a third of total electric revenue and controlled just over a fifth (22 percent) of electric capacity. The Canada Census of Industry did not comment on the percentage of power controlled by large corporations.[45]

Predictably, the private utilities and municipally owned utilities fought each other for customers, for service areas, and for the moral high ground. Since population concentrations made the best markets, profit-seeking companies often neglected small towns but fought strenuously to keep municipal systems out of the larger cities.[46] Although municipally owned systems have never supplied more than about 5 percent of the total electricity consumed in the United States, they have generated controversy far out of proportion to their actual power-supply role. Advocates of nonprofit public ownership attacked privately owned for-profit utilities as self-serving corporations overcharging captive customers to line the pockets of bankers on Wall Street. While often exaggerated, the rhetoric hit home, since many private utilities were controlled by financial holding companies, often associated with the J. P. Morgan syndicate after the utility mergers of the 1890s.[47] Private utilities replied that public ownership was anti-American socialism and argued that their own rates were reasonable and fair. When these two systems struggled over control of the most lucrative markets, the debate often reached a fever pitch.

In 1908, municipal administration scholar John Fairlie characterized these opposing positions with remarkable clarity:

> In the constantly accelerating public discussion on the question of municipal ownership in [the United States], the two radically opposing views have been becoming more and more predominant. On the one hand is the propaganda for immediate and universal municipalization, as the next great step in social and political reform. On the other side is the conservative or reactionary attitude that opposes any movement in the direction of collectivism.
>
> In spite of their antagonism, both of these views have this in common: they assume that the question is one which can be dogmatically answered either in the affirmative or the negative; that the problems can be settled by a universal rule expressed in a brief formula. There is also a tendency on the part of the supporters of each dogma to assume that the view they adopt has an ethical basis; and their views therefore are expressed in the categorical imperative; with little or no qualifications.[48]

Fairlie understood that municipal electric service could be adequately provided under either system, but this thoughtful middle ground did not make headlines or sell newspapers.

CHALLENGING PRIVATE POWER

This initial contest between public ownership and private ownership of utilities drew upon several contemporaneous events: the rise of national corporations, popular fear of monopoly power, highly erratic economic cycles, the growth of government regulation of big business, and the promotion of public ownership of public service corporations.[49] Contemporaneous with the rise of the electric power and electrified transit in the 1880s and 1890s in the United States was the rise of the People's Party (the "Populists"), one of the more successful third parties in American political history. The Populists railed against the wealth and political influence of industrial corporations during the Gilded Age and sought to displace pandering, corrupt politicians and return political power "to the people." Their fiery rhetoric inspired both loyalty and fear. Populists in many ways were progressive for their time, seeking to address problems associated with the rise of a new urban industrial order. At the same time, the party could be reactionary, resisting the changes confronting the nation and seeking naive solutions. One component of the People's Party platform included national ownership of the railroad, telegraph, and postal services. Populists felt that the transportation and communications systems linking the nation were too important to the welfare of all Americans to tolerate their control by profit-seeking corporations. The more electricity penetrated daily life and economic activities, the more reformers advocated public ownership of that service, too.

Besides the movement to establish municipally owned electric systems, Canada and some European nations were moving toward provincial or national ownership of electric systems at this time, providing a model for reformers in America.[50] In 1902, Canadians in the province of Ontario began advocating for a provincial-owned electric company to be ensconced at Niagara Falls to supplement or replace the two slow-moving private companies that had already been granted franchises to utilize the falls. Americans to the south had jump-started the economy of Buffalo, New York, through rapid development of hydropower on the U.S. side of the falls, and many Canadians wanted a similar result and were tired of waiting for the private companies to provide it. But more than a practical concern for development, advocates of public power in Ontario objected on ideological grounds to the whole for-profit corporate-dominated electric system developing in the province. In 1902 and 1903 in Berlin, Ontario, there were several well-publicized meetings of public power advocates from cities within transmission distance of Niagara. Ontario newspapers blessed the proceedings, endorsing what historian H. V. Nelles called a "Populist revolt." The Toronto *World* proclaimed,

> Capitalists get together and create monopolies. The municipal representatives at Berlin yesterday gave practical effect to the belief that the people should get together and create monopolies. . . . Corporate oppression has been possible largely because of the diffusion of the strength of the oppressed. The Berlin convention is a sign of an awakening.[51]

An Ontario Hydro-Electric Power Commission was created in 1903 with representatives from interested municipalities, resulting by 1910 in a public power authority delivering electricity to Berlin, Hamilton, and London. By 1913, Ontario Hydro delivered power at cost to fifty-six communities and had bought out all the private power companies in the region except for those in the larger urban centers of Toronto, Hamilton, London, and Ottawa.[52]

Other forces besides the Populists pushed for public ownership. Inspired by the utopian futurist novel *Looking Backward: 2000–1887*, by Edward Bellamy, "Nationalist Clubs" sprouted up all over America in the 1890s and early 1900s, promoting, among other things, public ownership of electric utilities.[53] Socialist philosophers, essentially calling for the elimination of all private concentrations of wealth, also gained popularity in Europe in the late nineteenth century and made some inroads on the political landscape of the United States and Canada at that time. But other than municipally owned water, sewer, electric, and trolley systems and the spectacular success of Ontario Hydro, advocates of public ownership saw few gains in these decades. Principally, their successes consisted of indirectly advancing a compromise arrangement whereby the private sector would be allowed to flourish but under increasing public oversight and regulation. The municipal franchise arrangement was a prime example of this at the local level.[54]

Many early forays into regulatory activism were generally popular with voters but were nevertheless opposed by a substantial segment of the business community, their political allies, and many voters. Thus, these regulatory efforts were compromised in their design, weak in financial resources and staffing, and erratic in implementation. But the long-term trend was toward stronger regulation of business practices when those practices had an inordinate influence on local and regional economies and daily life. By the turn of the century, the electric power industry came under increasing scrutiny by reformers, much as the railroads had a couple of decades earlier. The same arguments would be used again: that the electric companies, like the railroads, were becoming interstate monopolies controlling a service so crucial to the public welfare that they should either be publicly owned or closely regulated to ensure financial stability and good public service.[55]

EARLY HYDROELECTRIC POWER DEVELOPMENT

In addition to the debate over municipal franchises and municipal ownership of utilities, another controversy arose over the question of who could use and alter rivers, for what purposes, and under what restrictions. Rivers had long been the nation's primary arteries of inland transportation and were therefore considered a kind of public property—a legal tradition dating back to English common law. Damming rivers was always controversial. When a private party dammed a river for personal profit to the detriment of others, the protests grew particularly vigorous. By the turn of the century, building or retrofitting dams with hydroelectric generators had become a new and popular means for generating power, especially in the Pacific Northwest. Nationwide, hydropower generation jumped 464 percent in the ten years between 1902 and 1912. By 1907 hydropower supplied fully one third of the electrical output of central electric stations—an impressive launch for a relatively new energy technology. Moreover, this power source was particularly capital intensive and attractive to large corporations. Two thirds of the overall increase in hydropower between 1907 and 1912 came from very large turbines (over 5,000 horsepower, the biggest units available at the time). Commercial power companies (as opposed to municipals) owned 96 percent of the hydropower generators, with more than 2,000-horsepower capacity, and produced 90 percent of all water power, revealing the relative weakness of municipal systems in the U.S. sector of this industry. California and the Northwest had more than their fair share of these large hydropower plants, about one fifth of the national total, as reported in the 1912 electrical census.[56]

This rapid expansion of hydropower propelled the debate over dams to new heights, focusing on how the public's interest in all the values provided by rivers would be protected. The contest for hydropower sites escalated as the technology for long-distance transmission of electric power improved. Historians often point to the tapping of Niagara Falls by a company affiliated with J. P. Morgan as the moment at which the electric industry perceived long-distance transmission of hydropower as an attractive alternative to steam generators.[57] While the Niagara project started construction in 1889, it did not come online until 1896 and sent power only twenty miles to Buffalo that year. The Willamette Falls project, mentioned earlier, came online in 1889 and sent power fourteen miles to Portland seven years before any Niagara power reached Buffalo. The Niagara project involved much more electricity and some famous corporations, including Westinghouse and GE, so it gained more notoriety due to its scale and associations, as well as being situated in the state of New York, but the precedent for long-distance transmission of hydropower was really set in the Northwest. In any case, by the turn of the century, hydropower was on the wish list of any utility with a good dam site in its vicinity. In the Pacific Northwest, this meant every utility, since rivers were as plentiful as mountains.

Among the early entrepreneurs tapping hydropower in the Northwest was Charles Baker, a railroad engineer with a degree from Cornell, who happened to be the fortunate son of W. T. Baker of Chicago, president of the Chicago Board of Trade and also president of the 1893 Chicago World's Fair. The younger Baker moved to Seattle in the late 1880s, worked for a railroad company for three years, then opened his own engineering firm and became deeply involved in the electric streetcar business. He and many of his associates went bankrupt in the Panic of 1893, along with all the Seattle streetcar companies. But, like other shrewd investors with deep pockets, Baker saw opportunity amid the financial chaos. In the mid-1890s he quietly decided to pursue a grand scheme: consolidating Seattle's eight or ten ailing electric streetcar companies and developing a huge power plant at Snoqualmie Falls, thirty-two miles east of Seattle, to supply the electricity. With investment capital from his father, Charles Baker bought "options" on most of the streetcar companies and purchased the land and water rights at Snoqualmie Falls. Unfortunately for Baker, S. Z. Mitchell of General Electric outbid him on the streetcar companies, but Baker went ahead with his hydropower scheme anyway. In 1897 he organized a power company and completed an engineering marvel of a powerhouse at Snoqualmie Falls in 1898. The original Westinghouse generators had, for the time, an amazing capacity of 6,000 kilowatts, and Westinghouse's alternating current technology facilitated the long-distance transmission to Seattle (thirty-two miles), Tacoma (forty-four miles), and Everett (thirty-six miles). The construction cost per kilowatt was reportedly one fourth of that of the U.S. power plants at Niagara Falls.[58]

The engineering of this project involved some ingenious tunneling through bedrock, a feat common in the railroad industry but new to the hydropower industry. About 500 feet upstream from the falls, Baker's workers carved a trench in the streambed, channeling water to a vertical shaft carved 270 feet straight down through bedrock. Two eight-foot-diameter iron penstocks (pipes) carried river water down through this vertical shaft to a hidden powerhouse under the falls. At the bottom of the vertical shaft, the penstocks turned 90 degrees to the horizontal into a huge room carved out of bedrock, 40 feet wide, 30 feet high, and 200 feet long, housing the power plant—purportedly the world's first completely underground power facility. After the gravity-pressurized water passed through the turbines, a 400-foot-long tunnel carried the water out and emptied it into the river at the foot of the waterfall.[59]

Towns around Puget Sound receiving electricity from Snoqualmie Falls got a boost of energy—electrical and fiscal—as the transmission lines reached them. But to Baker's disappointment, cutthroat competition with the S. Z. Mitchell / General Electric consortium stymied his dreams of personal wealth. When Charles's father in Chicago died, the probate lawyers sold the Snoqualmie Falls electric company, which W. T. Baker had owned outright, to a General Electric affiliate. Charles left the Northwest to seek better fortunes elsewhere. Stone & Webster soon acquired the Snoqualmie Falls property and made it part of the Puget Sound Power & Light conglomerate. Renewed population and economic growth after the turn of the century required more power, so another generator was added to the powerhouse in 1905 and a second power plant built downstream in 1910.[60]

In British Columbia, West Kootenay Power and Light achieved a similar precedent-setting feat in the same year. The company incorporated in 1897 and by the next year finished a 4,000-horsepower hydroelectric plant at Bonnington Falls on the Kootenay River to supply electricity to the town of Rossland and its nearby mines. Although the power plant had about half the horsepower capacity of the Snoqualmie Falls plant, its electricity was sent the same distance of thirty-two miles from the falls to Rossland, pushed at an impressive voltage for the time, 20,000 volts. In only five years, this distance and voltage would seem inconsequential, but in 1898 skeptical technicians predicted that neither the Snoqualmie nor the Kootenay system would succeed. But they did.[61]

The Kootenay project did not involve the extraordinary rock excavation that Snoqualmie and Niagara Falls did, but a few years later and a few hundred miles to the west, BC Electric undertook an even more ambitious tunneling effort than its predecessors. About nineteen miles northeast of Vancouver, at a picturesque mountain lake named Lake Beautiful, BC Electric completed Vancouver's first hydroelectric project in 1903. A dam at the north end of the lake diverted water into penstocks that carried it downhill to a power plant at tidewater on Burrard Inlet. To increase the amount of water available for generating hydropower, the company undertook an extraordinary cross-basin transfer—the first of its kind in British Columbia, but not the last. The company drilled a 2.25-mile-long tunnel underneath Eagle Mountain to siphon off water from Coquitlam Lake and deliver it to Lake Beautiful, where the penstocks carried it to the power plant. Thus, water that would have flowed southeast from Coquitlam Lake into the Fraser River ended up to the north emptying into Burrard Inlet. In true imperial style, Lake Beautiful was renamed Lake Buntzen in honor of the company's general manager, Johannes Buntzen.[62]

Hydropower projects like these required building dams and diversions that raised lake levels and altered river flows. These activities affected people and ecosystems, especially as hydropower projects grew more ambitious. For projects on private lands, developers needed to buy the necessary property and water rights. BC Electric, for example, had to purchase the Coquitlam Water Works Company in order to get water rights at Coquitlam Lake. But even ownership of the requisite parcels of land and quantifiable water rights did not free a company to dam rivers and divert water as it pleased. Rivers were viewed as public property, necessitating formal permission to build "obstructions," to

Figure 2.3. Tunnel Connecting Coquitlam Lake and Buntzen Lake, B.C. Source: Leo Denis, *Water Powers of Canada* (Ottawa: Commission of Conservation of Canada, 1911), 306, 307.

alter the river's flow, or to create a reservoir. Moreover, many good dam sites were on government-owned lands rather than on private property. The upper reaches of rivers in the mountain West often flowed through federal, Crown, state, or provincial lands, which contained many valuable hydropower sites. In the United States, many of these dam sites were located on federal lands in the newly created national forest system. Although Snoqualmie Falls was on private property, the Snoqualmie National Forest surrounded it. Long Lake Dam on the Spokane River was located partly on the Spokane Indian Reservation held in trust by the U.S. government.

Whether on private land or government land, damming rivers posed an interesting challenge. Should valuable dam sites be available for unregulated private ownership and development on a first-come, first-served basis? Or should a government agency carefully administer development rights to ensure the best use of rivers and avoid conflicts? Or should these dam sites on public

Figure 2.4. Power Plant at Burrard Inlet. Source: Leo Denis, *Water Powers of Canada* (Ottawa: Commission of Conservation of Canada, 1911), 306, 307.

rivers and public lands be retained in public ownership and reserved for non-profit projects dedicated to the greatest good for the greatest number? Progressives squarely confronted these questions after the turn of the century.

THE U.S. FEDERAL GOVERNMENT AWAKENS TO ITS ASSETS

As the private sector accelerated its development of profitable enterprises on public waterways and public lands, the U.S. Congress moved cautiously toward asserting additional regulatory oversight of these projects. Going back all the way to the eighteenth century, the federal government, through the Army Corps of Engineers (CoE), had invested considerable sums of public money improving and maintaining the navigability of rivers and harbors. By the late nineteenth century, no one doubted that the protection and enhancement of river navigation were within the authority of the federal government, and most people who engaged in commerce in the pre-automobile era, when rivers were America's main arteries of commerce, appreciated these efforts. The government thus had a vested interest in what happened to the streams and rivers it had "improved" or might be asked to improve in the future. So the U.S. Congress passed the River and Harbor Act of 1884, which required congressional approval before anyone could build a dam or other obstruction

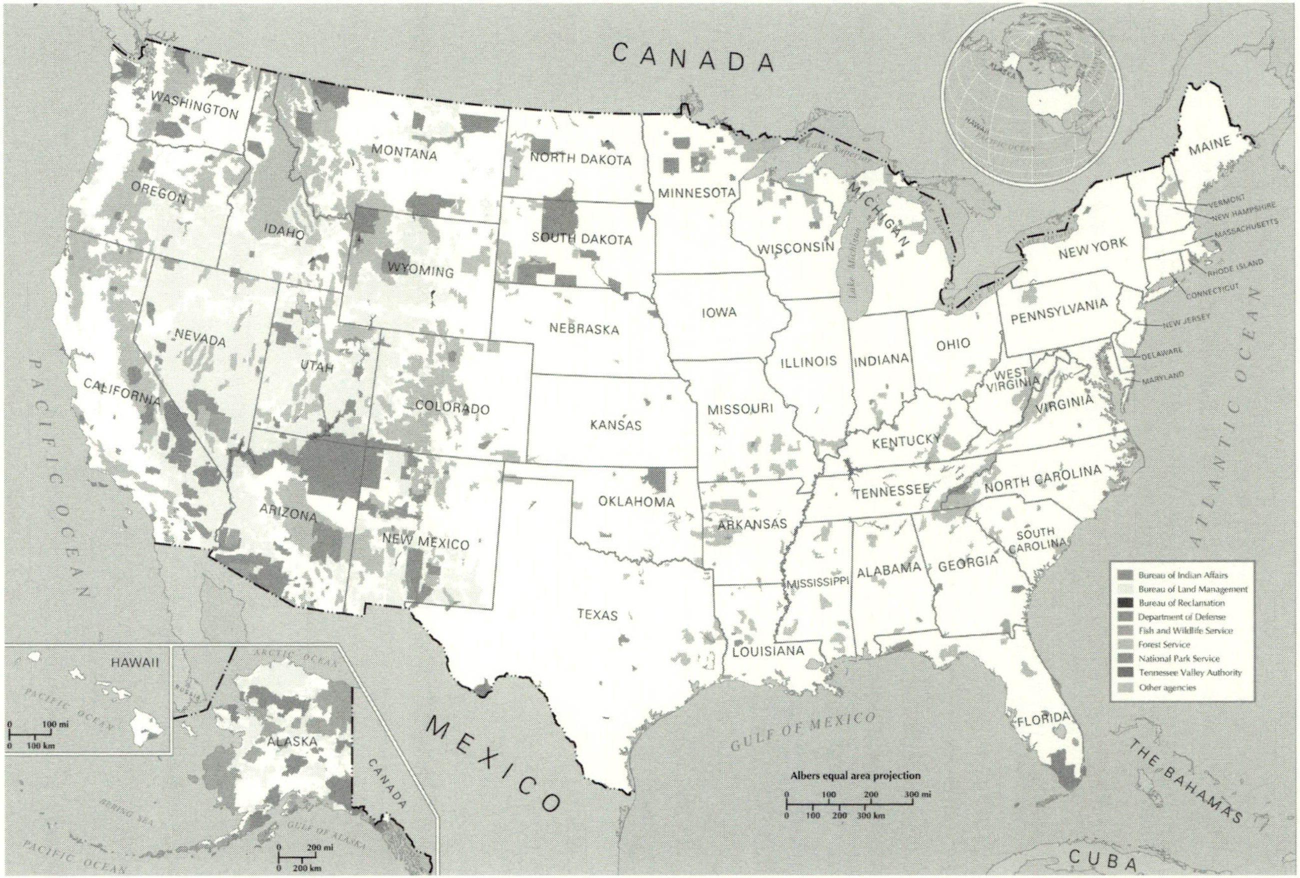

Figure 2.5. Federal Lands and Indian Reservations. Source: U.S. Department of the Interior, http://nationalatlas.gov/printable/fedlands.html#us.

on any navigable body of water. Congress refined this law in 1899, requiring congressional approval for dams on interstate rivers and state approval for dams on rivers that flowed through a single state, plus approval, by the chief of the CoE, of construction plans to ensure that navigability would not be impaired.[63] Interestingly, the topic of hydropower scarcely entered these debates since that technology was still in its infancy. This changed rapidly after the turn of the century and Progressive Era politicians in the United States revisited the issue during President Theodore Roosevelt's administration.

Another effort to beef up federal oversight and ensure responsible development was expressed through tightened regulations regarding the use of federal lands by private interests, specifically the granting of rights-of-way and the permitting of dams on public lands. At the time, 70 percent of the nation's hydropower potential lay in just nine western states,[64] and much of that on federal lands. Today, 49 percent of the land in Washington, Oregon, and Idaho is still in federal ownership.[65] These lands are now managed as national forests, national parks, Bureau of Land Management lands, wildlife refuges, and military reservations. One hundred years ago many of these designations did not exist; the lands were simply "public domain" property held by the Department of the Interior.

For most of the nineteenth century, the federal government aggressively pursued a policy of "disposing"—that is, privatizing—its extensive landholdings acquired by treaty or conquest over two centuries. Federal lands could be privatized through purchase, homesteads, or grants. There were even laws allowing settlers or businesses to use resources on the public lands—timber, minerals, water, etc.—without having to own the land itself. The General Land Office administered most of the privatization effort, and it was understaffed and overwhelmed by applicants. Most of the time the federal government exercised very little oversight, or none at all, of the use of public lands or their transference to private or state ownership.

In 1872, this policy began to change. That year, Congress established the nation's first national park, Yellowstone, to be held in perpetuity for the enjoyment of all Americans. Other national parks followed. About the same time, a group of influential scientists began pushing Congress to place tracts of federally owned forestlands into a "reserved" status, closed to homesteading or sale, to protect navigable streams and rivers. Forest exploitation by timber companies and slash-and-burn agriculture had severely damaged watersheds, causing destructive floods and troublesome river sedimentation that clogged waterborne commerce. Congress finally responded in 1891 with an act allowing the president to set aside "forest reservations" out of federally owned lands in the West. A succession of presidents designated tens of millions of acres of these forest reserves during the next ten years. The reserves were renamed "national forests" in 1907. Teddy Roosevelt tripled the system to nearly 170 million acres before he left office. The 1891 act thus represented a precedent-setting policy switch from federal land disposal to federal land retention. While a significant proportion of federal lands remained open to privatization for many more decades, the federal government had turned a corner and entered the land-management business.

By the first decade of the twentieth century, hundreds of private companies sought permission to build dams on federal lands, often for irrigation purposes in the arid West but increasingly for hydropower. Besides approval to build a dam, they needed permission to flood land behind the dam where the reservoir would lie, and they needed permission to build roads and run irrigation canals and power lines across federal land to reach farms and towns. Congress passed laws approving rights-of-way across federal lands for canals and reservoirs in 1891 and 1895 and then passed a bill in 1896 dealing with rights-of-way for electric power dams and transmission lines. Before this, private citizens acquired water-power sites on public lands through homestead laws or purchase, and the federal government paid no special attention to them. Many prime dam sites passed into private ownership this way.[66] But by the 1890s, with the government more interested in retaining forest- and parklands in the West, right-of-way permissions replaced privatization of these valuable sites. All of these right-of-way laws were combined and refined in the Right of Way Act of 1901, authorizing the Secretary of the Interior to issue permits for such purposes so Congress no longer had to approve every application.[67]

Initially, the Secretary of the Interior rubber-stamped these applications, exercising little oversight and making no effort to reserve any dam sites for federal development. But soon this laissez-faire policy changed as Progressives sought to retain more federal options and exercise more regulatory influence. Leaders in the conservation movement, like Gifford Pinchot, first chief of the U.S. Forest Service, promoted the idea that these federal properties were "public assets," not just plums for private-sector picking, and that the public should get the best value from them.[68]

CONCLUSION

Corporate consolidation of the electric light, power, and traction companies led to rapid expansion of the industry, improved service, technological innovation, greater economies of scale, and increased profits. At the same time, the public grew increasingly concerned about monopoly power, excess profits, and absentee ownership, leading to pressure to increase local, state, and federal government oversight. Corporations holding franchises to provide lighting, power, and transportation services to municipalities made up a special category of business called a "public service corporation" or, increasingly, a "public utility." (The nomenclature can be confusing because these are usually privately owned, for-profit companies. A "public utility" does not necessarily denote a publicly owned utility.) Since these businesses profited by holding special monopoly privileges and using public thoroughfares and public resources, many citizens, city managers, and other political leaders felt these privileges should be contingent upon the corporations charging fair rates and providing adequate and safe services. They consequently advocated for independent, professionally staffed commissions and government agencies with adequate resources and expertise to provide the oversight and regulation necessary to ensure corporate responsibility. Motivated by the increasing importance of electric power in urban and industrial life, the Progressives took on the task of developing those commissions and agencies (this is elaborated in chapter 4), as well as promoting publicly owned utilities as an alternative to the profit-driven capitalist model of public service. While enthusiastically promoting modernization in all its forms—material, social, political, and cultural—Progressives simultaneously became both advocates and critics of the electrical industries that made modernization possible yet also seemed to limit to the more privileged classes the good life it promised.

CHAPTER 3

Becoming Modern

FOR PEOPLE WHO DON'T USE ELECTRICITY:
Don't expect ironing to be a pleasure
Don't knock the kerosene lamp over
Don't imagine your home looks modern
Don't kick because your business is "bad"
Don't scold your workers because you can't increase your production[1]

—Bulletin of the Portland General Electric Company, 1906

After the turn of the twentieth century, electrification rapidly penetrated and fundamentally altered businesses, households, and daily life. For countless generations, people had moved about on foot or by horse, accomplished work by muscle power or fire, illuminated interiors with windows or flames, delivered irrigation water by gravity through hand-built canals, and made products piecemeal by hand and tool. In half a human lifetime these time-honored practices increasingly gave way to incandescent light bulbs, electric streetcars, precisely controlled motors, automatic washers, giant water pumps, and machine-made products—all made possible by electricity. The changes spawned by this new technology penetrated society rapidly, compared to the pace of technological change prior to the electrical age. The new world ushered in by electricity was so different, so novel, that it acquired a moniker: electrification became the symbol of "modernity." Pacific Power & Light's official motto at the turn of the century was "The Modern Way."[2] Of course, modernization is much more than mere electrification, but the two became inseparable. Those striving for a modern life or modern methods of production or a modern identity adopted the new technology.

The best analogy for understanding this electrification revolution is a more recent electrical technology revolution: the rapid penetration of the personal computer and Internet into our lives. While the analogy is not perfect, it is highly instructive. Both revolutions started with a specialized technology developed by well-funded researchers, some of whom became household names (Thomas Edison and George Westinghouse, Bill Gates and Steve Jobs). Early battles over system design (AC or DC, MS DOS or Mac) led to patent wars, hybrids, and eventual compatibility among competing systems. Governments and businesses made first use of the new technologies, but then both quickly diffused throughout society.

The initial diffusion of both technological revolutions was accomplished mainly by the private for-profit sector with some government assistance, and was pushed along by aggressive marketing campaigns. In both, the technologies started off almost prohibitively expensive, then steadily grew more affordable as usage increased. In both, quality, performance, and capacity to do work steadily increased as prices steadily decreased. Both progressed from techno-gimmicks to indispensable tools, with the haves seen as modern and the have-nots seen as unfortunates stuck in the past. Reformers during both revolutions called for more government oversight, argued that access should be a universal right, and pushed for government assistance in providing universal access (compare the rural electrification movement with the movement advocating computers in every school and library). Both revolutions fundamentally altered much of the structure and function of daily life, business, and leisure and did so in unpredictable and sometimes surprising ways.

While people quickly came to understand the utilitarian benefits of the new technologies, most did not understand how they worked. In both cases, people were absorbed into a technological system they did not create, control, or comprehend. When the system worked well, they were complacent; when it did not, they demanded intervention. In both revolutions, consequently, the private sector maintained its for-profit privileges by keeping customers relatively happy and acquiescing to government regulations designed to regularize and stabilize the systems through controlled service monopolies with protections for certain aspects of the public interest. Both attracted huge amounts of capital and generated astounding profits and concentrations of wealth, which then led to complaints about monopoly, followed by government antitrust actions and a slow but steady development of regulatory institutions.

The promises and the perils of both technological revolutions generated utopian and dystopian dreams and futuristic novels and films. Even before electricity had penetrated everyday life, Kenneth Roemer found it to be one of the three most common "scientific marvels" mentioned in 160 utopian novels published between 1888 (the year Edward Bellamy's *Looking Backward* appeared) and 1900. Many feminist utopian authors predicted that electricity would relieve women of drudgework and liberate them for intellectual development and civic activism. Thomas Edison thought electrical conveniences in the home would eventually make women and men equal (a radical idea in the nineteenth century). But there was a dark side of moral ambiguity and fear, as expressed in Eugene O'Neill's 1929 drama *Dynamo*, which featured an obsessed young man who makes electricity his god, turns his love from his spouse to his dynamo, and ends up electrocuted in a final deadly embrace.[3] For the more recent revolution, Jeremy Rifkin's book *The Hydrogen Economy* (2002)[4] was a best-selling utopian vision of a hydrogen fuel-cell-generated, electricity-driven, computer- and Internet-administered future that he predicted will weaken petroleum conglomerates and utility monopolies, strengthen democracy, and improve the environment. This hopeful vision contrasts starkly with the dark vision of the future in the technological dystopia portrayed in the 1999 film *The Matrix*, in which humans have become wired, somnolent slaves of a computer system that constructs a virtual reality dreamworld to lull people into thinking they are living real lives when in fact their unconscious physical bodies lie hooked up to a network providing energy to the master system. As computers reshaped the world of Generation Xers going into the twenty-first century, electricity reshaped the world of "progressives" going into the twentieth century. Everything was affected—economics, politics, science, education, art, personal identity, even religion. Naturally, both generated excitement and anxiety about the future.

THE BOOM IN ELECTRICAL USAGE 1902–1922

For a new technology, electricity spread remarkably rapidly throughout industry, retail businesses, and urban homes in America in the early decades of the twentieth century. As electric generators and electrical transmission became cheaper, safer, and more efficient, and as G.E. and Westinghouse made electrical machines and appliances more widely available, consumers with a little money to spare responded enthusiastically. After an initial hesitation, electric lighting quickly made the transition from a novelty to a necessity. Electric motors replaced bulky, noisy, inefficient steam engines with all their dangerous shafts and pulleys in manufacturing plants. Electric irons, refrigerators, vacuum cleaners, and electric ranges replaced flat irons, iceboxes, brooms, and woodstoves in homes. And, to a lesser extent, electric lights, pumps, and milking machines started replacing kerosene lamps, windmills, and muscle power on the farm. Certainly, none of this happened overnight. Only businesses with surplus capital or families with expendable income purchased these innovations, while people in rural areas, Indian reservations, and poor sections of town often continued to live without electricity for decades. But those with the means and the opportunity adopted lighting and electric tools relatively quickly. The data are revealing. Every five years starting in 1902, the U.S. Bureau of the Census collected statistics on the electric light, power, and streetcar industries. In 1902, central electric light and power stations generated 2.5 billion kilowatt-hours (kwh) of electricity for sale to consumers, while industries, mines, and electric railroads generated another 3.46 billion kwh for their own use. That equaled an average of 75 kwh per capita in the U.S. Both electrical generation and per capita consumption more than doubled in the next five years and nearly doubled again in the subsequent five years. By 1922, total electrical generation from all sources had increased tenfold over 1902, and per capita consumption had risen more than sevenfold.[5]

The separate figures for central electric light and power stations are even more revealing of the penetration of electricity into people's daily lives. The census records of central stations left out most electric railways and isolated industrial power plants, concentrating on the electric companies that served cities, retail businesses, homeowners, and irrigators. In every five-year census after 1902, the power produced by central stations increased dramatically. While total U.S. electric generation increased tenfold between 1902 and 1922, output from central stations increased sixteenfold—a rate of increase twenty-two times higher than the rate of population growth. Electricity was clearly penetrating American society at an impressive rate.[6]

The U.S. Northwest mirrored and in many ways exceeded these national trends. In the two decades after 1902, central electric light and power stations in the three states of Washington, Oregon, and Idaho increased their output by forty-six times—three times the national average. Per capita consumption of electricity from central stations in the Northwest states likewise increased at almost three times the national rate.[7]

Consolidating data for the three states obscures some interesting regional differences. Oregon's rate of increase in kilowatt-hour generation nearly matched the national rate, while Washington's rate of increase was more than three times the national rate and Idaho's was more than six times the national rate.

Much of this is explained by the fact that Oregon developed earlier than the other two states, so that its post-1902 growth rate was more moderate than the later-blooming states of Washington and Idaho. Also, Washington's growth greatly outpaced Oregon's starting in the 1890s because Puget Sound offered better ports than the Columbia and Willamette rivers and was situated nearer to the Fraser and Klondike gold rushes. Washington's population had just surpassed Oregon's for the first time in the 1900 census. By 1920 it had nearly dou-

Table 3.1. Electric Power Production and Consumption, 1902–1922*

	Kilowatt-Hours Generated			Number of Customers			Kilowatt-Hours per Customer			Increase in Kilowatt-Hours
	1902	1912	1922	1902	1912	1922	1902	1912	1922	1902–1922
WA	19,722	71,414	1,144,068	NA	73,421	260,610	NA	0.97	4.39	52-fold
OR	17,532	58,789	266,861	NA	33,627	78,534	NA	1.75	3.4	15-fold
ID	5,018	115,812	531,134	NA	26,078	65,298	NA	4.44	8.13	106-fold
US	2.5 billion	11.5 billion	40.3 billion	NA	3.8 million	12.7 million	NA	30	317.32	16-fold

*Central station light and power data only

Sources: U.S. Census, *Central Electric Light and Power Stations, 1915*, 19, 50, 82–83, 86–87; and *Central Electric and Power Stations, 1930*, pp. 12–13, 44.

ble the population of Oregon. Idaho experienced a high population growth rate very similar to Washington's but had fewer people—one third as many as Washington. The very rapid development of irrigation in Idaho after 1902 and the hydropower that accompanied irrigation dams led to Idaho's impressive growth in kwh output for a state with such a small population. Thus, more rapid population growth, combined with aggressive investment and abundant hydropower resources in the early twentieth century, made the Pacific Northwest (as well as California) a national leader in electrification.

All this electrification in the early twentieth century made its way primarily into municipal transportation systems, city lighting, commercial business, and heavy industry. The household electric market did not really take off until the second and third decades of the twentieth century. Even when it did, the amount of electricity consumed for domestic use paled in comparison to these other uses. The 1927 Electrical Census initiated new reporting procedures that distinguished for the first time among different classes of consumers, including domestic, commercial (broken into small and large businesses), farm, municipal street lighting, railways, and a couple of catch-all "other" categories. In the previous censuses, only "lighting" and "power" were distinguished. The 1927 statistics reveal that while domestic electric service had greatly expanded to 83 percent of all customers, it represented only 13 percent of total energy consumption. In contrast, the class of "large light and power" consumers included less than 2 percent of customers but consumed 52.6 percent of the power. This national pattern prevailed in the Northwest. For example, PGE (called Portland Electric Power Company in the 1920s) served seventy-two communities in Oregon by the late 1920s, yet its biggest customer was still one single business: Hawley Pulp and Paper, which consumed 100 million kilowatt-hours of electricity in 1929.[8]

Economic logic dictated this outcome. Mining operations, for example, consumed prodigious amounts of electricity in one location, in contrast to the small amounts of electricity consumed by widely distributed households in residential areas of cities; consequently, private-sector electricity entrepreneurs looked first for opportunities to sell the most power at the least cost. Transmission lines were expensive investments, and company executives built them only where they expected to garner large enough power loads to pay off the investment and earn profits. In contrast, the high cost and long payback of distributed service to homeowners and farmers led investor-owned utilities to delay extending electric lines to those areas and to charge higher rates than they charged big consumers. As mentioned earlier, this was the major motivation for small towns to launch nonprofit municipally owned electric systems. Recall also that in the 1890s, electric lines extended for streetcars offered the opportunity to electrify nearby neighborhoods. Likewise, neighborhoods near downtown commercial districts got lighting and power before outlying neighborhoods, and small towns in the path of transmission lines going from hydropower plants to load centers got power before other towns remote from these distribution lines.

To encourage more high-load customers, utilities used graduated rate structures: the more you consumed, the less you paid per kwh. For example, Washington Water Power's rates for commercial and residential lighting from 1900 to 1909 are shown in Table 3.2.

In 1910, WWP lowered these rates by about 25 percent and separated commercial and residential lighting, with the higher-load commercial lighting enjoying lower rates than residential lighting.[9] The industry as a whole lowered electric rates repeatedly for the first four or five decades of the twentieth century in large part due to economies of scale and improved power production and distribution efficiencies, but also because of state regulation of rates and services. Municipally owned power companies also offered graduated rate structures. Eugene's Water and Electric Board completed its first power plant in 1911 and approved a sliding rate scale according to the type and amount of use: lighting was 9 cents/kwh for the first 20 kwh and dropped in increments down to 3 cents/kwh for loads over 1,300 kwh. For power, the rate was even cheaper because customers buying power usually consumed much more electricity than customers buying lighting services: the highest rate was 5 cents/kwh for up to 100 kwh; the rate dropped to 2 cents for loads over 10,000 kwh.[10]

Table 3.2. Washington Water Power's Rates for Commercial and Residential Lighting, 1900–1909

Kilowatts per Hour	Price (cents/kw)	Kilowatts per Hour	Price (cents/kw)
0–25	13	151–250	9
26–40	12	251–625	8
41–60	11	626–1,000	7
61–150	10	Over 1,000	6

Source: Edward J. Crosby, *The Story of the Washington Water Power Company and Its Part in the History of Electric Service in the Inland Empire, 1889–1930* (Spokane: Washington Water Power, 1931), 26–27.

Sliding scales and lower rates for bulk customers were standard practice for electric companies, both private and public, and were endorsed by the state public utility commissions that began regulating utility rates in the 1910s.[11] Thus, economic logic inevitably led to inequities in the availability and price of electric service and complaints about favoritism and unfair practices.

As Americans came to view electricity as a central component of modern life, many who were dissatisfied with their electric service questioned the adequacy of a purely profit-driven system and advocated for government regulation of electric utilities or the expansion of nonprofit municipally owned electric systems committed to universal service at fair rates. Ironically, as electrical manufacturers and utility companies succeeded in making their products universally desired, they sowed the seeds of a political movement that restrained their ability to operate freely and maximize profits. In the first three decades of the twentieth century, both electrification and electric system regulation developed roughly in parallel. The former will be examined in this chapter and the latter examined in the next two.

ELECTRICITY IN INDUSTRY

Historian David Nye notes that there were two periods of electrification in American factories. In the first, from 1880 to 1900, factory owners adopted electric lighting to improve visibility and extend production hours. In the second, from 1900 to 1930, they adopted electric motors and experienced tremendous gains in productivity as a result. Even when factories were still consuming electricity mainly for lighting, they were already among the most important customers of electric utilities. In 1900, only 5 percent of factory production was powered by electricity, yet these same factories consumed more than half of all electrical generation in the United States—mainly for lighting.[12] As industrial production processes electrified, the importance of factories, mines, and mills to the electric power industry increased.

The Washington Water Power Company, for example, went to precedent-setting lengths to send its electricity to a booming mining district far off in north Idaho, called the Silver Valley. From its power plant at Spokane Falls, the company built in 1902 what was then the world's longest transmission line—100 miles long—pushing electrons at 60,000 volts, only the second time such high voltages had ever been used anywhere. WWP justified such an expense with the expectation that the Silver Valley mining industry would easily consume more than enough power to recoup the investment. At the same time, WWP also bought the rights to develop a dam at Post Falls on the upper Spokane River, halfway between Spokane and the mining district. The dam site had previously been acquired by wealthy mine owners in nearby Coeur d'Alene, Idaho, who planned to develop hydropower to serve the mining industry. Instead, they negotiated a deal to let WWP develop the site and provide the power. WWP completed the dam and power plant at Post Falls in 1906, connecting it to their 100-mile-long transmission line. Hydroelectricity from Post Falls greatly boosted WWP's power supply for both urban uses in Spokane and industrial production in Silver Valley.[13]

By 1910, the rapid shift from steam to electricity in mining and manufacturing was under way. Statistics help provide a sense of the scale and rapidity of this shift. In 1890 less than 1 percent of mechanical power in manufacturing came from electricity; steam engines or moving water supplied the remainder. By 1910 electricity supplied 25 percent of mechanical power in manufacturing, and by 1920 it provided the majority, at 53 percent. Steam still provided almost 40 percent of mechanical power in 1920, but by 1930 the transition was nearly complete, with electricity supplying 78 percent of motive power, steam 16 percent, and mechanical water power 1 percent.[14]

The reason for this rapid transition from steam-driven shaft-and-belt mechanical power to electric motors is partially explained by the unique nature of electricity. Unlike other forms of mechanical power, electricity can be controlled with remarkable precision and adjusted instantaneously, reducing the inefficiencies that come from having too much or too little power for the task, inefficiencies that often accompanied steam engines. The rapid adoption of electric motors in factories is also explained by the fact that it allowed for a complete reorganization of the factory workspace, vastly increasing the flexibility and efficiency of production processes. To fully appreciate the advantages of electricity one must understand how the older steam-shaft-belt-driven production system of the nineteenth century worked. Warren Devine describes the old system as follows:

> In most factories, a single centrally located prime mover, such as a water wheel or steam engine, turned iron or steel "line shafts" via pulleys and leather belts. These line shafts—usually three inches in diameter—were suspended from the ceiling and extended the entire length of each floor in a factory. . . . The line shafts turned "countershafts"—shorter ceiling-mounted shafts parallel to the line shafts. Production machinery was belted to the countershafts and was arranged, of necessity, in rows parallel to the line shafts.[15]

The entire network of line shafts and countershafts rotated continuously—from the time the steam engine was started up in the morning until it was shut down at night—no matter how many machines were actually being used. If a line shaft or the steam engine broke down, production ceased in the whole room of machines or even in the entire factory until mechanics repaired the system. Shafts had to be oiled continuously by oil drippers suspended above the shaft hangers, and belts constantly had to be tightened, repaired, or replaced. This required a substantial amount of maintenance labor since a typical factory might contain thousands of feet of shafts and belts and hundreds of pulleys and oil drippers—all of which created opportunities for something to go awry. With every machine directly or indirectly powered by one central spinning shaft attached to a steam engine, the only way to vary speeds for different machines was to attach a series of pulleys and belts to increase or decrease the spin speed of a secondary shaft. This whole centrally driven mass of moving parts was noisy, dangerous, temperamental, and inefficient. The moving belts and dripping oil also presented fire and health hazards. Moreover, since all the machines had to be connected to the shafts, everything lined up in parallel along the floors of the factory with no flexibility in workspace design. With no real alternative available, people simply got used to this system and adjusted as best they could to its limitations.

In the 1880s, electric motors appeared on the scene. At first, most factory owners who experimented with electric motors simply substituted them for the steam engines, making the electric motor turn the same inefficient central shaft-belt system. Eventually, engineers and factory designers caught on to the opportunities inherent in electric motors. The steam engine might remain the "prime mover," but it soon spun generators to create electricity instead of spinning a central floor-length shaft. This electricity would be distributed throughout the factory by electric wiring to any location that a motor might

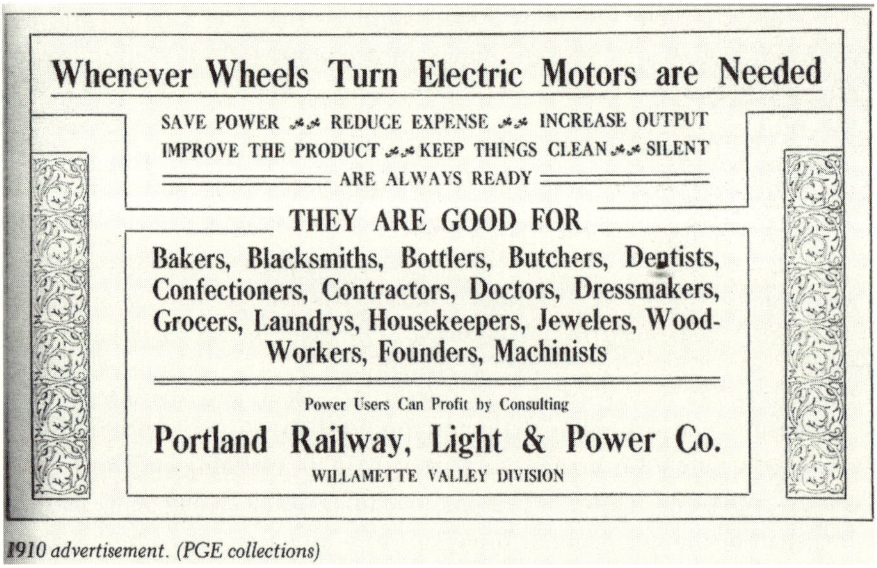

Figure 3.1. "Whenever Wheels Turn Electric Motors Are Needed." Source: Craig Wollner, *Electrifying Eden: Portland General Electric, 1889–1965* (Portland: Oregon Historical Society, 1990), 99. From Portland General Electric Collections.

be placed, unencumbered by the need to line up parallel to a shaft. Individual motors could run a group of machines all doing the same kind of work or all requiring the same speed of shaft spin, or small motors could run individual machines. This flexibility significantly reduced the amount of energy required for production work by reducing waste, especially as it allowed managers to instantaneously start certain machines when needed or stop machines when not needed without affecting the driving mechanisms or motors in the rest of the factory. As Devine pointed out, "The ability to shut down or start up selected equipment is taken for granted today, but such ability represented technical and organizational innovation in the 1890s."[16] Electric utility marketing personnel were well acquainted with the benefits of electricity in the factory, and they peppered their ads and bulletins with appropriate allures.

In a 1905 article titled "The Wasteful Steam Plant," PGE told its business customers that with electric motors "you use what you need, when you need it, and where you need it. There is neither waste of power nor of money."[17]

Commercial and manufacturing enterprises in the Northwest, large and small, quickly adopted electric-driven machines for production processes. Newspaper presses were among the first, as mentioned in chapter 2. By the 1910s Northwest businesses widely adopted electric motors and machines for elevators, ice-making, lumber-planing, flour-milling, illuminated signs, heating and cooking, slicing and canning, refrigeration, mining, smelting, processing, and much more. Examples of electrical services to industry can be found in company bulletins, annual reports, and public utility commission records. For example, the annual report of the Public Service Commission of Oregon for 1915 (the first year the commission had authority to regulate electric companies) included a table showing sales revenues from various sources for Eastern Oregon Light & Power Co., including many of the business and manufacturing uses described above. The table graphically illustrates the dominance of big industrial consumers, with mining consuming more than half the company's electric load. But because of the lower rates the electric company charged for mining uses, this largest consumer of power contributed only 24 percent of the electric company's revenue. Residential customers paid twelve times more per kwh than did the mines.[18]

The *Puget Sound Electric Journal*, published monthly beginning in 1911 by Puget Sound Power & Light, regularly profiled the company's service areas and important customers. The Renton coal mines, some of which were owned outright by the power company itself, appeared often in the early journals.[19] Washington at the time mined more coal than any other Pacific coast state: 3.5 million tons per year in 1915, the majority taken from mines near Tacoma and Seattle, including the Renton mines. Before 1915, electric lights and fans and some motors were employed at the mines, but most of the power for mining and processing the coal was still steam-engine-driven belts and pulleys. The development of hydroelectric power in the Cascade Mountains just east of Puget Sound brought plenty of power as well as transmission lines running from the dams past the coal fields to the cities at tidewater. The power company offered the mine owners low rates for wholesale electricity, including some very attractive off-peak rates to encourage the complete electrification of the mining operations, which was accomplished by 1915. The mine owners used electricity to dig, transport, wash, and process the coal, to pump water, compress air, light the facilities, and ventilate the mine shafts. The electricity came in large part from the power company's new hydroelectric stations, including Snoqualmie Falls, the Electron hydropower plant completed on the Puyallup River at the foot of Mount Rainier in 1904, and the White River/Lake Tapps project near Sumner completed in 1911.[20]

Mining was not the only heavy industry served by Puget Sound Power & Light. The Pacific Northwest contained the greatest temperate rainforest in North America and consequently supported a lumber industry of astounding size. The first major boost to the lumber business came with the arrival of railroads in the late nineteenth century—both narrow-gauge railroads that opened up remote mountain valleys to logging and the transcontinental railroads that brought the region's lumber to distant markets. One of the centers of lumber manufacturing and shipping was Everett, Washington, on Puget Sound, north of Seattle. Everett boomed in the 1890s after the Great Northern Railroad connected it to Chicago and Weyerhaeuser opened a mill employing 2,000 workers. The lumber business grew to such proportions that by 1903 the town was hailed as the lumber capital of the world—or at least the cedar shingle-making capital of the world, with the capacity to produce six million red cedar shingles per day.[21]

Puget Sound Power & Light served this community. Slowly but surely the company both promoted the electrification of mill operations and acquired contracts to provide the power for the lumber, shingle, and pulp and paper companies that were the mainstay of Everett's economy. In the nineteenth century, most lumber mills provided their own electricity by installing on-site generators and burning fuel, often wood chips and other mill waste. But as centralized power companies grew in size, efficiency, and reliability, they could often provide electricity at competitive rates and free manufacturers from the trouble of buying fuel and maintaining furnaces and generators.

The Everett Pulp and Paper Company provides an illustrative example of this shift. Originally, the company's sawing, chipping, loading, cooking, pulping, pressing, drying, and cutting machines were all driven by Corliss steam engines and cone pulleys. One by one, the company purchased electric motors to run various parts of the production process. By 1912, the company had seventy-seven electric motors with a combined total of over 2,000 horsepower, which drove all the machines except the pulp screens and paper rollers. Since the paper mill ran twenty-four hours a day, six days a week, indoor illumination was necessary, too. All the electricity was supplied by an on-site 500-kilowatt generator that burned fuel oil. Puget Sound Power & Light (PSP&L) convinced the company to run an experiment comparing energy costs from three weeks of self-generated power against three weeks of purchased power. At the end of the experiment, Everett Pulp and Paper contracted to buy all their electricity from PSP&L.[22]

The power company's new hydroelectric project at Lake Tapps, near Sumner, Washington, provided a good deal of surplus power after it came online in 1911, and that power needed a market. Fortuitously, the Fleischmann Yeast Company opened a large manufacturing facility on eleven acres in Sumner at about the same time. Fleischmann's found Sumner attractive because it had plenty of electric power and an artesian spring that produced 1.5 million gallons of pure water daily, and it was located near three transcontinental rail lines. Fleischmann's wanted an external source of electricity because the processes for making quality yeast required very clean air and water. Fuel-burning electric generators at that time spewed out smoke, ash, and soot (this was fifty years before Clean Air Act emissions controls). The yeast-making process required boxcars of grains to be milled, washed, cooked, and fermented. Centrifugal spinners separated out the bubbling yeast, which was hydraulically pressed into eighty-pound boxes and refrigerated. The factory operated every day, night and day, and required illumination. In fact, a reporter in 1916 claimed it had shut down for only four hours in the previous four years. This Sumner plant was one of eleven Fleischmann plants in North America. The yeast was shipped express throughout the United States, Canada, Mexico, and Cuba. PSP&L supplied its electricity.[23]

In southeast British Columbia, the region's major electric company and its largest mining corporation developed in tandem. Copper, silver, lead, and zinc ores abounded in southeast British Columbia, and a thriving mining district developed around Rossland, Trail, and Nelson in the second half of the nineteenth century. Seeing good profit potential from an industry that required moving large amounts of heavy ores, the Canadian Pacific Railway purchased one of the larger mines and lead smelters in the region at Trail near the international border in 1898. That same year, the provincial government authorized the incorporation of the West Kootenay Power and Light Company (WKP&L), whose owners saw an opportunity to use Kootenay River hydropower to electrify the mining operations in the region. From the very start, before power was even available, the Canadian Pacific Railway contracted to purchase a large block of wholesale power from the electric company for the lead smelter it owned at Trail. When the power company's facility at Bonnington Falls came online in 1898, it brought electricity and lights to several local mines and mining towns. By 1899, newspapers were touting both the new power plant and the huge electric hoist it powered at the War Eagle Mine in Rossland—purportedly the largest electric hoist in the world.[24]

Fortuitously for the mining industry, the Canadian government wanted to encourage domestic lead production and initiated financial incentives in 1902 to spur production. Capitalizing on the opportunity, the Canadian Pacific Railway opened another lead smelter in Trail that year. As mining and smelting operations grew, the railroad created a subsidiary company to manage that end of their business: the Consolidated Mining and Smelting Company, or COMINCO. Because electricity grew increasingly important to production processes, in 1905 the Canadian Pacific Railway also bought a controlling interest in WKP&L. Power company managers and consultants had been advocating for more electric load diversification since 1901, but mining companies ruled the roost in that region. Service towns and mining camps in the region were electrified, but they consumed only a pittance of the electric load. The power company's fortunes were wedded to the mining industry.[25]

The mining industry's fortunes were also wedded to a cheap and abundant supply of electricity, especially since a new process involving electrolysis to refine zinc from the region's low-grade ores, requiring extraordinary amounts of electricity, was developed in 1911. Several experimental zinc processing plants located in the region after 1911 to take advantage of WKP&L's cheap and plentiful hydropower. With the advent of World War I in Europe, Canada became a major supplier of war materials to Great Britain. Copper and zinc were critical war materials, and other than some small experimental smelters, Canada had no real production facilities for refining either metal. With encouragement and assistance from the national Zinc Commission, COMINCO expanded their electrolytic zinc refining operations in 1914 (see chapter 6 for more on the effect of the war on the region's mining and power industries). To ensure a ready supply and favorable rates for electric power, COMINCO purchased WKP&L in 1916. The relationship between these two companies paralleled the relationship that developed at the same time between streetcar companies and electric companies in the cities: the consumer bought the supplier.[26]

Another important segment of the metals industry developed interdependently with the electric industry: aluminum processing. Aluminum is the most abundant metal in the earth's crust, but it is very difficult to isolate and concentrate into a pure form. Before the invention of electrolysis, aluminum processors concentrated the metal chemically in meager amounts and used it only for jewelry and other specialty applications. When the Washington Monument was completed in 1884 it was ceremoniously capped not with gold but with a 100-ounce pyramid of aluminum, the largest piece of cast aluminum in the Western Hemisphere.[27] Aluminum was so rare in its pure form in the mid-nineteenth century that it sold for about $100 per pound (the price of gold in the United States at that time was just over $300 per pound). New forms of chemical processing eventually brought that price down to $8–$12 per pound by the late 1880s. Then electrical processing of aluminum surged after 1895 as the Pittsburgh Reduction Company, using cheap abundant hydropower from Niagara Falls, built two plants for processing aluminum in 1895–1896 with a capacity of 100,000 pounds of output per month. Aluminum prices subsequently fell to 20–30 cents per pound.[28] Aluminum thereafter found its way into a wide variety of products and applications, including the electric transmission lines used to bring power from Snoqualmie Falls to Seattle in 1898. Because the Northwest had large amounts of cheap hydropower, an aluminum-processing industry developed in the region during World War I and grew exponentially during World War II, fed by federal hydropower and federal defense contracts (see chapters 6 and 11).

All these factories, mills, and mines did not run just on steam and electricity. Hundreds, even thousands, of laborers were employed at such businesses, and electrification profoundly affected their environment and experience. As David Nye observed, electricity made the workplace "safer, quieter, cleaner, and brighter." But that was only part of the story. Evolving hand in hand with an increasing concentration of wealth, industrial electrification contributed to the exploitation of the working class and the weakening of worker control over production. Nye reminds us that electricity facilitated assembly-line manufacturing with its mind-numbing repetitive tasks; supported the "tyranny of the clock"—the regimentation of hours of labor and increased control over worker productivity; extended work hours, with indoor lighting allowing round-the-clock shifts that disrupted family and community life and worker health; brought new dangers of electrocution (a serious hazard for most of the twentieth century, until occupational safety and health regulations mitigated the worst hazards); and contributed significantly to job loss and de-skilling in factories that electrified production processes.[29] Since there was no turning back the clock, these problems and hazards would be dealt with directly through labor unions and government regulations in the new "modern" age.

THE DECLINE OF ELECTRIC RAILWAYS

While some electric companies primarily served natural resource industries outside the cities—mining, lumbering, and irrigation agriculture—most primarily served urban areas. For urban utilities, electric railways, factories, mills, and street lighting consumed the great majority of their power load, usually 90 percent or more. Electric streetcars in particular were seen as modernizing agents, and most trolley systems flourished under city franchises until the second decade of the twentieth century. Then, a series of interrelated physical, technological, economic, and cultural factors led to the near-universal demise of these systems, causing temporary financial problems for urban utilities and leading to a significant shift in the marketing and financing of electricity.

The physical limitations of trolleys, surmountable at first, would cripple them in the age of the automobile. Trolleys ran on limited and fixed tracks

Figure 3.2. Portland Electric Train, Circa 1915. Source: Craig Wollner, *Electrifying Eden: Portland General Electric, 1889–1965* (Portland: Oregon Historical Society, 1990), 82. Original image from Oregon Historical Society, OrHi 1604.

with fixed schedules and depended for their success on municipal cooperation, affordable fares, and willing riders. Tracks were difficult to install, expensive to maintain, and impractical to relocate. As cities continued to sprawl, electric streetcar companies, under regulated franchises from city governments, were required to expand service to new residential districts, which required constant investment. The capital for expansion as well as for maintenance came from rider fares or investment firms or both. Until the 1910s, conditions encouraged investment, profits, city cooperation, and abundant ridership. Cities with municipally owned electric systems often owned the trolley system, but cities served by privately owned electric utilities usually had privately owned trolley systems regulated by city government franchises. These private trolley companies in turn were usually affiliated with a company that supplied electricity to the city, such as the Portland Railway, Light and Power Company discussed in chapter 2. I use the Portland experience in detail below to exemplify the typical challenges facing this industry in the 1910s and 1920s.[30]

During the trolley heyday of the first decade of the twentieth century, Portland Railway, Light and Power (PRL&P) continually expanded, riders enjoyed good service and low fares, the company owned or promoted fourteen different recreation sites to encourage weekend riders, city government was reasonably satisfied with the company's performance, and eastern investors controlled 80 percent of the profitable company's stock. But this happy association soon

Figure 3.3. PEPCO Motor Bus. Source: Craig Wollner, *Electrifying Eden: Portland General Electric, 1889–1965* (Portland: Oregon Historical Society, 1990), 122. From Portland General Electric Collections.

disintegrated. The first problem was physical: urban growth slowly made the city less compact, decreasing the density of trolley lines and increasing the distance people had to walk to get to a trolley stop. The second problem was economic: internal combustion engines, Henry Ford's mass-produced Model T, petroleum, gasoline stations, and paved roads all rapidly penetrated urban life in the 1910s, creating the first competitive alternative to streetcars. The third problem was political and institutional: the conflicting demands of profit-seeking owners, wage-seeking workers, cost-conscious customers, city leaders, and downtown businesses simply could not be reconciled, especially as ridership declined.[31]

Besides those who bought their own car or rode with acquaintances, there were cars for hire. As early as 1914, Portland's trolley system lost riders to these unregulated "jitneys," large private vehicles that accepted paying passengers and often followed trolley lines looking for customers (the first incarnation of taxicabs!). In a bitter campaign, the owner of the trolley system, PRL&P, successfully convinced the city council to severely restrict these private jitneys in 1916, but by then another competitor had already entered the scene—motorbuses. Since trolleys could run only on their tracks and since the city grew more rapidly than the company could expand its track system, a growing percentage of Portlanders needed alternative transportation, and a regulated bus system was born in 1915. Unable to beat this competition, PRL&P eventually decided to join them. In 1924 it started the Oregon City Motor Bus Company, and the next year it had twenty-two buses running in Portland and nine buses running between Portland and nearby towns. By 1927 it had forty buses operating in Portland alone.[32] But this investment strategy did nothing to improve the financial situation of the electric streetcar system.

Another problem that led to the decline of streetcars in the 1910s was caused by a financial structure dependent on economic prosperity and continued growth in ridership. In Portland, ridership declined, starting in 1913. The decline in passenger traffic from 1913 through 1916 affected the company's ability to invest in expansion and maintenance, which further discouraged passengers. Since the system's inception, growing passenger traffic had guaranteed enough revenue to continue expanding and renovating the system as well as pay dividends to investors. The financial decisions of the company and the rates that the Portland city government approved for the trolleys were based on assumptions of continued passenger growth. When that growth stopped, something had to give. Although the company continued to operate in the black, dividend payments to investors were suspended in 1914 and 1915 and planned expansions stalled.[33]

Worse, electricity sales slumped and the company's interurban railway lost revenue because of a decline in the lumber business, exacerbated by a troubling economic recession from 1913 to 1915, which affected all parts of PRL&P's businesses. Unable to shift funds from other parts of its business and unable to attract capital from its investors, PRL&P had to beg the city council for a fare increase at the very time ridership was falling and 20 percent of Portland's male population was unemployed. World War I caused a temporary economic boom in Portland, with aluminum and shipyard facilities springing up along the Willamette and Columbia rivers and ridership hitting a new record high, but by 1920 this temporary boost was over. A postwar economic slump brought back the prewar financial challenges. A continued decline in passengers dogged the company for the rest of the decade. Trolley revenues declined every year after 1921, and by 1931 the company began operating in the red. World War II boosted revenues for a while, but the handwriting was on the wall. Portland's last electric streetcar ended operation in 1950.[34]

The story of Spokane's streetcar system is much the same as Portland's. Ridership peaked in 1910 with Washington Water Power's rail business serving 24.7 million passengers that year. A steady decline started in 1912, and passenger traffic fell more than 90 percent to 2 million riders by 1915, mainly as a result of competition from jitneys and automobiles. WWP shut down the interurban rail line to Cheney in 1922, invested in motorbuses in 1931, and finally closed down its last streetcar line in 1936.[35] Much the same pattern repeated itself across the United States. Net income for electric railway and streetcar companies (revenues minus expenses) rose until 1912, then declined thereafter.

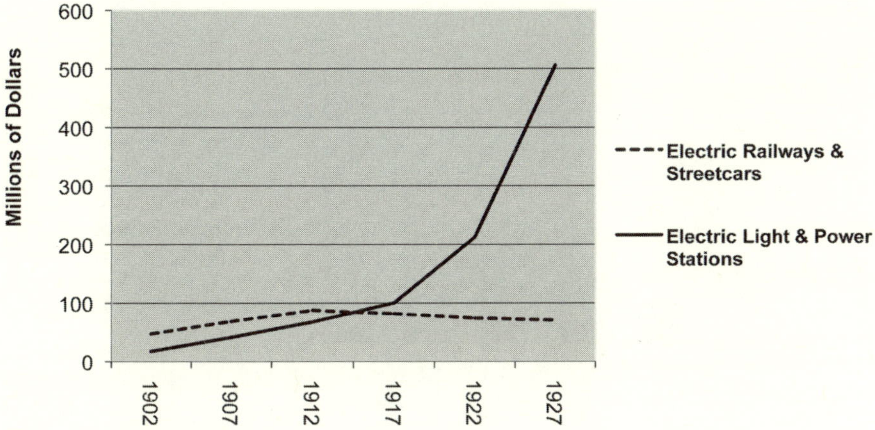

Figure 3.4. Net Income for Electric Railways and Central Electric Light and Power Stations in the United States. Source: U.S. Census Bureau, *Central Electric Light and Power Stations* (various years). Graph by Paul Hirt.

In contrast, net income for electric light and power companies not serving the streetcar industry rose steadily and steeply.

Needless to say, there was another, cultural reason for the decline of the trolleys: the American predilection for the private automobile and the personal mobility it allowed. Increasingly in the 1910s and 1920s, Americans shifted away from using mass transit and toward using motorcars. Only the middle and upper classes were able to take advantage of this new transportation technology at first, so the wage-labor class suffered most from the decline of the trolleys, but even they hoped to acquire a motorcar someday and supported the proliferation of public roads, without always realizing that trolleys were in competition with roads for public funds. Government transportation subsidies could have sustained the trolleys, but they went instead to building and maintaining roads.

Many people have argued that city ownership of streetcar systems could have sustained this form of mass transit, since earning a profit was secondary to providing a public service. But many public streetcar systems suffered similar problems: burdensome debt, rising costs, declining ridership, lack of investment capital, poor management, and competition from motorbuses, cars, and publicly funded roads. Seattle bought a formerly private streetcar company in 1918 but suffered financial woes over it for decades. Portland looked into buying the streetcar system from PRL&P in 1920 but demurred as the cost and legal complexity of doing so became clear. Cities were acquiring private streetcar assets all over the country at that time due to the general debilita-

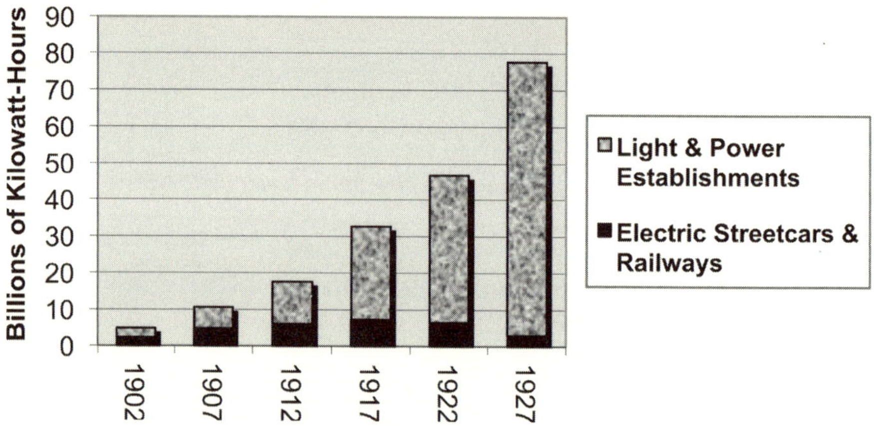

Figure 3.5. U.S. Electric Power Generated by Central Light and Power Stations versus Electric Streetcar and Railway Companies, 1917–1927. Data source: U.S. Census Bureau, *Census of Electrical Industries, 1912*, 19, 186 and *1927*, 12. Graph by Paul Hirt.

tion of the business and the exit of investors from this field. Indeed, ninety-one streetcar companies controlling 11 percent of all the track mileage in the United States were in receivership or out of business by 1920. As historian E. Kimbark MacColl concludes, "municipal ownership in itself did not necessarily lead to a more viable or cost-efficient transit system."[36] Nor did it necessarily lead to greater longevity. Portland's trolley system, privately owned to the bitter end, was one of America's last functioning electric streetcar systems when it shut down in 1950.

As the private sector pulled out of the electric trolley business and city governments invested heavily in roads and pavement, this late-nineteenth-century harbinger of the modern city became a relic of the past, replaced by an even greater symbol of modernity—the private automobile. Interestingly, many cities have recently brought back the trolleys on a limited scale as an expression of political support for organized mass transit, cleaner air, less traffic congestion, and a sense of historical appreciation for what has become a romantic symbol of a bygone era.

By the 1920s, following the decline of streetcar systems, utilities serving urban areas were exiting the streetcar business and shifting to a compensatory strategy of pushing electric "load growth" in households and retail establishments.

ELECTRICITY IN RETAIL BUSINESSES AND HOMES

Electric lighting, together with trolley transportation, transformed commercial businesses in downtown areas and stimulated the rise of a consumer culture. Commercial businesses installed lighted signs to attract customers, bright indoor illumination to accentuate products, electric heating and air-conditioning systems to keep shoppers comfortable, elevators and escalators to lessen the burden of using stairways in multi-story buildings, motorized conveyor belts to move freight from storage to display areas, and electric adding machines to total up customers' purchases. Compared to the dark and often dingy mercantile stores in the candle and kerosene age, opulently lit, multilevel department stores modernized by electricity created what David Nye refers to as "emporiums of desire" for consumers, especially women of the burgeoning new middle class. As a consumer culture arose, middle- and upper-class women's roles in the family economy shifted from being a major producer to being the manager of household consumption. In other words, women increasingly purchased food, clothing, and candles in stores rather than making them at home or bartering them with neighbors. Moreover, since more and more of their customers were women, most department stores hired a large number of female sales clerks, thus opening up a new employment option for women.[37]

National name-brand products and giant department stores became symbols of the time. "This is the era of Big Business," proclaimed the *Puget Sound Electric Journal* in 1915, and Seattle was privileged to have a branch of "the biggest 'mail order' house in the country—Sears, Roebuck and Company."[38] The Seattle mail order branch of Sears maintained additional warehouses in Spokane; Portland; and Stockton, California, which distributed products to all Pacific coast states as well as Idaho, Montana, and Nevada. In one of its important customer profiles, the journal effused about Sears's state-of-the-art, nine-story building that occupied an entire square block in Seattle's downtown. All the electricity was supplied by what was then Puget Sound Traction, Light, and Power from two steam-driven generators housed in the building's basement that burned unneeded wooden crates and boxes for fuel. Emphasizing its modern features, the journal praised the facility's 13,000 light sockets, eleven high-speed passenger and freight elevators, extensive telephone system, pneumatic vacuum tube network for delivering paper messages (in leather cylinders) throughout the building, and a rooftop electric sign 160 feet long with seven-foot letters lit up by 310 lamps spelling out the company's name.[39] These impressive electrical features awed workers, visitors, and shoppers with a sense of being witness to cutting-edge technological progress.

After the turn of the century, electricity found its way far beyond the commercial districts and into suburban and small-town residences. In 1904, residents of Emmett, Idaho, received electricity for the first time. The power company provided electricity only from sundown to midnight and for three daylight hours each on Tuesday and Thursday. The nighttime electricity supported indoor illumination after dark, allowing householders to retire their kerosene lamps. The Tuesday and Thursday hours of service allowed women to use their new electric irons after wash days.[40] The electric iron was one of the first household electric appliances invented and marketed to homeowners and consequently was often the first electric appliance most women acquired and used.

The arrival of electric lights and household electric appliances—irons, cooking stoves, toasters, butter churns, clothes washers—heralded a transformation in household labor. Incandescent lighting meant no more daily kerosene wick-trimming and lamp-oil filling, less smoke to breathe (although stoves still burned coal or wood), and less soot to clean off walls, ceilings, curtains, and clothes. The electric iron and electric clothes washer made an even greater impact on domestic chores, as the following passage about a pre-electric home reveals:

> At their farm in eastern Idaho, Mildred Jensen helped her mother on wash days. . . . The work required priming the hand pump and hauling the water by bucket into the house. One wash needed 50 gallons of water, which weighed 400 pounds. To heat it, Mildred filled the hot water reservoir in the stove. She rubbed the clothes on a washboard to knock out the grime and dirt. Whites and baby clothes were boiled. Then there was lifting the waterlogged articles, including heavy work clothes, wools and linens, wringing them out, lugging them outside in baskets or tubs, stooping and lifting to hang them on the line, where in the winter the wind froze her damp hands. Once the clothes were dry, there was taking them down and hauling them once more back to the house. . . . Heat for the irons came from a hot stove, so it was necessary to stand near it, comfortable only in the winter. To prevent a hot iron from scorching the cloth, each iron fresh from the stove had to be tried on a piece of scrap cloth or paper.[41]

Numerous versions of hand-cranked washers and wringers were invented and sold in the nineteenth century to assist in this labor-intensive task, but the electric motor really improved the usefulness of the machines and boosted sales. The first rudimentary electric washing machine was invented in 1908. By 1925 dealers sold nearly a million electric washers at an average price of $150. Nine years later, in the midst of the Depression, 1.4 million washers were sold at an average price of $60.[42]

Always keen to acquire new customers and encourage them to consume

Figure 3.6. "Electric Ranges Make Happy Housewives." Source: Craig Wollner, *Electrifying Eden: Portland General Electric, 1889–1965* (Portland: Oregon Historical Society, 1990), 109. From Portland General Electric Collections.

more electricity, utility companies regularly sponsored newspaper advertisements for electrical appliances while their company bulletins cheerfully recounted stories of happy customers with their newfangled electric gadgets (see Figure 3.6). Two newspapers in The Dalles, Oregon, joined a campaign with Pacific Power & Light (PP&L) offering customers a year's subscription to either newspaper plus a Hotpoint electric iron, all for the bargain price of

The Most Popular

Electric Appliances

A PRICE LIST

For Your Convenience

	Price	Cost per Hour at 5c K.W.H.
Air Heaters for small rooms	$7.50	4c
Boudoir Sets: 3-lb. iron, curler and stand	5.00	1½c
Curling Irons	4.50	1/12c
Fans, 8-in., 12-in., 16-in	19.65, 26.55, 33.00	¾c
Grills and Toaster Stoves	4.00 to 7.50	3½c
Hair Dryers	16.00 to 25.00	¼c
Irons: 6-lb. and 61/2-lb., all standard makes	5.50 to 6.50	3c
Percolators: many styles and sizes	7.50 to 20.00	1¾c
Plates: single, for light cooking	8.00 to 12.50	4c to 7½c
Plates: double, almost equal to a range	18.00 to 20.00	8c to 11½c
Sewing Motors: attach to any machine	15.00	1/8c
Sewing Machines with motors attached	39.50 to 51.00	1/8c
Toasters	5.00 to 7.00	2½c
Vibrators	18.00 to 25.00	1/8c
Vacuum Cleaners: a great variety	30.00 to 65.00	1c
Warming Pads	8.00	1/25c
Washing Machines	115.00 to 145.00	5/8c

Figure 3.7. "The Most Popular Electric Appliances." Source: *Puget Sound Electric Journal* (August 1918): 37. Facsimile prepared by Paul Hirt.

$3.50. The local manager of the PP&L facility in Pomeroy, Washington, was renowned for his success in selling washing machines to utility customers. In 1913, 20 percent of the company's electric customers had electric washers, with an average of four washers sold each month. The Sunnyside and Dayton offices of the company had window displays with electric egg incubators, and Dayton even sponsored a contest to see who could guess how many of the 120 incubated eggs would actually hatch. The company bulletin proudly reported that thirty Dayton-area farmers used incubators.[43]

PP&L participated in a traveling demonstration of electric cooking stoves; sponsored an exhibit at the Oregon State Fair in 1915 with electric washing machines, milking machines, grinders, freezers, cutters, pumps, and ranges; and negotiated with architects and builders to place electrical wiring and outlets in new homes. In Astoria, Oregon, in 1916 the Hyland Homes Company agreed to build eight homes at the end of the utility company's streetcar line, all outfitted with special circuits for plugging in electric ranges. Several all-electric homes (sans chimneys) appeared in Hood River, Oregon, also in 1916. For the entire year of 1916, PP&L sold the following to its customers: 84 electric ranges, 20 electric water heaters, 504 toasters, 45 curling irons, 50 sewing machines, and 2,027 electric flat irons.[44] Figure 3.7 shows a facsimile of a household appliance advertisement that appeared in the August 1918 issue of the *Puget Sound Electric Journal*. All of the appliances were for sale by the Puget Sound Traction, Light, and Power Company.

A relatively new innovation in financing facilitated the purchase of all these appliances: consumer credit. Normally, a customer had to pay full price cash for retail products, but in the early twentieth century, businesses increasingly extended credit and term loans, especially for more expensive durable goods.[45] In fact, it was the only way to boost consumption. Power companies regularly offered their own forms of credit and what we now call package deals. As subsidiaries of the big electric appliance manufacturers like G.E., many local power companies sold both appliances and electricity, often as a single unit. Commonly in the early years of the business, for instance, a customer would get an electric iron or a cookstove and all the electricity to run that appliance for a set annual or monthly price. Sometimes power companies installed electric appliances, free, into new homes through arrangements with builders and then recouped their costs through subsequent electric bills. This practice still exists today in a few places where utility companies give away free electric water heaters to their customers with the only stipulation that the customer actually install and use it for at least two years. Heating water with electricity consumes, on average, 30 percent of a typical homeowner's monthly electric usage today, so it is easy to see why giving away electric water heaters makes good business sense.

In the early years, metering of electric usage was inconsistent and sometimes complicated. When utilities sold customers an appliance and the current to run it as a unit for a set fee, no meters were installed at all. When customers bought electric appliances separately and then needed to purchase electricity for them, utilities would often meter each appliance as well as the lights, so that a home might have multiple electric meters. By the 1920s, many of these inefficient practices were replaced by the system we are familiar with today: a single electric meter registering all the electricity used by a home or business. Because of these innovations and incentives, between the turn of the century and the 1920s, electricity and electrical appliances transitioned from novelty to standard equipment in homes, particularly urban homes, bringing to bear economies of scale that led to a steady decline in prices for both. In a self-reinforcing cycle, the decline in prices stimulated more purchases and more electricity use, which allowed companies to further lower prices. Already by 1912, one sixth of American homes were electrified; fifteen years later two thirds were electrified.[46]

These modern, electrified homes with their newfangled conveniences were supposed to be cleaner, brighter, healthier, and happier. And the labor required to maintain these homes was supposed to be less onerous and time-consuming. Compare the earlier passage about women's work in the pre-electric home with this poem from a utility company journal in 1929, titled "My Mother's Happy Now":

> My mother says she's never tired
> Since father had the house all wired.
> First thing she percolates and toasts,
> And afterwards she boils and roasts.
> The clothes are washed and on the line
> Before the kitchen clock strikes nine.
> The ironing's done and put away
> Lots quicker than the sad iron way.
> Her vacuum makes the rugs like new
> And cleans the dust from corners too.
> She sews for miles on her machine
> And sharpens knives to edges keen.
> She turns on switches everywhere
> To light the lamps and curl her hair.
> And when the weather's hotter than—
> My mother has her electric fan.[47]

Of course this advertising ploy was more hopeful than real. If happiness could be secured at the end of an electric plug, we would all be in perpetual bliss. Moreover, the idea that electricity reduced household labor, while a logical assumption, in fact turns out not to be the case. Historian Ruth Schwartz Cowen thoroughly dissected and debunked this myth in her 1983 book *More Work for Mother*,[48] which shows that as certain tasks became easier or quicker to perform with the aid of household technologies, new tasks were added to women's work routines—maintaining the same overall hours of labor and sometimes increasing them. Even more ironically, as vacuums and washing machines made it easier to clean houses, social expectations of cleanliness rose, and women ended up spending more time cleaning house in the twentieth century than they did in the nineteenth century without these conveniences. With better indoor lighting, women worked in the home for longer hours before dawn and after dusk. With better cooking equipment, women were subject to higher culinary standards that required more time in the kitchen. In short, Cowan found that the modern (electrified) home rarely made the lives of housewives more leisurely as many ads implied. This does not deny that certain tasks involved less drudgery and that electrified homes were often more pleasant and healthy places to live, but it is nevertheless important to distinguish between the alluring claims of advertising and the realities of life.

Similarly, an important distinction should be made between electrification and electrical modernization, as historian Ronald Tobey has argued. It is one thing to be hooked up to a power pole and quite another to own and use electric appliances on a regular basis. Many homes and farms were "electrified" long before their owners were able to purchase and fully integrate electrical products into daily life—to "modernize." While electrification was widespread by the 1910s, electrical modernization did not become common until the 1920s in urban areas and the 1930s in most rural areas.[49]

Historians also differ in how much they portray this phenomenal diffusion of electric appliances as a product of consumer demand versus a product of sharp marketing. Obviously both factors were at play. But the aggressive advertising and astute marketing strategies of electric manufacturers and power companies led at least one scholar to opine that electric appliances in these early years were "sold" more than "bought."[50] This was big business, consistently profitable enough to attract big investors via savvy marketing campaigns. But consumers were never passive and never afraid to demand better products, prices, services, and social benefits when companies performed poorly or failed to meet expectations (see next chapter).

ELECTRICITY AND AGRICULTURE

Isolated farmers and small farm communities rarely got power until the New Deal's Rural Electrification Administration started stringing power lines to the unelectrified countryside (see chapter 9). As late as 1930, less than 10 percent of U.S. farms had electricity service from central station electric companies, another 5 percent had their own generators, and the remaining 85 percent lived and worked without electric light and power. In fact, many farms remained off the grid until the 1950s. But there were some important exceptions, especially in the arid West, where successful agriculture requires artificial irrigation. Wherever farmers needed dams, canals, and pumps to bring water from rivers or aquifers to croplands, electricity found a niche.

Although the popular image of the American Northwest is a wet one—green forests, snow-clad mountains, rushing rivers, misty coasts, and ubiquitous drizzle—much of the region is actually quite arid due to the "rain shadow" effects of its mountain ranges, especially the Cascade Mountains. It is indeed wet on the Pacific slope of the mountains, but the low-elevation interior Columbia Basin of eastern Washington and Oregon is very dry, with the Tri-Cities area getting only six to eight inches of rainfall annually. That is less precipitation than Tucson, Arizona, gets in the Sonoran desert (eleven inches). Southern Idaho is the other main desert region in the Northwest, with most of the area along the Snake River Plain getting less than ten inches of rainfall annually (see Figure 3.8). These arid regions of the Northwest attracted irrigation crusades that accelerated after federal and state incentives and assistance arrived following passage of the Carey Act in 1894 and the Newlands Reclamation Act of 1902.[51] Idaho took full advantage of both laws. Indeed, Idaho irrigated more acres than any other state under the Carey Act. These reclamation projects developed at the same time that electrification first developed, so the two often grew up in tandem. For example, two hydropower/irrigation dams installed in arid Ada County in southwest Idaho, Arrowrock and Deer Flat dams, led to a tripling of the number of farms there (from 1,600 to 5,000 farms) between the 1900 and 1920 censuses.

Geography influenced the system. Despite Idaho's famous aridity, upland forest and rugged mountains prevail in the areas north and east of the Snake River Plain, capturing enormous amounts of winter snow, which is released in the spring and summer down raging tributaries that feed into the Snake River. From the Wyoming border near Grand Teton National Park (where the Snake River originates) westward to the Oregon border, the Snake River drops a vertical half-mile in elevation. Along its westerly journey, the river plunges over Idaho Falls, Twin Falls, American Falls, Shoshone Falls, and Swan Falls before leaving the irrigable plain and coursing through Hells Canyon, the deepest river gorge in North America. This river is a hydropower developer's dream. Both public and private efforts to exploit this resource proliferated after 1900.[52]

The very specific geography of the Snake River itself also influenced the development of irrigation and electric power in the region. Some irrigation projects needed only a dam and gravity to deliver water to farms, but the Snake

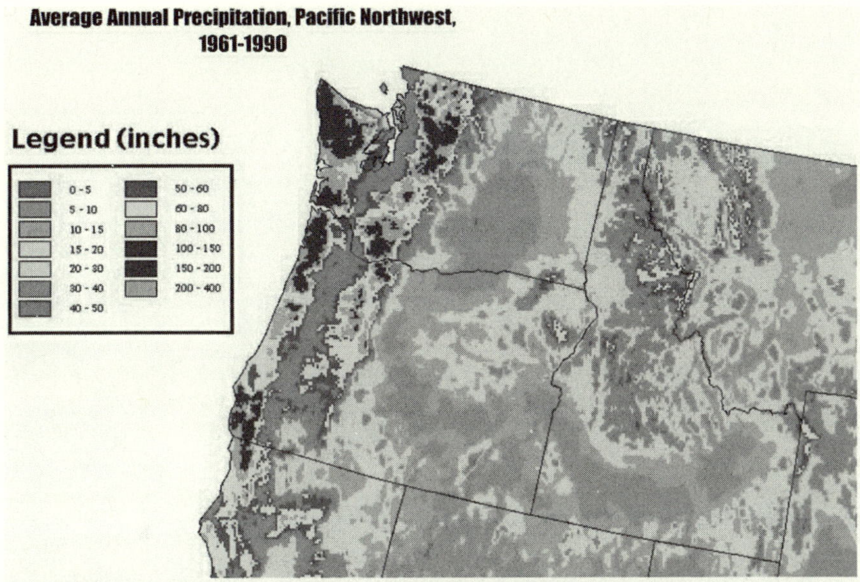

The Cascade mountains divide the wetter west from the drier east. Source: Mapping by C. Daıy, graphic by G. Taylor and J. Aiken, copyright ' 2000, Oregon State University.

Figure 3.8. Average Annual Precipitation, Northwest States, 1961-1990. Source: Oregon State University, Spatial Climate Analysis Service. Copyright © 2000, PRISM Climate Group, Oregon State University, http://prism.oregonstate.edu. Map created by C. Daly, graphic by G. Taylor and J. Aiken.

River flowed at the bottom of a steep-walled canyon, so pumps were usually needed to get the water from the reservoir up to the benchlands where most of the arable land was located.

Water pumps could be powered by an engine burning oil, kerosene, or other fuels, but since dams provided excellent opportunities for generating hydropower, electric pumps were most common in irrigation districts after the turn of the century. Electricity literally and figuratively energized irrigated agriculture in the Northwest, contributing significantly to the founding of farm communities, population growth, and economic development in these regions.

In 1907, the Bureau of Reclamation (BoR) completed Minidoka Dam in southeast Idaho, one of its first irrigation projects. The dam included turbines to generate power, which was used to pump irrigation water into a distribution system. But much more power was available than the BoR needed, so starting in 1910 it delivered surplus power to the various farm communities served by the reclamation project. Towns such as Burley, Declo, Heyburn, Rupert, and Minidoka created municipal electric systems to receive the low-

Figure 3.9. Snake River, Southern Idaho. Photograph by Paul Hirt.

cost federal power.[53] Unincorporated farm communities also organized nonprofit electric companies to receive cheap federal power from Minidoka Dam to provide electric service to scattered farms. These electric cooperatives, or "mutuals," flourished in the Minidoka project area in part because the farmers were familiar with this type of organization. Cooperative water users' associations established under state incorporation laws were the primary means for receiving and distributing the federal irrigation water coming from Minidoka. It took just a little more effort to organize a second cooperative to receive and distribute electricity. One of the Northwest's first electric co-ops, the Farmers Electric Company, formed at Heyburn in 1914. At least five others followed its lead in the next seven years. By 1920, power from the Minidoka project delivered through these various municipals and mutuals served 1,100 families. Seventy percent of the project farmers had wired their homes and farm buildings, and more than 6,000 electric appliances had been purchased, including irons, washing machines, vacuum cleaners, and cooking stoves.[54] Besides the thousands of electric home appliances farmers bought, they also acquired electric egg incubators, butter churns, grinding stones, water pumps, and other electrical tools, as well as heating and lighting for livestock facilities.

The 1912 Census of Electrical Industries noted that for those farms with access to power, new uses for electricity developed very rapidly. It cited a National Electric Light Association report enumerating fifty different applications for electricity in farmhouses, thirty different electric motor applications for barns and fields, and twenty applications for dairies, not to mention fruit and vegetable processing and canning operations.[55]

Schools, too, adopted electric power for heating, lighting, and even cooking. The three-story red-brick schoolhouse in the town of Rupert, served by Minidoka power, claimed to be the first school in the United States to be heated entirely by electricity (1914). In Boise in 1913, two private power companies competed for a contract to provide Boise High School with sixty electric stoves for the school's "domestic science" lab. The school board enthused that "Boise school girls will learn to cook on electric stoves in an electric kitchen in the future."[56]

As noted above in the case of Boise, private power was also available in southern Idaho, often competing for customers with publicly owned systems. Before Minidoka came online, small, independent, isolated systems sprang up all along the Snake River and its tributaries. Most served small areas with poorly maintained, low-budget facilities. But the federal government's commitment to irrigation at Minidoka eventually attracted a financier from Pittsburgh, named William Kuhn, to invest in the region. The same year the Bureau of Reclamation completed Minidoka Dam, Kuhn started buying electric companies, water rights, and dam sites, intending to consolidate and improve electric power services and provide irrigation water to farmers in southern Idaho. In the next few years, he and his brother built a central station power plant at Shoshone Falls and bought a hydropower and irrigation project at Twin Falls (the North Side project). Hydropower from the falls fed into high-voltage transmission lines linked to other power plants Kuhn purchased and to a network of customers up and down the valley.[57]

Eventually, the Kuhns overextended themselves and suffered from bad financial timing. In an attempt to monopolize the electric power business in the region, they borrowed millions of dollars to buy out or squeeze out competitors and open new irrigation projects on lands they bought. Rather than servicing only existing populations, they often extended power lines and irrigation infrastructure ahead of settlement into areas they controlled, in hopes of future profits. Not content to dominate an irrigation empire, they ran power lines to Pocatello and Boise, and bought the Boise and Interurban Railway Company. All this aggressive investment occurred too quickly and recklessly. At American Falls in 1913, a reservoir the Kuhns built flooded land they did not own, embroiling them in litigation and stalling the project. Many irrigation projects built by private, for-profit companies in southern Idaho, including some of the Kuhns' projects, were disintegrating from poor planning, faulty construction, and inadequate financing. Farmers refused to pay water bills, profits slipped away, and investment capital grew scarce. When the 1913 economic recession hit, a Kuhn bank in Pittsburgh collapsed, and the Kuhns' financially strained business ventures fell like dominoes. In 1916, S. Z. Mitchell of Electric Bond and Share Co. intervened and created the Idaho Power Company out of the mess left by the Kuhns and some of their smaller competitors.[58] Thereafter, both power and irrigation expanded steadily and in tandem in southern Idaho, pushed by the Idaho Power Company on one hand and the federal government on the other.

In southern Idaho farm country, electric space-heating was the next greatest consumptive use of electricity after powering irrigation pumps. Electric heating is very energy-intensive, much more so than lighting, traction, or powering motors. Nevertheless, it got a solid foothold in irrigation communities in the early twentieth century, starting with the Minidoka project in 1910. Since irrigation pumps were used only during the growing season, and since Idaho winters are long and cold, the Bureau of Reclamation decided to offer cheap electricity for home heating in the winter, when the water pumps were shut down. All the towns and families served by power from Minidoka Dam quickly got used to affordable electric heat to mitigate the chill of winter. In those days buildings had no insulation. Because this off-season power was so affordable, farmers adopted home heating with great enthusiasm.[59]

Farms not served by the Minidoka project wanted the same opportunity, but Idaho Power and other private companies were in business to earn profits, and most did not offer winter electric rates as cheap as the BoR's. Some companies with plenty of surplus winter power did contract to sell cheap electricity for winter heating, but after S. Z. Mitchell consolidated private utilities into the new Idaho Power Company in 1916, the company stopped renewing those contracts. A dry winter and low water in 1915 had highlighted the potential conflict between selling hydropower cheaply to farmers for heating and storing that water in reservoirs for use during the more profitable spring and summer irrigation season. So, while Minidoka project farmers enjoyed heated homes, customers of Idaho Power experienced higher rates and curtailed heating service. Consumers organized a movement promoting winter home heating and fought the Idaho Power Company's effort to cancel home-heating contracts, arguing that they had invested money wiring their homes and purchasing heaters and other appliances with the expectation of receiving low-cost winter power. Idaho electric consumers took the battle to the Idaho Public Utilities Commission (PUC) in 1918, which had the authority to regulate electric rates. But the PUC disappointed the farmers and sided with the Idaho Power Company, allowing it to raise home-heating rates to "profitable" levels. The commissioners struggled over this decision, however, because their investigation revealed that representatives of the power companies had convinced more than 500 customers to adopt electric heating technology in new buildings by promising that the electricity would be abundantly available and cheaper than coal heating. Changing to a coal heat source in many cases would require large investments.[60] According to Idaho Power's historian, Susan Stacy, this decision endorsed Idaho Power's desire to prioritize the use of electricity for pumping irrigation water and closed the subject of electric home heating for investor-owned power company customers in Idaho until 1958.[61] Apparently, electric heat would not be a part of most modern homes in southern Idaho until after midcentury.

CONCLUSION

For many people in the early twentieth century, modernity meant having new opportunities made possible by new technologies. In southern Idaho and other parts of the arid Northwest, electricity lifted life-giving waters out of deep canyons to irrigate lands previously incapable of growing commercial crops. It warmed without smoke or soot stone-cold houses during long, gray winters. Electricity offered a new and ostensibly better life, a "modern" life. Whether one lived on a remote farm, in a small town, a streetcar suburb, an industrial tenement, or a hilltop mansion, the adoption of electricity promised (though did not always deliver) greater productivity and profit, comfort and safety, cleanliness, satisfaction, and status. At the same time, electricity accelerated trends toward the increasing concentration of wealth and power in the hands of an elite class of national and international business leaders and financiers, what Alan Trachtenberg has called "the incorporation of America."[62] Electrical manufacturers like G.E. and Westinghouse, and the entrepreneurs behind electric light, power, and traction companies, spawned revolutions in factory production, transportation, communication (telephones and telegraphs are electrical technologies), consumer products, entertainment (radio and motion pictures are electrical technologies), education, and culture. Nearly everyone was affected by this new, "modern" electrical technology.

In the beginning, the electric light and power business suffered growing pains, technological failures, cutthroat competition, and booms and busts. But eventually, the "search for order," in Robert Wiebe's memorable phrasing,[63] led to the systematization and stabilization of the industry into a reliably profitable capitalist enterprise. But as electrical technology penetrated ever deeper into economic, public, and private life, the body of citizens as a whole became impatient with the gaps and inequities inherent in this system designed primarily for profit. Consequently, a parallel search for order and equity unfolded through the political process, energized by a progressive ideology that Samuel Hays appropriately called "the gospel of efficiency."[64] Increased public ownership of electric systems and public regulation of private systems resulted. Starting out as a series of obscure inventions, the harnessing of electricity for lighting and motive power became in succession a technological spectacle, then a small business opportunity, then a regional commercial enterprise backed by national and international financiers, then an integral part of modern life, and finally an object of intense political debate and public regulation. That last stage is the subject of the next chapter.

CHAPTER 4

Profit and Welfare

The man of great wealth owes a peculiar obligation to the State, because he derives special advantages from the mere existence of government.

—Theodore Roosevelt, 1906

Americans and Canadians entertained grand visions of electrical modernization in the first three decades of the twentieth century, but their technological optimism was tempered by some disappointment in the performance of electric utility companies, anxiety over growing monopoly power, inequities in the price and availability of electrical service, and business instability. Aggrieved consumers, large and small, demanded that government intervene to secure more reliable service and more affordable rates. Community leaders saw electricity as integral to economic growth and social welfare. At the same time, investors with property rights in the electrical facilities saw this profitable enterprise as appropriately under their control. Initially, customers negotiated with suppliers in a free market, but the interests of consumers often conflicted with the interests of producers, so consumers sought an external advocate for their interests. Moreover, in this era of rapid expansion, the producers themselves were often at odds with each other. Many sought an arbitrator to settle conflicts and create a more secure investment climate. The unfettered market worked to a degree, but not well enough to the liking of either producers or consumers. Consequently, a diverse coalition of citizens, business leaders, and politicians pushed for laws designed to make the market work better by structuring the relationship between consumer and supplier in such a way that both got a steady stream of the things they wanted: affordable service for the one and reliable profits for the other. Throughout the United States and Canada in the early decades of the twentieth century, businesses sacrificed a measure of freedom in exchange for reliable profits. Governments legitimized and protected for-profit corporations as monopoly providers of public services, requiring in exchange basic standards of service and fairness.[1]

Cities and states have long sought to secure essential public services either by providing them directly or by contracting with for-profit companies to provide them under specified conditions. If a private company could be induced to provide the desired service at a secure profit, then the government avoided the trouble of providing the service itself. This kind of public-private partnership, exemplified by the municipal franchise discussed in chapter 2, has provided many essential public services over the last century and a half. This special relationship between governments and privileged business interests providing public services was most common at the municipal level but involved higher levels of government as the scale of enterprise expanded beyond discrete urban areas. In the United States and Canada, for example, state, provincial, and federal governments had set up civil commissions to regulate railroads and other "common carriers" since the late nineteenth century. Over time a suite of companies deemed "public utilities" or "public service corporations" came under the regulatory authority of these commissions, providing transportation, water, natural gas, municipal lighting, electric power, telephone, telegraph, and other similar public services.

The United States pioneered state regulation of electric companies in the early 1910s, with Canada shortly following suit. From 1911 to 1916, all the states in the U.S. Northwest classified electric light and power companies as "public service corporations" or "public utilities" and placed them under the authority of state commissions that regulated their rates, services, service territories, facility standards, accounting procedures, and public safety. In 1917, Québec became one of the first provinces to adopt the American model of utility regulation. British Columbia briefly created a Public Utility Commission in 1919 but abolished it a year later before it had accomplished anything. British Columbia then lagged behind its counterparts in both the United States and Canada, postponing the establishment of a functional and lasting public utility commission until 1938. This left to the discretion of B.C. electric utilities the right to set electric rates and to determine where and how to expand electric service. Only contract stipulations with municipal authorities or large industrial customers constrained the liberty of British Columbia's electric utilities.[2]

Along with the rise of state and provincial regulatory agencies in the early twentieth century, a vocal contingent of reformers in both the United States and Canada fought to have electricity provided by nonprofit publicly owned utilities selling power at cost and offering universal service as a measure against which to judge the fairness and adequacy of private utility performance. Advocates of the public interest at the turn of the century railed against the increasingly powerful and consolidated private-sector utilities earning record profits from what reformers considered inflated charges to customers. Ontario led the way at the provincial level, creating its own provincial electric power authority in 1910, developing hydropower facilities at Niagara Falls, building extensive transmission facilities, and directly marketing cheap power throughout the province. No state in the Northwest, then or now, produced or produces power for its citizens through a state-owned electric utility. Apparently unimpressed with Ontario's achievement, British Columbia did not establish a provincially owned power system until 1945. Municipally owned utilities did thrive to some extent in British Columbia in the early years of the century and they will be discussed below, but because British Columbia came late to both regulatory activism and public power production, this chapter focuses predominantly on developments on the U.S. side of the border with a particular emphasis on the 1910s, when state legislatures first began regulating electric utilities.

Publicly owned power systems provided a threatening alternative to regulated private ownership of public utilities. Defenders of capitalism and free enterprise attacked the "creeping socialism" they feared in public ownership, while their opponents attacked the profiteering of wealthy corporations at the expense of the humble ratepayer. In between these two ideological poles, government policy trod a crooked path. While the two sides alternately won different battles, neither won the war, forcing them into an uneasy accommodation. As a result, both countries developed a curious, hybrid system of publicly owned and privately owned utilities sharing the electricity supply sector, with private utilities embedded in a strongly regulated environment.

MERGERMANIA

Selling electricity and electrical machines and appliances became so lucrative in the early decades of the twentieth century that America's banking and investment elite poured their pocketbooks, their hearts, and their souls into the business. By 1910, electric utilities were the most heavily capitalized industry in the nation, as well as one of the most consistently profitable.[3] This promise of profitability attracted steady investment, system expansion, technological improvement, and constant mergers and acquisitions. It also spawned shady schemes for wealth aggrandizement, political corruption, and a very large measure of public concern and distrust. Despite many early consolidations, so many new companies were launched that the total number of electric utilities in the United States continued to increase steadily until World War I. Thereafter, however, corporate mergers significantly reduced the number of utilities, even while power production and numbers of customers continued to grow, as shown in Table 4.1.[4] Many small utilities lacked the capital to expand

Table 4.1. Number of Utilities, Generator Sizes, and Total U.S. Kilowatt Capacity*

Date	Number of Utilities	Average Kilowatt Output per Generator	Total Kilowatt Capacity in U.S.
1902	3,620	380.2	1,212,235
1907	4,714	648.5	2,709,225
1912	5,221	1,075.9	5,165,439
1917	6,542	1,475.9	8,994,407
1922	6,355	2,330.3	14,313,438
1927	4,335	6,145.3	25,811,305
1932	3,429	10,425.5	34,662,554

* Data source: U.S. Census Bureau, *Census of Electrical Industries, 1932, Central Electric Light and Power Stations* (Washington, DC: U.S. Census Bureau, 1934), Table 30, and adapted from tables provided in Brigham, *Empowering the West: Electrical Politics before FDR* (Lawrence: University Press of Kansas, 1998), 10, and Edison Electric Institute, *The Electric Light and Power Industry in the United States*, Statistical Bulletin no. 4 (New York: EEI, January 1937), Table VI, p. 10.

or purchase new technology, and they became prime targets for takeover by larger companies. In the Pacific Northwest and throughout the country, an elite, nationally organized collection of mega-corporations acquired control over electricity services for more and more customers. This would be the trend throughout the twentieth century, most especially during the 1910s and 1920s (see chapter 7 for more on this).

Electric companies based on hydropower experienced this consolidation even earlier. While steam-driven power plants can be located almost anywhere, there are a limited number of good hydropower sites on rivers. Consequently, sharp investors quickly scoped out the best opportunities and staked their claims. The U.S. Bureau of Corporations reported in 1909 that thirteen utility holding companies controlled one third of the nation's commercial water power. Consolidation came so quickly that only three years later the same federal agency reported that a mere ten companies controlled three fifths of the nation's water power.[5] Among northwestern utilities, which depended heavily on hydropower, seven national holding companies controlled virtually all of the investor-owned (for-profit) electric companies by 1930. These financial conglomerates engaged in a turf war with each other to dominate the best dam sites and the most lucrative service areas, absorbing dozens of struggling, isolated, little electric companies that had sprouted up in the previous two decades.[6]

Technological factors contributed significantly to this consolidation trend.

Table 4.2. Increase in Size and Voltage of Generating Units in U.S., 1900–1930

Date	Maximum Size of Generating Units in U.S. (kilowatts)	Date	Maximum Transmission Voltages in U.S. (volts)
1900	1,500	1892	10,000
1920	60,000	1908	110,000
1930	75,000*	1923	220,000

* In 1930 there was actually one unit generating 208,000 kilowatts, but the typical maximum then was 75,000 kw. The unit generating 208,000 kw remained the single largest electric generator in the U.S. until 1956.

Source: Federal Power Commission, *National Power Survey, 1964* (Washington, DC: U.S. Government Printing Office, 1964), 14.

First, the development of high-voltage, long-distance transmission made it possible for one company to serve very large areas. Second, power production technology continually increased in scale and cost, advantaging companies with ready access to investment capital. Companies constantly sought ways to produce more power more cheaply and they accomplished this primarily through economies of scale. Throughout this era, the size and efficiency of turbines increased, temperatures and pressures in steam-driven generators increased, transmission voltages and distances increased, and central power plants grew ever larger (see Table 4.2).[7] All of this lowered the average price per kilowatt-hour of electricity, and the savings were partly captured as profits and partly passed on as lower electric rates for consumers.[8] To successfully compete in this economy, utilities had to invest continually in bigger and better electrical generation and transmission technology. Companies tied to national financial houses were in the best position to upgrade and expand their operations.

The logic of profit-seeking also led to consolidations. Utilities followed three main strategies for increasing their revenue. They encouraged existing customers to use more electricity, usually by marketing electrical appliances; they expanded their distribution lines within their service area to reach new customers; and they bought out or drove out competitors to absorb additional systems and service areas. This elimination of competition appeared efficient and rational from one perspective but ominously monopolistic from another perspective. Eight electric utilities accomplished most of the consolidations in the U.S. Northwest. Pacific Power & Light originally centered in Astoria, Oregon, is a good example. Incorporated in the state of Maine in 1910 as a subsidiary of the gigantic eastern holding company Electric Bond and Share, it bought out in quick succession utilities throughout the Columbia Basin,

Figure 4.1. Electron Power House, Puyallup River, Washington. Source: Puget Sound Power and Light, *Hydro-Electric Development: An Illustrated Story of the Power Properties of the Puget Sound Power and Light Company* (Seattle, WA: Puget Sound Power & Light, 1920), 15.

including systems in Yakima, Kennewick, and Pasco in 1910; Dayton, Waitsburg, White Salmon, Goldendale, Prosser, Hood River, and Toppenish in 1911; and Pomeroy in 1912.[9] The same man who created Idaho Power, as described in the previous chapter, S. Z. Mitchell, orchestrated the PP&L acquisitions, as well as mergers in Utah, Idaho, and Wyoming to create the Utah Power & Light Company, incorporated in 1912. Mitchell's successful consolidation of the Idaho Power Company came four years later, in 1916. Nineteen electric utilities served Idaho at the time, and most were experiencing financial difficulties. A Boise newspaper editor named Calvin Cobb pushed for Mitchell and his eastern financial backers to intervene. The companies that Mitchell consolidated into Idaho Power in 1915–1916 owned about a dozen hydroelectric plants on the Snake River and its tributaries, including those at American Falls and Shoshone Falls. Mitchell planned to integrate the entire 300-mile-long Snake River Valley of southern Idaho into a single electric grid tied together by a central high-voltage transmission line. While his mergers were successful, his electric grid integration dream did not become a reality until 1937, with the intervention of the federal government's Bonneville Power Administration.[10]

In the populous Puget Sound area of western Washington, another national heavyweight, Stone & Webster, consolidated five local utilities into Puget Sound Power & Light Company in 1912. This new company controlled the three biggest hydropower facilities in the Puget Sound region: the dams at Snoqualmie Falls, Lake Tapps on the White River, and the Electron Project on the Puyallup River.[11]

Puget Sound Power & Light had only one real competitor, the municipally owned Seattle City Light, and their rivalry over rates and over control of valuable hydropower sites enlivened local politics for many decades.[12]

Four utilities serving the four major population centers of Montana all merged in 1912 to form the Montana Power Company. Fed almost entirely by hydropower, this system generated more electricity than any other system in the Northwest and northern Rockies because of the huge electric demand of the mines in Butte, Montana. The Butte electricity market was so lucrative that it paid for the construction of unusually long high-voltage transmission lines quite early in the century. In 1910, there were already 720 miles of high-voltage transmission lines developed by the companies that would soon merge to become Montana Power.[13]

Washington Water Power (WWP), with headquarters in Spokane, followed its own path of expansion and consolidation in these years too. After 1907, electricity from its hydropower plants at Spokane Falls and Post Falls was distributed throughout eastern Washington and North Idaho via 600 miles of 60,000-volt transmission lines. WWP expanded its Post Falls plant in 1913 and built new hydropower plants at Little Falls, Upper Falls, and Long Lake on the Spokane River in that decade. Until the 1910s, WWP sold wholesale power to dozens of local utilities serving small populations in this largely agricultural region. Thereafter, the company aggressively acquired the small utilities to which it had previously sold power. By 1930, after forty separate acquisitions involving over 100 small utilities, WWP controlled a unified electric system covering 20,000 square miles and serving 200,000 customers.[14]

As small electric companies consolidated into well-financed regional corporations, economies of scale resulted in better service, lower rates, and higher profits. Of course, rate reductions and improved service also resulted from the regulatory interventions of state public utility commissions (more on this below). Electrical generating technology improved constantly as bugs in the early equipment were worked out regularly by engineers employed by the electrical industry. Newer equipment was more reliable, lasted longer, ran more efficiently, and met the needs of increased loads and higher voltages.[15] Pacific Power & Light's mergers, for example, absorbed many small family-owned businesses that were providing electricity in a haphazard manner with unreliable equipment, unsophisticated controls, and poor maintenance. For the outdated system serving Waitsburg, which PP&L absorbed, the operator purportedly used a light bulb to determine when to adjust the voltage: the bulb was connected to the electrical system, and when it went dim, the operator would send more water through the penstock to the turbine; when it burned too brightly, the operator reduced the flow of water. The bulb reportedly hung over a cot where the operator napped.[16] In another example, the system PP&L acquired serving Goldendale, Washington, ran on a small hydroelectric generator placed in a precipitous 600-foot-deep canyon, where erratic water flows provided erratic service. A poorly maintained power line from the canyon to Goldendale and problems with the generator resulted in electrical outages an average of five nights every month.[17] According to the minutes of PP&L's board meetings, during the first twelve months of the company's existence (1910–1911), it invested just under $1 million in improvements to these newly acquired systems and new transmission lines—a substantial amount of money in 1911.[18]

PP&L not only absorbed its competitors' systems, but simultaneously initiated a construction program that brought new electrical service to dozens of communities in the next decade. Ten communities in Washington and Oregon received their first electricity from PP&L in 1911, thirteen more in 1912, three more in 1913, eight more in 1915–1916, and five more in 1917–1918. These service expansions continued at a similar pace in the 1920s.[19] In addition to serving communities, utility companies in Washington and Oregon finally recognized, as Idaho had earlier, that irrigation farming offered a great opportunity for electricity sales, as farmers often needed to pump irrigation water up from rivers or out of deep wells. To tap this market, PP&L built transmission lines to six new irrigation districts in 1912. By 1915, PP&L supplied power to 125 irrigation systems, watering 25,000 acres of farmland in the Columbia, Snake, and Yakima river valleys.[20]

The more compact municipally owned systems expanded their customer base at this time, too. Seattle's population skyrocketed from 90,000 to 240,000 in the single remarkable decade between 1900 and 1910—an astounding 250 percent increase in ten years. Seattle City Light (the municipally owned utility) captured about 20 percent of the market, so it had to constantly expand its generation and its transmission to new customers. By World War I, Seattle's utility was planning to construct three new dams on the Skagit River, which it eventually completed in 1924, 1927, and 1940. A little further to the south, the municipally owned utility Tacoma City Light constructed a $2.3

million hydroelectric project in 1910, tapping water from the Nisqually Glacier some thirty miles from the city. The utility then sold its excess power to more than a dozen nearby rural communities that formed their own publicly owned systems to enable them to purchase the power wholesale.[21] These municipal systems, however, were legally restricted from selling electricity beyond their municipal boundaries, so they did not grow large and dominate extensive territories the way private companies did.

ELECTRICITY AND SOCIAL RESPONSIBILITY

Amid the booster fanfare of expanded electrical services, lower prices, and improved reliability arose complaints from those who were not yet on the electrical grid, those who felt their utility was overcharging (rates were by no means uniform, especially between public and private utilities), those whose service was inadequate or intermittent, and those who feared the growing monopoly power of the utility companies. These concerns arose coincident with the broad political movement in North America we now refer to as "progressivism." During the Progressive Era (1890s–WWI) Americans struggled to adapt their political systems to the new challenges of a modern world. Since at least the Civil War, economic opportunities and the businesses that pursued them evolved much faster than did political institutions in America. As a result, the nation faced the complex problems of modernization with the political tools and social values of an agrarian age.[22] This provided capitalists and entrepreneurial corporations a unique window of opportunity: a moment in American history when they enjoyed more free political rein and more unfettered liberty to expand their operations and accumulate property and profits than at probably any other time in the nation's history. The results were mixed. While industrial capitalism brought considerable wealth, power, and new products to the nation, the distribution of the first two—wealth and power—remained highly concentrated in the hands of a small portion of the population in the late nineteenth century. Moreover, business interests in a world of cutthroat competition pursued self-interest almost as single-mindedly as capitalism's founding philosopher Adam Smith predicted they would. Unfortunately, Smith's "invisible hand" often failed to guide these thousands of selfish decisions in directions beneficial to the public weal as a whole. The entrepreneurial power and creative forces of market capitalism many times floundered in cutthroat competition, market saturation, resource depletion, pollution, public health problems, labor conflict, social approbation, and corrupt governance.

A large but fragmented business elite and its political allies essentially governed the U.S. economy in the late nineteenth century, overseeing the conversion to industrial capitalism while tolerating a mixed and sometimes troubling record of economic and social changes.

By the turn of the century, many Americans were convinced that a more rational, equitable, and sustainable path to modernization was both possible and necessary. Dozens of disparate reform organizations at the local and state levels finally coalesced into a national movement to reshape social, political, and economic institutions. One of their key tools for reform, of course, was government. At all levels progressives sought to expand the power and reach and professionalism of governing institutions to enable them to guide, assist, and when necessary restrain the power of corporations—all in pursuit of their vision of responsible economic and social development. These efforts were sometimes intelligent, sometimes naive, always compromised, and never fully successful at solving targeted problems. Nevertheless, they represented incremental advances in directing the complicated process of modernization in more beneficial paths.

The Progressives loved electricity and the modernization that it represented, but they had their own ideas about how electrical services should be organized so as to provide the greatest good to the greatest number. Investor-owned, for-profit utilities—which accounted for the great majority of electrical supply in the United States then and now—developed with one primary guiding purpose: to make money. Americans got a system designed primarily to generate profits and build stock values for large corporations. Often enough, the profit motive and market-based behaviors yielded socially desirable results, but many times the reverse was true. To Progressives, electricity seemed too precious, too useful, too essential to the modern age to deny it to anyone or to allow monopolistic corporations to extract excessive profits at the expense of relatively powerless consumers. The unregulated, privately controlled, for-profit electric system seemed a major design flaw needing correction. According to historian Jay Brigham, those engaged in this debate after the turn of the century broke down into three general camps: (1) those who opposed any government intervention, (2) those who favored government regulation of privately owned utilities, and (3) those who favored a wholly publicly owned electric supply system. As Brigham describes it,

> Private power proponents, usually political conservatives and utility representatives, argued that laissez-faire should apply: government should not be involved in the production and distribution of electricity. The regulatory group, which included many

Progressives, believed that electrical production and distribution should remain in private hands but be subject to state or federal regulation. That group believed private utilities should receive a fair return on their investment but should not overcharge consumers. Public ownership supporters insisted that only government ownership could guarantee fair electrical rates. This group was convinced that greed would prevail if laissez-faire economics prevailed. They also considered regulation inadequate; selfish and unscrupulous utility executives could easily buy off regulators and alter account books.[23]

Reformers launched state regulation of electric utilities in this unsettled political climate. Although Progressives railed against government corruption, they usually trusted government more than they trusted unrestrained corporations, for the main reason that government was accountable to voters. Progressives abhorred inefficiencies, too—plowing a significant percentage of a private utility's revenue back into stock dividends for distant investors and paybacks to pyramided holding companies seemed inefficient at best when that money could be better spent improving service or lowering electric rates for customers. Public ownership of electric utilities simply made sense to many people.[24] Of course, investor-owned utility executives opposed the publicly owned utilities for both ideological and financial reasons. But because so few laypeople sympathized with the profit woes of big out-of-state corporations, the defense of private utilities usually centered on the ideological: public ownership was "un-American," it smacked of socialism, it deprived entrepreneurs of their right to earn a profit, and, they constantly reiterated, publicly owned utilities did not pay taxes to support local and state government services. Even though municipalities could provide more public benefit by owning the electric system than by taxing the profits of private utility owners, this political rhetoric sold well among certain segments of the electorate. Advocates of public ownership had both financial and ideological replies: public ownership almost invariably offered lower electric rates, cheaper power attracted business, a nonprofit utility could accommodate considerations of equity and fairness rather than simply profitability, revenue from city-owned electric utilities lowered the overall tax burden of residents, and it was more "democratic"—in effect, all the utility customers were owners.

To the above-mentioned three political camps in the electric system debate should be added a fourth camp: those who felt a hybrid system of private and public utilities was acceptable. These advocates sought to build up the publicly owned utilities to the point that they would adequately serve as competition for the investor-owned systems, keeping a downward pressure on electric rates and providing a benchmark for evaluating fair rates. This group also supported a regulatory framework for privately owned utilities. The well-known Progressive economist Richard T. Ely was perhaps most prominent in this group of policy advocates.[25] At the time, however, these innovations met stiff resistance from many quarters, including from advocates of local control, such as special interest groups who stood to lose free access to resources or other privileges, politicians who preferred "log-rolling" democratic negotiations to decision-making by bureaucratic experts, federal administrators and judges who believed the government lacked legal or constitutional authority to engage in many innovative regulatory activities, and finally the industries targeted by the reformers. Despite resistance, increasing regulatory power in government institutions emerged as the clear trend from the Progressive Era onward throughout the twentieth century, paralleling the consolidating, centralizing tendency in corporate financing and industrial organization.

MUNICIPAL-OWNERSHIP ADVOCACY BEFORE WORLD WAR I

As discussed in chapter 2, cities finally got involved in the electric service business, either by granting franchises to specific companies to provide these services under specified conditions or by developing their own city-owned light and power systems. The franchise process was susceptible to corruption, however, and small towns were no match for the growing power of the regional electric companies. Only large cities could bargain to their advantage and enforce franchise terms effectively. The alternative was to form a municipally owned system. Such systems experienced a growth spurt between 1897 and 1907, during which time 60–120 new municipal utilities formed each year in the United States, until by 1912 they represented about 30 percent of all utilities. But the growth of new municipal systems slowed after 1912, when many states began regulating investor-owned utilities. Richard Rudolph and Scott Ridley, in their 1986 book *Power Struggle*, argue that state regulation actually hurt the public power movement, as many reformers came to believe that the state regulatory framework would protect consumers enough that publicly owned systems were no longer necessary.[26]

Even though the growth of city-owned light and power systems slowed in the second decade of the twentieth century, the hostile rhetoric between public power and private power advocates accelerated. Both were competing for the lucrative urban markets, desirable dam sites, the support of policy makers, and the loyalty of customers. This debate played out on the national and international stages as well as in the Northwest. The larger debate drew in the United States, Canada, and several western European nations. In England and Germany, municipal ownership of electric utilities had advanced quite a bit by the second decade of the twentieth century, and in Ontario the provincially owned hydropower system based at Niagara Falls had also expanded greatly. These examples energized advocates of public power as they launched campaigns to "municipalize" electric power systems in the United States. The debate grew increasingly polarized and ideologically driven.

The National Civic Federation, a prominent group of business and labor leaders, academics, and politicians in the United States, which sponsored policy debates on important topics of the day, in 1915 published a book featuring pro and con arguments. They had studied the issue for two years a decade earlier and voted in 1907 to endorse state regulation of private utilities rather than public ownership. But the issue persisted, so the federation took up the debate again in 1914–1915, publishing two sets of essays, one group in the affirmative and one in the negative, on the question "Shall the Government Own and Operate the Railroads, the Telegraph and Telephone Systems? Shall the Municipalities Own Their Utilities?" The redoubtable Progressive reformer Frederic C. Howe and the information director for the National Socialist Party, Carl D. Thompson, were among those arguing in the affirmative. Howe provided testimony on the success of public ownership in England and Germany, with information from a book he had written on the topic. He argued that the transition to public ownership in Europe had led to better service, lower prices, increased revenue to the city from fares and electric rates, an easing of the tax burden for citizens, better conditions and wages for city workers, less labor strife, and less bribery and political corruption related to franchises. It sounded almost utopian. Carl Thompson took up the ideological argument, stating unequivocally that "private ownership of a public utility is fundamentally hostile to and inconsistent with the public welfare." He provided pages of anecdotal evidence of private ownership leading to excessive rates, poor service, labor exploitation and strikes, extortionate profits, overcapitalization, the concentration of wealth and power, and political corruption. He then argued that state and federal railroad and utility regulation had failed, citing a study by Daniel Hoan on the topic, and concluded that only public ownership protected the public interest.[27]

Reading the arguments of those on the negative side of the debate shows just how confusing this could be to the average citizen. Using a similar mix of statistics, anecdotes, and appeals to fundamental principles, the opponents contradicted nearly every argument of the advocates. J. W. Sullivan (whose expertise is not identified in the book) contributed an essay titled "Municipalization or a Just Regulation? A Plea for the Facts," in which he countered many of Howe's claims about municipal-ownership success. He argued that municipalization in Europe had slowed and in some cases reversed because of financial and managerial problems and that municipal ownership of utilities in the United States was rife with failures. He challenged Howe's economic data asserting that municipal systems were less efficient. Revealing interesting biases, he complained that the improved wages, hours, and conditions offered in public employment only increased labor strife outside public employment. He also claimed that a large body of public employees beholden to a politically appointed administrator threatened both democracy and the rights of labor to collective bargaining. The ironic implication was that laborers were better off bargaining for wages and conditions with a company unaccountable to voters and focused on profits than with a company accountable to voters and focused on social welfare. Sullivan's closing peroration indicates just how emotionally charged the debate was and how an "appeal for the facts" could degrade into personal recriminations and ideological posturing. Denigrating Howe's "infantile guesswork observations, his unbolted statistics, and his blissful ignoring of the recorded facts that have disproved his dreamy theorizing," Sullivan attacked public ownership at its core:

> What is the matter with municipalism? The reply is that it brings with it bureaucracy—inelasticity, "overstaffing," sinecurism, the evils in general of political management; and the menace of a million united municipal employees; and a tyranny over the million by the superior officials . . . and withal the blundering, inertness, costliness, and injustice of misapplied government.[28]

Clearly, the advocates of public ownership trusted government more than corporations, while their opponents just as clearly trusted business more than government. Citizens joined whatever side of the debate most conformed to their own views on the matter and did not usually bother to analyze the claims and statistics offered by these dueling experts. Those who clove to neither ideological position could be swayed by personal experience or simply favored whichever system promised them the best chance for lower rates and better service. Consequently, some efforts to form municipal utilities at this time succeeded, especially when a positive experience of municipal ownership existed nearby. Starting up a municipal system required municipal bond funding, which had to be approved by voters, so advocates had to struggle in the public arena to win the hearts and minds of citizens.

Once urban electrification was widespread and service monopolies were established within each city—essentially accomplished by the 1910s—choosing between a municipal and a privately owned system became a zero-sum game: one's gain was the other's loss. If public power advocates wanted to switch to a municipal system, they had to buy out the existing private system, a very fraught political action involving heated negotiations over fair market value, franchise rights, and more. Likewise, if a private utility wanted to sell power in a city with a municipal system, it had to convince the city to sell its system or carve out specific monopoly service areas for the company. These factors exacerbated the hostility between public power and private power advocates.

Choosing between the two types of ownership was only the first stage in the rivalry between public and private power. Struggling over who would serve new suburbs as cities expanded, competing over electric rates, and battling for control of water-power sites comprised the second stage. One of the most prominent and long-term conflicts of this type was the competition between the municipally owned Seattle City Light and the privately owned Puget Sound Power & Light (called Puget Sound Traction, Light, and Power in the early 1910s). Seattle began producing its own power to provide street lighting as early as 1905 but did not create a city-owned electric utility until 1911, then called the Seattle Lighting Department. The first annual report of the department opened with a statement revealing the underlying conflict at play:

> The Seattle plant has been written up frequently by private lighting companies in both western and eastern states and figures have been twisted out of recognition to prove that Seattle's plant is a "monumental failure." The fight against municipal ownership of lighting plants has been a strong and persistent fight. . . . The Seattle Lighting Department has taken the attitude that energetic work and consequent success will do more to offset unjust criticism than all of the answers to critics that it could have written.[29]

From 1910 to 1912, voters in Seattle quickly approved a series of bonds to expand the city's hydropower facility on the Cedar River and to purchase rights to hydropower sites on Lake Cushman in the Olympic Mountains and White River in the Cascade Mountains. The 1911 report notes proudly that "the people of Seattle plainly and clearly demonstrated their determination to keep the lighting and power situation of the City of Seattle in their own hands."[30] The energetic work and public support paid off. By 1913, Seattle's municipal utility claimed that with only one or two possible exceptions, it had the lowest electric rates in America and had deservedly earned its reputation as "America's best lighted city."[31]

The willingness of voters to authorize bonds to expand the municipal light and power facilities derived in large part from the immediate, positive effects of city ownership on lighting rates. The 1911 Seattle Lighting Department report laid it all out in black and white. In 1902, before state utility regulation was in place, the two private electric companies serving Seattle charged 20 cents per kwh for residential lighting. This high rate led to citizen agitation to establish a city-owned lighting plant. In response, the private companies immediately dropped their rates to 12 cents per kwh, but the city power plant was authorized nonetheless. When the city-owned electric plant came online in 1905, the city set residential lighting rates on a descending scale starting at 8.5 cents per kwh. To retain their customers, the private companies lowered their rates again to 10 cents per kwh. In 1911 the Seattle Lighting Department lowered its rates again to 7 cents per kwh and then to 6 cents in 1912. Puget Sound Traction, Light, and Power, the only remaining private utility serving Seattle in 1912, responded by matching the city rate—but only where the private company was in direct competition with Seattle City Light's lines. Outside areas of direct competition, the company's rates remained about double the rates of the Seattle Lighting Department.[32] The antagonistic relationship between these two utilities drew regional and national attention for decades. They battled over control of prime hydropower sites, fought over the construction of transmission lines, undercut each other in the political arena, and tried many times to eliminate the other from the Seattle service area.[33]

This rate disparity was not unique to Seattle. In a 1912 study of electric rates in twenty-seven Washington communities, the Northwest Electric Light and Power Association (ironically, a private industry association) found that electric rates in towns with municipally owned utilities (e.g., Seattle and Tacoma) were on average half as much as those in towns served by private utilities. While the Seattle and Tacoma systems charged 6 to 7 cents per kwh, the twenty-five investor-owned utilities charged an average of 12.7 cents—and this was three years after Washington had started regulating its electric utilities.[34] Similar situations occurred across the region as Seattle's experience emboldened other communities to take similar actions. In one interesting case in 1915, the city of Baker forced Eastern Oregon Light and Power to significantly reduce its rates for commercial lighting by threatening to expand its municipally owned street-lighting facilities to serve downtown businesses. Commercial customers in Baker were satisfied with the lower rates subsequently charged by the private electric company, but folks in neighboring La Grande, Oregon, were not. La Grande did not have a city-owned light and power plant and therefore lacked the leverage needed to convince the power company to lower its rates to match what it charged in Baker. Oregon had recently given its Public Service Commission authority to regulate electric companies, so the La Grande

Commercial Club filed a complaint against Eastern Oregon Light and Power for charging "discriminatory rates." The commission investigated and found that the lower rates in Baker provided the company a reasonable profit and ordered the company to reduce its commercial lighting charges for La Grande and other communities in eastern Oregon to the rates charged in Baker.[35]

All this competition had predictable political repercussions. Citizens in towns with successful municipal utilities grew enthusiastic about public power, while private companies watched their profits dwindle and markets siphon away. The for-profit utilities responded vigorously, claiming the competition unfair, the city rates artificially low, the municipal lighting plants uneconomic, and the principle of public ownership un-American. They lobbied state legislatures to limit the spread of municipal utilities and limit the areas where city-owned utilities could sell power. Some of these efforts were successful and others not. Washington legislators, for example, passed a law in 1915 prohibiting municipal utilities from selling power outside their city limits. Seattle and Tacoma, each with a successful public power system, had attempted to interconnect their utilities with a high-voltage transmission line to provide low-cost power to neighboring communities. Private utility lobbyists blocked it through the legislature.[36]

Private electric companies also fought the municipals by trying to monopolize the best hydropower sites on the region's rivers. An intriguing struggle during World War I between the Stone & Webster holding company and Seattle City Light over the rights to a dam site on the Skagit River situated on national forest land in the North Cascades is profiled in chapter 6. This struggle and many other similar competitive battles over good dam sites during the first half of the twentieth century highlight several important characteristics of the evolving electric power system in the Northwest. First, throughout the twentieth century, the primary source of energy for electric systems in the region was falling water. The region's rivers were at the very center of power production—physically, economically, politically, even emotionally. In British Columbia in the 1910s, hydroelectric generators comprised 84 percent of the province's total electric generating capacity.[37] South of the border in that decade, the figures were similar. Hydropower represented 81 percent of Washington's generating capacity, 65 percent of Oregon's, 89 percent of Montana's, and a whopping 92 percent of Idaho's capacity. While hydropower has always dominated Canada's power system—86 percent of it in 1918—it has never represented more than about a third of the power supply in the United States—exactly 33 percent in the 1912 electrical census.[38] The U.S. Northwest is thus particularly well-endowed with hydropower capacity. Second, the Skagit controversy underscores how the extensive presence of federal lands in the region facilitated a powerful role for the federal government in shaping the electric system. These two regional characteristics—the influence of rivers and of federal authorities—are the main subject of the next chapter.

Municipally owned utilities did flourish in British Columbia, but their numbers exaggerated their importance. At the time of the first Canadian Census of Electrical Industries in 1917, there were forty-eight central electric stations in British Columbia, twenty-one of them municipally owned. At first glance this appears to indicate a significant public power sector, but in fact those municipal stations controlled only 6 percent of the province's power supply. Comparisons with the U.S. Northwest are instructive. While Oregon's municipally owned providers also controlled about 6 percent of the power supply (in the 1912 electrical census), Idaho's municipals controlled only 1 percent, and Washington's municipals controlled 48 percent.[39] Clearly Washington was the public power leader.

From a social standpoint, comparing the customer base of the municipal utilities to the private utilities may be a more useful exercise than comparing with that of horsepower or kilowatt capacity because at this time industry consumed much more power than did domestic use. One mining operation or lumber mill could easily consume as much or more power than all the homes and businesses in all the communities adjacent to the mine or mill. The big private electric companies sought out these high-use customers first and only secondarily went after scattered residential customers. Municipal utilities, on the other hand, primarily electrified streetlights, homes, and businesses, putting them in direct contact with a larger percentage of total customers than their small amount of generated power suggests. While municipal utilities in British Columbia controlled only 6 percent of the power supply, they served 17 percent of all electric customers. The U.S. Northwest again offers a wide divergence in the percentage of customers served by municipal systems, with Oregon close to British Columbia at 10 percent, Idaho at the bottom with only 5 percent, and Washington scoring a rather amazing 60 percent of customers served by municipal utilities in 1912.[40] Washington's enthusiasm for public power persisted as it led the charge for regional public power in the 1930s (chapter 10), and retains today the largest public power sector in the U.S. Northwest.

This corrective of looking at the customer base, however, may once again exaggerate the importance of the public power sector in British Columbia, since many of these municipal systems did not produce their own electricity but instead bought it at bulk rates from private electric companies or local industries with surplus power to sell. So, even though the municipalities controlled distribution and retail rates, the wholesale power came from commercial for-profit enterprises. This occurred in the United States too, but more often the successful Northwest municipal utilities generated their own power (as in Seattle and Tacoma) or bought it from a federal agency (e.g., the Bureau of Reclamation's Minidoka Project or, after 1937, the Bonneville Power Administration).

A comparison of electric rates charged by private and municipal utilities in British Columbia further challenges the significance of municipal ownership there. Table 4.3 displays the data I have been discussing here regarding electric power systems in British Columbia. It shows that electric lighting rates in British Columbia were highly erratic from community to community, just as in the United States, ranging from 5 cents per kwh to 25 cents per kwh. But unlike in the U.S. Northwest, there was no clear price advantage to being served by a municipal utility. Among the thirteen cities with the lowest rates, ten were served by private electric companies, including the most populous service areas around Vancouver and Victoria. Among those with the highest rates, half were municipal utilities. Only economies of scale in hydropower provided clear price advantages. The communities served by utilities with large hydroelectric facilities paid the lowest rates. This likely explains why British Columbians took so long to embrace a provincial public power system. In fact, the main impetus for the development of a public power system in Ontario and then later in British Columbia was frustration with the slow pace of private development.

STATE INTERVENTIONS IN THE U.S. NORTHWEST

While the battle between publicly owned and privately owned utilities raged on in scattered cities across the region, the great majority of businesses and residences got their power from private electric companies. As these private electric systems grew larger and more regionally integrated, effective regulation required a higher level of authority beyond the municipal franchise. Cities could not control these large utilities, nor did they exercise any protective jurisdiction over utility customers outside their city limits. At the same time, the utility industry itself began searching for a means to slow the movement for public ownership and to blunt the voices of those calling for federal antitrust proceedings against utility monopolies.

One of the more farsighted leaders of the electric industry, Samuel Insull, who owned one of the nation's largest utility empires, which was centered

Table 4.3. Electric Generation and Distribution in British Columbia, 1918

Town/City	Town/City Population	Number of Customers	Municipal or Private	Hydro	Heat Fuel	Lighting Rates
Alberni & Port Alberni (combined)	1,440	333	Municipal	No	Oil	.11/kwh
Anyox	2,320	250	Private	Yes	Oil	—
Armstrong	950	250	Municipal	Yes	Oil	.10–.15/kwh
Ashcroft	—	81	Private	No	Oil	.20–.25/kwh
Britannia Beach (Britannia Mining & Smelting Co.)	—	275	Private	Yes	Oil	.5/kwh
Chase (Adams River Lumber Co.)	600	100	Private	No	Wood waste	—
Chilliwack*	1,600		Private	Yes	Oil	.4–.11/kwh
Clayburn** (Clayburn Co. brick & tile works)	200	25	Private	Yes	No	—
Coal Creek—mines (Crow's Nest Pass Lt & Power)	500	102	Private	No	Coal	.50/lamp/mo.
Copper Mountain—copper mines	200	Mines only	Private	No	Coal	—
Courtenay	500	110	Private	Yes	No	.7–.14/kwh
Cranbrook	2,500	550	Private	No	Coal	.18/kwh
Cumberland—coal mining (Canadian Collieries Ltd.)	1,200	550	Private	Yes	No	.7–.14/kwh
Duncan	800	165	Municipal	No	Oil	.15/kwh
Enderby (Okanagan Saw Mills Ltd.)	950	169	Private	No	Wood waste	.12–.14/kwh
Fernie	3,146	600	Municipal	No	Coal	.13/kwh
Fraser Mills (Canadian Western Lumber Co.)	900	118	Private	No	Wood waste	—

Golden (Columbia River Lumber Co.)	900	120	Private	No	Wood waste	.70–$1.00/ 50-watt lamp
Grand Forks***	1,700	400	Municipal	Yes	No	.5–.13/kwh
Greenwood	778	150	Private	Yes	No	.9–.13/kwh
Hedley (Hedley Gold Mining Co.)	500	90	Private	Yes	No	.15/kwh
Kamloops	4,500	1,150	Municipal	Yes	Coal	.7–.13/kwh
Kaslo	722	225	Municipal	Yes	No	.15/kwh
Kelowna	2,500	525	Municipal	No	Wood & coal	.18/kwh
Ladysmith	3,000	425	Municipal	No	Coal	.13/kwh
Merritt	703	300	Municipal	No	Coal	.15/kwh
Michel—mining town (Crow's Nest Pass Lt & Power)	1,100	81	Private	No	Coal	.13/kwh
Mill Creek (Whalen Pulp and Paper Mills)	—	50	Private	Yes	No	Incl. in dwelling rent
Mission**	797	60	Private	Yes	No	.10/kwh
Nanaimo	8,000	1,800	Private	Yes	Coal	.7–.14/kwh
Naramata	200	15	Private	Yes	No	.50/lamp/mo.
Nelson	4,476	1,400	Municipal	Yes	No	.7–.9/kwh
New Denver	300	90	Private	Yes	No	.20/kwh
New Westminster*	15,000	3,297	Municipal	Yes	Oil	.4–.9/kwh
Ocean Falls (Pacific Mills Limited)	2,000	135	Private	Yes	No	No charge to employees
Peachland	385	60	Municipal	Yes	No	.10–.50/lamp
Penticton	3,000	564	Municipal	No	Oil	.10–.15/kwh
Phoenix***	662	—	Private	Yes	No	.6–.15/kwh

(*continued*)

Table 4.3. *Continued*

Town/City	Town/City Population	Number of Customers	Municipal or Private	Hydro	Heat Fuel	Lighting Rates
Powell River (Powell River Co. mills)	—	225	Private	Yes	No	.5/kwh
Prince George	1,500	205	Municipal	No	Oil	.18–.22/kwh
Prince Rupert	6,505	1,000	Municipal	Yes	Coal	.75–.105 / kwh
Princeton (Princeton Coal Co.)	400	100	Private	No	Oil	.15/kwh
Quatsino (Whalen Pulp and Paper Mills)	—	55	Private	No	Wood & coal	Incl. in dwelling rent
Revelstoke	3,150	796	Municipal	Yes	Gas	.8–.13/kwh
Rossland (incl. Cascade)***	3,500	800	Private	Yes	No	.4–.10/kwh
Salmon Arm	500	105	Municipal	No	Oil	.14–.16/kwh
Sandon	300	75	Private	Yes	No	—
Spence Bridge	105	20	Private	Yes	No	.50/lamp/mo.
Summerland	2,500	150	Municipal	Yes	No	.125/kwh
Swanson Bay (Whalen Pulp and Paper Mills)	—	35	Private	Yes	No	Incl. in dwelling rent
Trail***	4,000	700	Private	Yes	No	.4–.10/kwh
Vancouver (incl. Burnaby, Cloverdale, Delta, Huntingdon, Langley, North Vancouver, Point Grey, Port Moody, South Vancouver)*	149,873 (Vancouver alone = 102,550)	40,600	Private (BC Electric Railway Co.)	Yes	Oil	.3–.8/kwh
Vernon	2,500	775	Municipal	No	Oil	.10/kwh

Victoria (incl. Esquimalt, Oak Bay, Saanich District, Tod Inlet, Uplands)*	45,000	13,254	Private (municipal street lights)	Yes	Oil	.4–.11/kwh

Data source: Leo Denis, *Electric Generation and Distribution in Canada* (Ottawa: Commission of Conservation of Canada, 1918), 212–230. Summary tables are also found on pages 281–283 of the report. Some summary totals in Denis's tables vary slightly from totals derived from my table above due to minor differences in how we lump and split communities. Also, I did not list one community found in the 1918 report because it had been "depopulated" due to the war and the electric system had been shut down.

*Denotes towns/companies that purchased bulk electricity from British Columbia Electric Railway Company.

**Denotes towns/companies that purchased bulk electricity from the Western Power Company of Canada (WPC), which owned the hydroelectric power station on Stave Lake. In 1918, WPC sold 63 percent of its power to the British Columbia Electric Railway Company, 21 percent to Puget Sound Traction, Light, and Power, and 16 percent to its own industrial customers.

***Denotes towns/companies that purchased bulk electricity from West Kootenay Power and Light Company.

in Chicago, saw the regulatory handwriting on the wall and proposed state regulation of utilities as early as 1898. At that time, Insull was president of the private industry-sponsored National Electric Light Association (NELA) and gave his policy speech as his presidential address to the association. To both industry representatives and policy makers, he argued that electrical service should be considered a "natural monopoly" in which competing systems were duplicative and wasteful. A utility should be granted a monopoly franchise, he argued, for the sake of efficiency. To appease the anti-monopolists and justify this special political privilege, these franchises would be authorized and overseen by state-level citizen commissions accountable to the people of the state. The commissions would guard against price-gouging and other undesirable business practices. The utility companies' financial books would be open to inspection, and commissions could approve or deny rate increase requests. In return, the utility would have no competition and a guaranteed "fair return" on its investments.[41]

Insull's own organization, NELA, did not endorse this proposal until 1907, but the idea caught on. Starting in 1907 with Wisconsin and New York, one state legislature after another passed enabling laws creating public utility commissions (PUCs) to regulate the electric industry in their states. By 1921, every state had a PUC except Delaware.[42] Though eventually bankrupt and disgraced for his role in the post-1929 financial collapse of his utility empire, Insull saw at least one aspect of the future of his industry presciently and helped to build the regulatory system that held sway in the United States for nearly the entire twentieth century.[43]

Most of the public utility commissions—also known as public service commissions—grew out of earlier citizen boards set up to regulate railroads. The commissions in the Northwest were no exception. Oregon's utility commission started out as a State Board of Railroad Commissioners in 1887 (established the same year as the Interstate Commerce Commission) regulating railroad rates, safety, and service. The legislature temporarily abolished it in 1898, creating a new Railroad Commission in 1907, expanding its duties to include the regulation of public utilities in 1911, and finally renaming it the Oregon Public Service Commission in 1915. Washington state created its railroad commission in 1905, and in 1911 the legislature expanded its duties to include the regulation of all common carrier transportation companies as well as public utilities, including telephones, telegraphs, ports, irrigation districts, and electric, gas, and water companies—renaming it the Washington Public Service Commission. Idaho and Montana created their PUCs in 1913, also out of older railroad commissions, with similar responsibilities that expanded over time.

Idaho currently calls its board a public utility commission, while Montana calls its board a public service commission.[44]

State laws extending the authority of public utility commissions over electric light and power companies issued from a convergence of interests. Consumers, businesses, and political leaders all lobbied on behalf of state regulation often for very different reasons. Moreover, these interest groups were not monolithic. Many powerful businesses themselves consumed large amounts of electricity, such as mines, lumber mills, and manufacturing plants, and they sought stability, fair rates, and reliable service just as homeowners did. In fact, power-consuming industries used the state commissions as leverage to force down electric rates to improve their own profits at the expense of electric company profits. The "Commercial Clubs" of many cities, precursors to today's Chambers of Commerce, went to the commissions seeking to force down electric rates and secure more reliable service in order to attract businesses and residents to their towns. City and state government leaders and the commissioners themselves (politically appointed) used state regulation to ensure more orderly development and encourage investment by creating a stable business-friendly infrastructure of affordable lighting, power, and transportation. Not all petitions to the commissions focused on business interests or self-interest, though. Some politicians and consumer advocates used the commissions to seek broader principles of fairness and equity in social and economic development. The PUCs themselves served all those interests in unpredictable and sometimes inconsistent ways.

A more or less common set of authorities and responsibilities structured the work of the commissions, providing a counterbalance to the diversity of pressures and special interests. First, to avoid the waste and inefficiency of redundant facilities and cannibalistic competition, they protected the monopoly status of electric companies within defined service areas by not permitting other companies to "invade" those territories. They accomplished this by requiring companies to obtain a "certificate of public convenience and necessity" from the commission before building power plants and transmission lines. The commissions simply denied these certificates to any company trying to encroach on another utility's service area. States hoped this policy would promote orderly, low-cost development of light and power systems.

Second, the commissions mitigated this monopoly privilege by controlling rates and service. Every electric company had to report its rate schedules, and any schedule could be reviewed and adjusted by the commission if it seemed unreasonable or if a consumer filed a complaint. Any subsequent changes in rates had to be financially justified and approved by the commission.

Third, to protect the interests of investors and keep the industry prosperous, the commissions allowed utilities to charge rates that provided a "reasonable return on investment"—usually 5–10 percent annually. In order to make defensible decisions regarding fair rates and remunerative return, commissions had the authority to review company financial records, inspect facilities, determine fair market value of a utility's properties, and distinguish between legitimate and illegitimate operating expenses and financial practices. Many companies were secretive or had poorly documented financial records, and the industry as a whole used highly inconsistent accounting methods. Consequently, one of the first tasks many commissions undertook was to prescribe uniform methods of accounting for the various utilities they regulated.

Beyond these financial authorities, the commissions had a number of public welfare responsibilities, too. They required companies to file annual reports of injuries and accidents, ordered facility upgrades to improve both service and safety, and penalized companies that acted against the public interest or in violation of laws or commission orders. The following survey of three Northwest PUCs during their early years of regulating electric utilities illustrates their work, challenges, and accomplishments in the second decade of the twentieth century.

In 1911, Washington's legislature transformed the state's Railroad Commission into a more expansive public service commission (PSC) with regulatory authority over 400 public utility companies in the transportation, communication, and electric power industries. The first report of the new commission summarized its legislative mandate: to ensure "reasonableness of rates, the sufficiency of service, [and] adequacy of facilities." In the introductory narrative of the report (a letter of transmittal to Governor M. E. Hay) the commissioners justified railroad and utility regulation as necessary to promote the orderly economic development of the state. High shipping rates hurt commerce and population growth, the report observed, while irresponsible fly-by-night irrigation companies cast a pall of "disrepute" on agricultural development in the arid regions of central Washington, inhibiting investment. At the same time, a distinct tone of public-interest advocacy is noticeable in the narrative: "Exact justice and equality to consumers and patrons is contemplated by the law and to that end the commission is directing its energies." To accomplish its extensive duties, the PSC, composed of three commissioners and a secretary, had the services of one engineer with four assistants.[45]

The number of Washington companies under the PSC's jurisdiction jumped to over 480 the next year, reflecting the rapid economic growth occurring at this time. These included railway, electric, water, gas, telephone, telegraph, and irrigation companies. The largest group, totaling 121, was the electric companies. The commission's staff had expanded to eight with the addition of an attorney, a rate expert, and several inspectors, but its responsibilities increased, too. In 1912, the PSC had 40,000 tariffs and rate schedules on file; launched efforts to prescribe three "uniform systems of accounting" for the gas, electric, and water utilities respectively; and required all regulated companies to file accident reports. State law required the commission to assess the value of all public utilities in the state to support rate evaluations, but limited staff and resources meant that the commission investigated only companies against whom complaints had been filed.[46]

Among the pro-consumer actions taken by the commission in 1912 were two orders establishing rules for how gas, water, and electric companies charged customers. In these early years, rate schedules were astonishingly complex and inconsistent. Utilities metered some electricity consumption (typically lighting and commercial power), while charging flat monthly rates for other items like illuminated signs, street lighting, irons, electric ranges, and water pumps. Irrigators were charged monthly rates for pumps based on horsepower rather than actual electrical consumption. Sometimes lighting companies charged a flat fee per bulb per month. Different kinds of uses were charged highly inconsistent rates: lighting cost a premium per kilowatt-hour, while cooking or heating or pumping enjoyed very low rates per kwh. Some customers were charged a minimum monthly fee; others were not. For metered customers, utilities invariably offered graduated rates, charging increasingly less per kwh the more a customer consumed. Commercial use rates were cheaper than residential use rates. These rate schedules were a complicated morass of marketing ploys vulnerable to charges of discrimination, and the utility commissions took them on. Washington's PSC held hearings in 1912 and issued orders simplifying these schedules and forbidding utility companies to collect large deposits from customers or to make them pay for electric meters. The commission also began publishing utility company valuations and prescribing rates for them whenever existing rates were deemed unreasonable.[47]

Over the next several years, the Washington PSC survived numerous court challenges to its authority, acquired greater expertise, especially in accounting, and was engaged in a multi-year struggle over rates with several of the big players on the scene: Pacific Power & Light, Washington Water Power, and Puget Sound Traction, Light, and Power. Following up on rate complaints, the commission embarked on an effort to assess the value of these companies in order to determine a fair return on investment and reasonable rates. Some of the rate complaints came from citizens, some from municipalities, some from large industrial customers. Occasionally, the commission itself would initiate an investigation against a utility. The valuations for these large companies were extremely complex, time-consuming, and contestable. Because of the scale of these companies, even minor downward rate adjustments could mean significant lost revenue. Consequently, valuations of the big companies were lengthy affairs involving hearings, site inspections, tedious financial analysis, appeals, and litigation. The commissions were generally careful and reasonable, however, and most lawsuits were settled in their favor. Often the utility and the complaining party would negotiate a settlement without lengthy hearings or prescriptions imposed by the PSC. For example, in 1914 the city of Spokane filed a formal complaint with the commission, alleging that Washington Water Power (WWP) charged exorbitant rates for lighting, heat, and power. The PSC had not finished its valuation of WWP's properties, so it could not prescribe a new rate schedule. Instead, it helped negotiate a compromise between the city and the utility, in which WWP agreed to reduce its residential lighting rates 20–40 percent, saving customers over $100,000 a year.[48]

A precedent-setting case in 1916 involving the Puget Sound International Railway and Power Company foreshadowed systemic problems in the electric streetcar industry and its eventual financial collapse. Typical of the time, this company ran several related enterprises for the city of Everett: a railway system involving electric interurban lines, an electric lighting system, and a water system. A formal complaint about lighting and water rates from the city of Everett led the commission to investigate. It found that the company was making a return of 15.8 percent on its water and lighting businesses and using the profits to subsidize its ailing electric railway business. The commission ruled that the return was excessive and set rates for water and lighting that would generate an 8 percent return instead. Importantly, it also stated:

> The users of light and water have a right to a supply of electrical energy or water at fair, just, reasonable and sufficient rates, independent of whether or no [sic] some other department operated by the same company is profitable or otherwise. If the rates of some other department are not fair, they should be made so, but the Public Service Commission has no right under the law to tax users of electrical energy or water in order that the patrons of a street railway may be permitted a low rate, or in order that the utility may be permitted to compete with some other form of transportation.[49]

Another precedent-setting case involved Pacific Power & Light, a regional utility serving a large number of towns, each with a different rate schedule. The principle of nondiscrimination was at stake, and the commission's action shows how difficult it could be to determine the best way to ensure equity among consumers. Washington state law forbade public utilities from providing "any undue or unreasonable preference or advantage" to any person, corporation, or locality. Likewise the law forbade subjecting anyone to "unreasonable prejudice or disadvantage." The commission therefore wrestled with the propriety of PP&L's charging different rates in different towns for the same kinds of electrical service. Electric light and power charges varied widely across PP&L's service area. Towns with higher rates filed complaints to get them lowered, claiming that PP&L was discriminating against them and favoring other towns with lower rates. Did the principle of nondiscrimination require the commissioners to equalize rates among the many towns served by one utility? Washington's commissioners decided not. In justifying their decision, they endorsed the principle of "segregation of political units" and the concept of "natural advantages." Many of PP&L's power plants and transmission lines were small operations serving local populations. The utility was consolidating and interconnecting these systems, but during valuation and rate hearings at this time, the facilities looked very much like a conglomeration of distinct units. So the commission ruled that "each political division . . . should be considered by the commission as a separate unit" with fair rates established independently for each. Moreover, different towns had different advantages inherent in their location or conditions. Some towns were close to water-power facilities, others more distant. Some towns consumed large amounts of power, offering economies of scale, while others did not. The commissioners ruled that uniform rate schedules penalized towns with natural advantages and subsidized those without them. Therefore, uniform rates would be discriminatory.[50]

This decision was a product of its time and place, making sense because PP&L's electric power system still functioned as a loosely organized collection of formerly discrete systems with separate power plants used to supply electricity for local consumption. As power plants increased in size and as production and distribution facilities grew to serve regional rather than local populations, the principle of segregation became harder to justify. In fact, the Idaho Public Utility Commission in 1914 took the opposite stance on this issue, in part because of the greater integration of the power systems there but also because the Idaho commissioners took a very different position on how to enforce the principle of nondiscrimination. Responding to numerous complaints about inconsistent electric rates throughout the towns of southern Idaho, the Idaho PUC initiated a comprehensive action involving all four major electric utilities serving the region (case F-33). After a brief investigation, the commissioners ruled that "there is no reason why the rates and charges . . . should not be uniform." The commission then prescribed a single rate schedule for all four southern Idaho utilities for commercial and residential lighting and for domestic and commercial cooking, saying those prescribed rates would be "reasonable and more satisfactory to the patrons." This case did not prescribe rates for irrigation farmers or industrial customers, so the uniform rates policy was aimed at domestic consumers, clearly a populist decision.[51]

This was the Idaho utility commission's first year in operation after a long tenure as a railroad commission, and it launched with reformist zeal. The introductory letter to the governor fumed against the 1913 enabling legislation, complaining that the utility industry had emasculated the bill in committee. As introduced in the legislature, the Idaho bill mirrored a California law, but key passages were altered at the behest of the electric utility industry before the Idaho bill became law. The commissioners made no secret of their contempt for this outcome and recommended ways to strengthen the law in the future. They also identified one passage in particular that appeared to exempt electric companies from having to get a certificate of public convenience and necessity from the commission to build new power plants and transmission lines. In a lengthy legal analysis, the commissioners argued that this clause contradicted the entire letter and spirit of the law and that they fully intended to ignore it.[52] Idaho courts later upheld the authority of the commission to require these certificates.

The authority to regulate the location and pace of electric power facility development proved crucial to the commission's early work. Since the commission apparently had no accountant yet on staff to begin detailed property value assessments, it focused its energies instead on settling complaints and defining the territories within which certain companies would enjoy monopoly service rights. The complaints driving commission activities came mainly from big businesses and utility company competitors. The commission noted in its letter of transmittal to the governor that protecting the public interest by maintaining free competition in this industry was futile because these utilities naturally consolidated into monopolies even without regulation and that competitive public utilities were "inefficient."[53]

Ironically, electric utilities called on the commission often in its early years to exercise this authority to regulate power system expansion to protect them from competitors. While electric utility lobbyists had inserted a clause in the law attempting to gut the commission's authority to control expansion, companies did not demur from requesting the commission to use this authority to their benefit. In fact, six out of eight of the electric utility "applications" to the commission in the 1914 report dealt with precisely this issue. In the first case, the commissioners protected a service area of Utah Power & Light in southern Idaho, near Ashton, Egin Bench, and St. Anthony, from a hostile company trying to "invade its territory," as the commission and the utility said.[54] A second case involved the mountain region of central Idaho surrounding McCall and New Meadows, where the commission forged a compromise allowing a new company to build a hydropower plant on the Little Salmon River to serve customers in the region but restricting it to areas not already served by a preexisting power company.[55] A third case again involved Utah Power & Light, this time in a controversy in which the Swan Creek Electric Company wanted to build a power line in Bear Lake County to two small communities served inadequately by Utah Power. Customers had complained about the service to the commission, and Swan Creek wanted to offer alternative service to the residents in a bid to acquire some of Utah Power's customers. The commission noted that inadequate service justified allowing a new company to enter another company's territory, so to protect itself and its monopoly Utah Power agreed to significantly upgrade its facilities. When the commission was satisfied that Utah Power was diligently engaged in improving service to the communities, it denied Swan Creek a certificate of public convenience and necessity for its power line.[56]

A final case worth mentioning was a complicated morass involving numerous competing utility companies serving the lucrative Snake River Plain, where most of Idaho's agricultural and urban development was located. The electric power system there was dominated by large hydropower dams on the Snake River and its tributaries. Rapidly expanding, heavily capitalized by outside investors, and engaged in ruthless competition over service areas, the companies made risky decisions, spawned irrational price wars, and caused general business instability. Both investors and state policy makers grew concerned. Rolling together a number of pending cases (F-15, F-31, F-40, F-43, and F-48), the commission stepped in and created a semblance of order. It protected utility monopolies in defined service areas and prescribed rates it felt were both fair to consumers and remunerative to investors. Several utilities challenged the commission's intervention, but the courts sustained it. In the ensuing several years, S. Z. Mitchell of Electric Bond and Share brought in deep-pocket financing and consolidated most of these competitors into one giant monopoly—Idaho Power Company—demonstrating the commission's prediction that competition was futile and monopoly inevitable and preferable if effectively regulated.[57]

The section of the Idaho PUC annual report dealing with complaints from customers reveals interesting political and economic dynamics of the industry at this time. Most of the formal complaints filed with the commission were not from homeowners or small businesses but from large, wealthy, electricity-consuming firms in the state, seeking lower rates: big business against big business. The mining industry in particular wasted no time filing complaints with the commission. In case F-6 the Federal Mining and Smelting Company in Wallace, Idaho, asked the commission to review the rates charged by Washington Water Power (WWP). WWP served much of the north Idaho panhandle, including the Coeur d'Alene Mining District and Silver Valley. The mining company had had a contracted rate with WWP since 1908, but when the contract expired in 1913, WWP raised electric rates by 10 percent. This occurred at a time when rates in general were declining from economies of scale and technological improvements. The mining company had hired accountants from New York to examine WWP's books to determine whether their rates were fair and reasonable, but WWP denied the accountants access. So the mining company asked the commission to determine fair and just rates. The commission launched an investigation and a property value assessment in 1914 while WWP launched court challenges. With deep pockets of its own, WWP dragged out the valuation and rate-making process for six years, simultaneously fighting the Washington Public Service Commission regarding its property values and rates in that state, too. Finally, after joint hearings held by the Washington PSC and the Idaho PUC in 1919, WWP's property assessment was completed in 1920, appeals ended, and a complete rate schedule was finalized, including many rate reductions. The Idaho PUC report of 1920 noted that the rates were acceptable to north Idaho customers. Interestingly, two WWP company histories both mentioned how rates declined during these years, but neither acknowledged the role state utility commissions played in bringing those rates down.[58]

A case from Oregon provides an especially illuminating example of how PUCs helped bring fiscal responsibility to the industry while guarding the interests of the customer. In 1912, a particularly bruising episode of competition took place in the Hood River area, where irrigated fruit orchards drove the economy. Hood River Gas and Electric exclusively served the area until the entry of the Hydro Electric Company in 1912. Immediately, a "bitter competitive war" ensued, according to the Oregon Public Service Commission. Neither succeeded in destroying the other, although they tried until the commission forced a stalemate in 1913. Because the territory did not have enough business to support two competing electric utilities, neither earned a sufficient profit, so their facilities and service suffered. By 1913, Wall Street financiers had stepped in and bought controlling interests in the stock of both companies as well as their larger regional neighbor, Pacific Power & Light. All three became subsidiaries of the giant holding company American Power and Light. At that point the financiers wanted sufficient profits to recover the debt from their purchases and the debt accumulated from operating losses incurred during the bruising period of competition. The Oregon commission spent 1912 to 1914 assessing the two Hood River companies' financial records to set fair rates. In their 1915 annual report, the commissioners came down hard on both companies for their irresponsible behavior in 1912 and 1913. They rejected expenses related to the two companies' hostile and "imprudent" competition, denying them the ability to recover those losses through electricity charges to customers.[59] Clearly, utility commissions could significantly advantage or disadvantage a company in the regulated marketplace, depending on how the company performed.

While British Columbia did not create a formal utility commission in these years, the provincial legislature did intervene to accomplish several crucial regulatory functions in lieu of a commission. Inflation pressures during World War I caused a financial crisis for public utilities across Canada. While their material and labor costs escalated rapidly after 1915, the rates they charged were locked in by long-term contracts with client cities. Electric streetcar enterprises suffered the worst financial troubles, while electric light and power companies continued to turn a profit. Like many other utility conglomerates, BC Electric controlled a mixed group of light, power, transit, and gas utilities. Rising costs, declining ridership, and wage demands from streetcar workers erased the profitability of the streetcar franchise by 1916, but BC Electric could not raise its fares until its contract with the city expired in 1919. For a while, BC Electric simply subsidized its ailing electric streetcar business with profits from its light and power sales. But general dissatisfaction with the high rates for light and power made this strategy increasingly untenable.[60]

The success of the public power authority in Ontario after 1910 made headlines in Canada and emboldened populists who believed nonprofit power systems would better serve the public interest. Vancouver's electric rates ranged on the high side compared to those of Ontario's cities served by the provincial power authority, so disgruntled reformers agitated for a municipally owned electric utility. But there was a stumbling block. The B.C. legislature, at the behest of BC Electric, had included in the city's charter of 1895 a clause forbidding Vancouver from establishing a municipal power authority that would compete with BC Electric. Consequently, many city boosters who felt lower electric rates would attract more industry joined with advocates of public power and the Vancouver City Council to petition the legislature to amend the charter. Historians Christopher Armstrong and H. V. Nelles argued that

Vancouver supported this more as leverage against the utility than as a genuine commitment to a municipal utility. Vancouver wanted a reduction in its street lighting rates, but BC Electric had resisted because it needed the revenue to shore up its unprofitable streetcar franchise.[61]

Normally, this agitation against the company would not have posed much of a threat, but in 1916 the Conservative government lost the election to the Liberal Party, weakening the company's long-established political connections with the provincial Conservatives. The legislature subsequently removed the offending clause from the Vancouver City Charter in November 1917, opening the way for Vancouver to establish a municipal utility, though it never did. Its risky position led BC Electric to lobby the legislature for assistance of the kind that a utility commission would normally provide. It asked lawmakers to appoint a binding arbitrator to assess its financial situation and determine reasonable streetcar fares, electric rates, and a fair return on investment for the company's utility franchises. The legislature agreed and appointed a Special Commissioner respected by both sides in the dispute, Dr. Adam Shortt. After a thorough financial evaluation, Shortt determined in November 1917 that streetcar fares should be raised, competition from jitneys restricted, and electric rates lowered. Shortt also called for creation of a public utility commission, which the legislature briefly created in 1919 but abolished a year later over a jurisdictional dispute with the federal government. In 1922, provincial lawmakers approved the BCER Passenger Rates Act, which codified most of Shortt's recommendations. The law stabilized the profitability of the company's streetcar business and set up a procedure to resolve franchise conflicts with the city of Vancouver without resorting to establishing a PUC.[62]

Armstrong and Nelles characterize this as a "corporate triumph" in which the legislature "legitimize[d] and stabilize[d] private ownership." Historian Patricia Roy similarly characterized this episode as a win for BC Electric. The company got a fare increase to make its streetcar franchise profitable, amended its contract with Vancouver to allow a fare re-evaluation every three years, kept its electric and gas franchises unregulated and profitable, and forestalled the creation of a public utility commission. While these scholars are correct that BC Electric sought and supported this provincial intervention, the final resolution also led to lower electric rates, a wage increase for streetcar workers, a limit on company profits, and a more empirical basis for determining a fair return on investment. City leaders in Vancouver were generally as pleased with the outcome as the owners of BC Electric. As Armstrong and Nelles acknowledge, while regulatory intervention stabilized private ownership and profit, it simultaneously inserted a principle of fairness into public utility administration.[63]

Scholars differ in their opinion about the performance of public utility commissions, in part based on which state and which decade are under scrutiny. Clearly some PUCs over time have been weak and dominated by the regulated utilities themselves, while others have been more successful at protecting the public interest—however they defined it.[64] Legislatures gave these commissions a mixed mandate: to ensure the strength, growth, and profitability of the regulated businesses while simultaneously ensuring fair rates, reliable service, and broad public benefits. Some commissioners leaned more toward profit while others leaned more toward social welfare in their regulatory approaches. In either case, the PUCs had broad authority and often exercised it assertively in the early years of their existence, significantly influencing the development of the industry and the welfare of its customers.

CONCLUSION

By World War I, the private electric utility industry in the United States had settled into a mostly comfortable milieu of state-licensed monopoly enterprises, with episodes of bitter competition between municipal utilities and public (nonprofit) power advocates. Regulators progressed from simply licensing utilities to actively structuring business practices. Just like the electrical industry, state regulatory agencies grew in size, scope, and financial resources at this time, enabling them to play an increasingly active role in encouraging and shaping development of the electricity sector. They served as boosters while negotiating citizen rights and corporate responsibilities. During these crucial first decades of the twentieth century, a novel technology had become an essential component of modern life, and a freewheeling entrepreneurialism had morphed into a business-government partnership to provide profits for investors while advancing the interests of consumers, cities, states, and the nation. Both businesses and governments, however, were imperfect human institutions, profits and privileges remained uncertain, and the public interest was a constantly moving target.

CHAPTER 5

"Let Every Inch of Water Do Its Duty"

If it be for the purpose of irrigation, let every inch of water do its duty, and, if it be for the purpose of power, let us see that the works are so carried out as to get from the investment and from the water conservation, the very best and most profitable results.

—Sir Richard McBride, Premier of British Columbia, 1914

Rivers formed the central nervous system of Northwest electric power in the early decades of the twentieth century. At that time, electric utilities in British Columbia, Washington, Idaho, and Montana extracted 80 to 90 percent of their electric power from the region's prodigious and ubiquitous rivers. Thus, the story of electricity in the region cannot be separated from the story of the Northwest's rivers, mountains, glaciers, and lakes. Historian Thomas Hughes, in *Networks of Power*, argues that physical geography shaped electric power systems almost as much as did human institutions (technology, politics, markets). Networks of power developed regional characteristics as well as a regional culture based in part on the location and material conditions of the region.[1] While the abundance of coal shaped the power systems of the Great Plains and Northeast, rainfall and rivers shaped the power system of the Northwest. The nature of rivers influenced the nature of power—both electrical and political, bringing to bear on the watersheds of the region precedent-setting debates over access, use, and control of rivers.

The U.S. and Canadian governments had long been committed to promoting economic development but left the nuts and bolts of production and consumption to the private sector. Governments shaped markets, encouraged manufacturing, provided transportation infrastructure to support trade and national defense, protected property rights, issued patents, and even encouraged low commodity prices to stimulate consumption. But rarely did government directly manipulate production activities or consumer behavior, except during times of war. One big exception was the establishment and regulation of public spaces and things considered public goods: roads and navigable waterways, urban water supply and waste management, public parks, and, by the twentieth century, energy—gas, light, and electricity. The previous chapter explored the rise of public regulation of electricity in the public interest. This chapter looks at the same struggle over the use and regulation of rivers. Were the region's waterways to be privately owned and dedicated to unrestrained private use, or were they public property to be used for public purposes adjudicated by governing institutions? Or, as we saw in the case of electric systems, would the region's movers and shakers elaborate a mixed political economy of private rights and public obligations for the region's rivers? Lacking any monolithic authority, a messy blend of capitalism and state regulation evolved to control river use. It was not a battle between government and corporations, though often it has been portrayed as such. Instead, the two negotiated how to achieve a common goal: maximizing the utility of the region's rivers.

Americans and Canadians had been ideologically obsessed in the twentieth century with dividing capitalism and socialism into two dichotomous and dueling systems, which has obscured how modern political economy actually functions. For-profit corporations and the regulatory state evolved together. They are and remain inseparable. The technological systems of the modern age are a product of this marriage. In some instances, private interests prevail in a more or less laissez-faire environment; in other instances, profit-seeking is restrained and directed by governing institutions; and in still other instances, state institutions adopt certain functions of business organizations, producing goods and services on a not-for-profit basis. The debate is not over private enterprise versus the state, but over the proportional influence of each in shaping our social and physical worlds.

This chapter starts with an examination of the political philosophy of conservation, a reform movement that guided natural resource development in the early twentieth century, and ends with the growing rift among conservationists over the "duty" of rivers to hydropower development versus the duty of rivers to the fisheries industry.

CONSERVATION IN THE UNITED STATES AND CANADA

The now-familiar term *conservation* was purportedly invented by political progressives in the United States in 1907.[2] Actually, the word itself already existed, but its full meaning as a comprehensive policy regarding natural resources developed in the fertile soil of Theodore Roosevelt's administration, according to Gifford Pinchot, one of the founders of the movement. Pinchot, at that time the nation's leading forester, had an epiphany one cold, misty day in February

1907 while riding his horse through Rock Creek Park near Washington, D.C. Observing the forest around him and meditating on its interdependence with the soil, streams, rainfall, rock formations, flora, and fauna, thinking deeply about how the welfare of the land was connected to the welfare of its people, Pinchot arrived back in the capital city with a newfound, holistic vision of nature and nation.

> *E Pluribus Unum* is the fundamental fact of our political affairs. *E Pluribus Unum* is and always must be the basis in dealing with the natural resources. Many problems fuse into one great policy, just as many states fuse into one great Union. . . . The very existence of our Nation, and of all the rest, depends on conserving the resources which are the foundations of its life. That is why conservation is the greatest material question of all.[3]

In the most general terms, the Progressive Era conservation movement advocated socially responsible and sustainable economic development. Taking a long-term view of natural resource use, progressives vigorously sought to minimize waste, inefficiency, and unregulated private monopoly and to maximize social benefits. The term *conservation* and its underlying principles spread remarkably rapidly throughout the United States and internationally, in part because one of its chief partisans was the flamboyantly influential American president Theodore Roosevelt. Later in the same year as Pinchot's epiphany, Roosevelt formally established conservation as a national policy by featuring it in his State of the Union message, saying, "the conservation of our natural resources and their proper use constitute the fundamental problem which underlies almost every other problem of our National life." Conservation to Roosevelt meant establishing "a planned and orderly development of our resources in place of a haphazard striving for immediate profit."[4] The next year, 1908, Roosevelt helped jump-start the movement by convening a national conference of governors to discuss and promote conservation policies at all levels of government and business. The convention was a rousing success and a fitting launch for the movement. After the conference, Roosevelt established a National Conservation Commission, with Pinchot as its chairman, to begin an inventory of the nation's lands, forests, minerals, and waters and to advocate for their rational, integrated development.[5]

A Canadian delegation attended the national governors' conference and returned to Canada determined to accomplish similar results. In May 1909 the Canadian Parliament established a Commission of Conservation, and on January 1, 1910, the commission launched its inventory of Canada's resources with the goal of promoting systematic, integrated, efficient development. In the next decade Canada's Conservation Commission published at least four important volumes on electric systems and water power in Canada, among many other reports and policy documents related to natural resources. One of the reports, titled *Water Powers of British Columbia*, published in 1919, revealed the existence of the term *conservation* in British Columbia's 1897 Water Clauses Consolidation Act. The act sought "to provide for the due conservation of all water and water power so vested in the Crown . . . and to provide means whereby such water and water power may be made available to the fullest extent in aid of the industrial development, and of the agricultural and mineral resources of the Province." This use of the term *conservation* specifically meant that the government of British Columbia "conserved" or retained for itself the rights to unappropriated water and water-power sites on rivers with the purpose of ensuring that they would be developed efficiently for the greatest good of the province. While this reflects the earlier meaning of the verb *conserve*, clearly it is not so far from the meaning of the policy concept as advanced by Pinchot, Roosevelt, and other conservation leaders in the United States ten years later, so it seems Pinchot may have exaggerated the originality of his invention.[6]

Conservation leaders tended to be highly educated modernists with a strong faith in science and technology. Pinchot, born of Pennsylvania wealth, made a lifelong career in public service promoting intelligent economic development guided foremost by concerns for economic efficiency, national welfare, and social equity. After leaving federal government service in 1910 (President Taft fired him for insubordination), Pinchot returned to Pennsylvania and stayed active in conservation campaigns. He served as president of the National Conservation Association from its founding until 1925 and was elected governor of Pennsylvania in 1922.[7] While he is mainly remembered for his role in promoting professional forestry and founding the U.S. Forest Service, Pinchot was also deeply involved in the public debate over electric power. The following passage on the "Water-Power Trust" from his 1910 book *The Fight for Conservation* reveals his perspective on conservation, corporate monopoly, and the good society:

> The imagination is staggered by the magnitude of the prize for which we work. If we succeed, there will exist upon this continent a sane, strong people, living through the centuries in a land subdued and controlled for the service of the people, its rightful masters, owned by the many and not by the few. If we fail, the great interests, increasing their control of our natural resources, will thereby control the country more and more, and the rights of the people will fade into the privileges of concentrated wealth. There

could be no better illustration of the eager, rapid, unwearied absorption by capital of the rights which belong to all the people than the waterpower trust.[8]

This classic class-warfare rhetoric (ironically coming from a patrician) was designed to justify more vigorous laws and public institutions to check and balance the power of corporations with the power of the state in pursuit of a broad vision of the public good. The fiery language in Pinchot's quote shows why the movement was sometimes called a conservation crusade. But this anti-monopoly rhetoric disguised the fact that most progressive conservationists did not oppose big business. In fact, many were themselves prominent business leaders who saw national corporations as the logical outcome of a national economy. Such anti-corporate rhetoric also gave the appearance that conservation was a movement of the People against the Moneyed Interests. But as Samuel Hays in the United States and H. V. Nelles in Canada point out, the movement was launched and sustained by a rather elite corps of scientists, technocrats, and business leaders. Nelles, for example, shows how the successful crusade for "public power" in Ontario during the 1910s was from the outset "a businessman's movement."[9]

Conservationists believed a strong government led by scientifically trained experts should partner with business to promote efficient economic development responsive to a broad array of social needs and national interests. Conservation leaders pioneered a host of new political institutions that became standard fare during the twentieth century: state, provincial, and federal regulatory institutions (such as the public utility commissions discussed in the previous chapter); government land and water managing agencies (e.g., reclamation, forestry, and fish and game agencies); retention rather than disposal of the public-domain lands (e.g., provincial and national forests and parks); multipurpose river basin development; and government sponsorship of irrigation and hydropower projects. A suite of cultural values that remain familiar today drove these initiatives: scientific management of lands and waters, sustainable resource use, nature protection, and long-term, integrated, multi-resource planning.[10]

The first report of the U.S. National Conservation Commission (1908) served as a conservation manifesto. It clearly revealed the conditions that spawned the movement and conservation's core values and agenda. The opening sentence of the report addresses what we now call the challenge of sustainability. "The duty of man to man, on which the integrity of nations must rest, is no higher than the duty of each generation to the next." The report then describes the rapid and wasteful exploitation of resources that occurred in the nineteenth century, supported by a perception of great abundance and a cultural climate of laissez-faire individualism, remarking that "the sense of responsibility to the future has been slow to awaken."[11] Revealing a bias for collective values over individualist values (and revealing some naive political idealism), the authors of the report suggest that Americans had evolved through three stages of historical development in the use of natural resources: a first stage of individual enterprise for personal and familial benefit, followed by a second stage of corporate enterprise for the benefit of specific groups and communities, followed by a third stage just unfolding, in which "enterprise is collective and largely cooperative, and should be directed toward the larger benefit of communities, States, and the people generally." Echoes of communitarian social reformers like Edward Bellamy are evident in this passage. Each of these stages held implications for prosperity and sustainability. "In the first stage the resources received little thought. In the second, they were wastefully used. In the stage which we are entering wise and beneficial uses are essential, and the checking of waste is absolutely demanded."[12] As history, this evolutionary scheme is little more than imagination, but its political implications are clear. Predictably, libertarians did not flock to the movement. But those who wished to join forces with others to solve large social problems benefiting the community and nation found conservationism attractive.

As evidenced in the above passages, conservationists battled incessantly against waste, while just as incessantly championing what Samuel Hays appropriately called "the gospel of efficiency." The commission report even set up a typology of waste. "The most reprehensible waste is that of destruction . . . attributable, for the most part, to ignorance, indifference, or false notions of economy." This kind of waste problem would be solved mainly by education—progressives characteristically believed information and moral suasion had the power to change behavior and improve society. "Nearly as reprehensible is the waste arising from misuse, as in the consumption of fuel in furnaces and engines of low efficiency." Here, innovation and a progressive-minded willingness to adopt what we now call state-of-the-art technology would provide solutions. The last type of waste, "the waste arising from nonuse,"[13] is a curious one, but just as critical to understanding the era and the movement. While modern environmentalism champions the value of leaving some areas in a purposeful condition of "nonuse"— codifying this concept in the Wilderness Act of 1964—conservationists at the turn of the century were less comfortable advocating nonuse. Productivity rather than pleasure or contemplation at that time drove the values of the middle class, still influenced by Victorian values. Development promised tangible benefits in that materialist age, while preservation appeared to be a policy of withholding resources from doing work for society. As Canadian premier McBride admonished, "let every inch of water do its duty . . . [for] the most profitable results."[14]

The final and key element of the conservation agenda is revealed in the Conservation Commission report's concluding statement on the means for eliminating waste: "the most effective means will be found in the increase and diffusion of knowledge, from which is sure to result an aroused public sentiment demanding prevention. The people have the matter in their own hands. They may prevent or limit the destruction of resources and restrain misuse through the enactment and enforcement of appropriate state and federal laws."[15] Ultimately, Progressive Era (and later) conservationists believed an informed public would demand action and the government would oblige by passing laws to restructure economic activity to make it more efficient, sustainable, and socially and environmentally responsible. To a large extent, this is exactly what happened.

The Commission of Conservation in Canada advocated a very similar agenda. In fact, its 1911 report on *Water Powers of Canada* quoted the report of its U.S. counterpart and cited conservation studies and policies in the United States as important precedents. The purpose of the report, its authors explained, was to assist in the intelligent development of Canada's water power. It advocated the need for thorough inventories to support that development, for coordinated planning to ensure that all the uses of rivers (navigation, irrigation, power, etc.) were properly balanced, for effective laws to encourage the full use of rivers and discourage waste, and for the government to reserve some valuable dam sites to ensure that private, single-use interests did not monopolize all power-development opportunities and foil efforts for coordinated development.[16]

Lest we succumb to the same naive idealism that some progressives did, it is helpful to remember that this laudable ideology provided comfort and cover for plenty of selfish interests. H. V. Nelles perceptively captured the blurring of business interests with "the public interest" in the fight to establish a public power authority in Ontario:

> Hydro made its deepest impression upon the businessmen and small manufacturers in the provincial towns. The thought of boundless, cheap hydro-electricity inspired these men of property and industry with evangelical zeal. . . . A deep and abiding local pride had long since blurred the distinction between their own interests and those of the communities around them. . . . Because they conceived the public interest as indistinguishable from their own private interests, the manufacturers and businessmen of

Ontario ardently embraced hydro-electric power as the materialization of their fondest hopes for their beloved towns. . . . By the phrase "the people's power," the businessmen meant cheap electricity for the manufacturer, and it was assumed that the entire community would benefit as a result.[17]

Conflict and inconsistency haunted the conservation movement. Putting every inch of water to work irrigating dry lands harmed river navigation. Efficiently developing hydropower harmed efforts to conserve fisheries. Turning natural forests into timber farms and croplands harmed soils and watersheds. Conservationists advocated all of these overlapping activities while maintaining faith that good science, honest government, and managerial expertise would harmonize these diverse uses to achieve the greatest good for the greatest number over the long term—Pinchot's aphorism borrowed from his colleague W J McGee, who borrowed it from the philosopher Jeremy Bentham. How to balance these various values? Some conservationists advocated dedicating each area of land and each river and stream to its "highest and best use," which usually meant the most economically productive use. Others advocated more of a blend of multiple uses for lands and waters to ensure that one use did not excessively dominate or completely preempt other legitimate uses. Others with more democratic leanings argued that public preferences, rather than efficiency experts, bureaucrats, or businessmen, should control land-use decisions even if that meant sacrificing efficiency, forgoing some development opportunities, or favoring the needs of the present over those of future generations. A few conservationists even challenged the rank utilitarianism of the movement's leadership, advocating for a moral perspective on nature that challenged the predominant economic thinking. John Muir's oft-quoted peroration against those seeking to dam Yosemite Park's Hetch Hetchy Valley is iconic: "These temple destroyers, devotees of ravaging commercialism, seem to have a perfect contempt for Nature, and, instead of lifting their eyes to the God of the mountains, lift them to the Almighty Dollar."[18]

The movement spoke with many voices. But not all exercised equal influence. Amid this cacophony, the most powerful economic interests usually prevailed. Still, both nations moved incrementally toward a more structured political economy that used government authority to promote profit-oriented economic development while attempting to secure some degree of integrated management, social responsibility, and sustainable use of natural resources.

RIVERS, DAMS, MONOPOLY, AND THE POLITICS OF DEVELOPMENT

Since the 1890s, control of water-power sites had been growing increasingly controversial—as attested to in Pinchot's attack on the water-power trust in his book *The Fight for Conservation.* Because water power required no fuel for companies to purchase and burn, it promised to be a virtual cash machine once owners paid off their initial investment in dams and generators. All over the United States and Canada, and especially in the water-rich Pacific Northwest, a speculative frenzy occurred as investors looking ahead to future profits and advantageous positions sought ownership or franchises for key dam sites, in many cases long before the power was actually needed. By 1912, the U.S. Bureau of Corporations found that ten organized groups of commercial interests controlled 60 percent of all commercial water-power sites in the United States. In Washington state, two private corporations and their subsidiaries controlled 70 percent of all developed water power; in Montana two corporations controlled 96 percent of developed water power.[19] In British Columbia, three corporations controlled 60 percent of developed water power.[20] This concentration of power made many people nervous.

Pinchot, who helped lead the charge against this growing water-power trust, testified before Congress in 1914 in support of a bill to extend federal regulatory control over water-power development on the nation's rivers. In his testimony he noted that the ten large corporate interests singled out by the U.S. Bureau of Corporations report in 1912 had increased their control of developed hydropower by over 30 percent in the subsequent two years and even more alarmingly had increased their control of undeveloped water-power sites by nearly 60 percent in the same two years. The troubling trend continued. By 1919 Canada's Commission of Conservation noted that "a survey in both Canada and the United States, of power sites most suitable for economic development, shows that most of the best sites either have already been developed or are being held by various [private] interests for future development."[21]

In Canada and in the U.S. West, where hydropower was the foundation of electric power production, this concentrated control of hydropower sites appeared to many people as yet another example of big corporations grasping public resources to enhance their power and profits with little thought to the larger public good. Consequently, in both nations, governmental oversight and intervention in the water-power arena increased at the state, provincial, and federal levels. In the United States, federal involvement in economic affairs came more quickly and forcefully in the West than in other regions simply because of the large proportion of federal lands in western states. Throughout the mountainous West, federal lands held many excellent dam sites, especially the national forests under Pinchot's control after 1905. By 1916, forty-two percent of the developed water power in the West came from dams located on national forests.[22] In Canada, a large percentage of the lands and waters were similarly owned or controlled by the Crown, facilitating government involvement in the use of those resources. Corporations, policy makers, and social reformers wrestled over critical questions regarding who should be allowed to develop rivers and for what purposes and how competing interests should be reconciled.

Government intervention to establish rights and responsibilities and to coordinate development advanced slowly but steadily in the early decades of the twentieth century. The U.S. government was already engaged in minor regulation of the nation's navigable rivers under the authority of the Interstate Commerce Clause of the Constitution. But not until 1899 did Congress pass a law requiring congressional approval to build dams on navigable rivers. The same law required the U.S. Army Corps of Engineers to approve construction plans. The subsequent rush to acquire rights to good dam sites for hydropower put quite a burden on Congress since it had to legislate approval for every application. As early as 1905 in a House debate over a new water-power bill, Georgia representative William Adamson complained: "Almost every week and each day somebody is . . . seeking permission to construct dams for the improvement of water power."[23] Adamson was pushing for a new law to deal with this flood of proposals, for which the 1899 statute had already proved to be inadequate. The 1899 law provided no clear criteria for evaluating dam proposals, no guidance for applicants seeking to dam a navigable river, and no policy for dealing with competing proposals. Moreover, the rights granted in the permit were essentially perpetual, with few strings attached, and free of charge. To get a permit, one had only to get approval from Congress (or from the affected state if the river was entirely within a single state), along with the CoE's approval of construction plans.

The situation was similar in British Columbia. Since the establishment of the province during the gold rush of the 1850s, rights to use water and to dam or divert rivers had been granted in an uncoordinated, piecemeal manner by provincial authorities, mostly to benefit the mining industry, which dominated the provincial economy. By the 1890s, thousands of water rights and permits had been granted—for mining, irrigation, domestic and municipal water supply, and hydropower—many of them overlapping and conflicting. In 1897, the B.C. Legislative Assembly passed a comprehensive law consolidating all previous laws relating to rivers, appropriately titled the Water Clauses Consolidation Act. Evincing a provincial commitment to rationalize and coordinate river development, this law confirmed that all unappropriated waters and water-power sites belonged to the Crown ("in right of the Province"), requiring government permission to dam or divert waters; it forbade any permanent diversions or exclusive uses of water unless authorized by act of Parliament; it authorized the Crown to charge rents, royalties, and tolls for the use of provincial waters; it required water permit holders to exercise their rights within a specific period of time (to discourage speculation); and it allowed provincial authorities to cancel a permit for unreasonable use or willful waste. Interestingly, the law also required electric power companies to obtain a certificate of approval from the Lieutenant-Governor in Council before beginning construction of any water-power projects. While this latter requirement sounds much like the certificate of public convenience and necessity mandated by state public utility commissions south of the border, in fact it was merely a means by which the province determined the amount of capital to be subscribed for the project (to ensure adequate financing) and fixed the time frame for initiation and completion of the project. The authority was not used at that time to control development or protect monopolies as in the United States. Canadian analysts working for the Commission of Conservation in the 1910s judged the act "a highly commendable conservation enactment," but noted that the province neglected to set up executive agencies to enforce the law and failed to cancel the water permits of those who abused their privileges.[24] This spate of legislation on both sides of the border essentially confirmed the right of the government to regulate private development of rivers, but both the authority and the will to intervene remained weak in these early years.

Some observers felt this system was inherently democratic: the people's representatives would decide in their wisdom whether a dam was in the public's interest and whether navigability and other public interests in rivers would be preserved, or at least not greatly impaired. Others felt this system was inherently politicized, uncoordinated, and inefficient since an act of Congress or the Provincial Parliament was required for every dam and since most legislators had no expertise in engineering or hydrology. Well-connected companies with capital and clout were most likely to get a permit authorizing their projects, and river development would proceed mainly at the behest of profit-seeking corporations with narrow pecuniary aims.

In the United States, the Theodore Roosevelt administration and some members of Congress grew increasingly dissatisfied with the 1899 dam permit system. When Roosevelt became president in 1901, the Army Corps of Engineers was the only federal bureau that constructed dams, and its only purpose was to assist navigation and flood control. But it would soon have a competitor: the U.S. Reclamation Service, created in 1902 by the Newlands Reclamation Act.[25] Later renamed the Bureau of Reclamation (BoR), this federal agency facilitated irrigation-based farming in the arid West. Revealingly, the 1902 Reclamation Act never mentioned electricity. But turbines were included in the BoR's first dams, and the agency quickly discovered the benefit of producing electricity, a by-product of its irrigation projects that generated much more revenue for the government than the sale of land and water to farmers ever would. Electricity was often needed to pump water out of reservoirs into distribution canals. It also provided lighting for the project and related administrative facilities. But BoR dams invariably produced a great deal more power than they needed for project purposes, so almost immediately the government authorized the agency to market its surplus power, passing a supplemental reclamation law in 1906 (the Town Sites and Power Development Act[26]) stipulating how hydropower from its dams would be sold and distributed. Electricity sales from federal reclamation dams quickly became essential to the financing and long-term sustainability of the associated irrigation projects. Without hydropower, federal reclamation would have either died an early death or evolved into an entirely different institution than it is today.[27]

The CoE got on the bandwagon too, putting generators in its new dams and incrementally retrofitting many older dams. This brought revenue back to the U.S. Treasury and helped subsidize the costs of the CoE's navigation and flood-control projects. Hydropower thus became a cash cow for federal river-development activities (see chapter 9). As a result, the government after 1902 grew keenly interested in securing for itself excellent dam sites for its various purposes and saw the electric utility industry as a potential competitor. Progressives therefore sought to tighten regulations on the private sector, reserve some dam sites for federal development, and articulate an innovative national policy of comprehensive river development that would ensure, ostensibly, the most efficient and beneficial multipurpose use of rivers.[28]

The reins began to tighten on private developers with the General Dam Act of 1906.[29] This law took the first step toward specifying the conditions under which hydropower dams would be approved. It still required congressional approval of projects and CoE approval of the construction plans, but it also stipulated the following:

- Dams must include navigation locks, lights, signals, and fish-passage facilities (the latter requirement was often ignored).

- The federal government retained the right to control water discharges and reservoir levels (to protect navigation needs).
- Construction had to begin within a year of permit approval and be completed in a timely fashion (to discourage speculation).
- The permit could be terminated for noncompliance.
- Any dam that hindered navigation could be ordered removed at the owner's expense.[30]

British Columbia moved in the same direction, though more conservatively. The number of water-use applications proliferated after 1897, with more than 6,000 licenses (permits) in force in the province by 1909. The granting of licenses proceeded ad hoc, with no central coordination or long-term planning. Progressive Era inclinations toward government-administered order finally came to some fruition in the B.C. Water Act of 1909, which established a water commission to grant water licenses and a board of investigation to adjudicate conflicts among water claimants. The Water Act granted broad authority to power companies, however. In addition to water rights, companies producing electric power had extended rights to occupy additional lands for power generation and distribution facilities. Bucking the trend in the United States, the B.C. Water Act also granted electric companies the right "to fix the rates and charges for the use of light, heat, and power, and the carriage of goods and passengers."[31]

British Columbia's Water Act also granted municipally owned utilities the same rights as private companies to apply for licenses to divert water or build hydropower dams. Municipal systems received no preference, although the law authorized the Lieutenant-Governor in Council to reserve any unallocated water "for the use of the Crown or municipal purposes." This authorization to reserve water or water-power sites was rarely used, however. Instead, the government encouraged private companies to develop Crown lands and resources. In fact, the Water Act explicitly directed companies to build waterworks on Crown lands instead of private lands whenever possible. Showing the same concern as in the United States that dams might obstruct water flow, harm navigation, or impinge on other water rights, the 1909 law stipulated that dams and reservoirs should not impinge on domestic water supplies or "obstruct any navigable stream" without specific authorization.[32] Of course, major obstructions—large hydropower dams—would be regularly authorized and built in the name of economic development in the coming decades regardless of their negative impacts on other users and resources. As rail and road transportation increasingly replaced boat and barge transport, the river navigation concern diminished. Instead, fisheries' interests became the critical factor in weighing the pros and cons of building obstructions in rivers (more on this below).

Like Canadian provinces, states in the United States control the waters within their boundaries and establish the legal framework for exercising private "rights" to waters. States often played a laissez-faire role in water rights administration and water development during the nineteenth century but became more active players over time, especially to arbitrate water conflicts and rationalize the development of rivers and streams in the state. But the Northwest states lagged behind both the federal government and British Columbia in watercourse regulation. Washington state, for example, in 1917 passed a Water Code setting up a procedure for determining water rights and requiring users to obtain a state permit to divert waters from streams or lakes, similar to the B.C. Water Code passed eight years earlier. In 1921, the state belatedly created a Department of Conservation and Development, with a Division of Water Resources, to administer water rights applications. In the early years, water rights administration in Washington and other northwestern states focused mostly on diversion rights for agriculture. Mining, municipal supplies, hydropower, and lumber mills constituted a minority of applications. By the mid-1920s, however, water-power applications increasingly dominated the field. For example, in the first biennial report of Washington's Division of Water Resources in 1921–1922, one out of every three water rights applications were for water power (requests to divert streams for hydropower production), with the majority being agricultural water rights applications. By the third biennial report (1925–1926), power applications far outnumbered all other applications combined.[33] The title of the agency, "Department of Conservation and Development," reveals the Progressive Era tendency to link the two terms closely. *Conservation* at the time meant coordinated, sustainable development. However, unlike more interventionist federal government initiatives at this time, this state agency did not regulate the actual use of water or judge the appropriateness of water-diversion proposals or reserve water rights or dam sites for the state to develop in the public interest. It did nothing to protect fisheries or navigation interests. It simply recorded rights and promoted development.

Political reform movements often start at the local level and move to state levels and then the federal level. But in this instance, the impetus for assertive government intervention was first federal and then state and then local. Washington's 1917 Water Code and its 1921 creation of the Department of Conservation followed on the heels of much more interventionist federal actions. The Washington Division of Water Resources' first report of 1921–1922 provided further illustration of this relationship. In a section discussing the 1920 Federal Power Act, the report acknowledges that this federal law provided the main stimulus for the large and growing number of water-power applications at the state level. The Federal Water Power Act stimulated hydropower development but also required applicants to comply with all state water laws, leading to an explosion in water-power filings at the state level. Thus, federal action elicited more regulatory activity at the state level and may have provided impetus to the state to create the Washington Department of Conservation and Development a year later. (The Federal Water Power Act is discussed in more detail in the next chapter.)

The more enthusiastic and visionary of the Washington, D.C., conservationists, those who hoped "good government" would usher in a new age of enlightened social and economic development, sought to go beyond mere oversight of private proposals for dams and promote government-sponsored multipurpose river basin development. W J McGee, Frederick Newell, James Garfield, and Gifford Pinchot, in particular, thought the American public would be best served by comprehensive river management administered by knowledgeable federal bureaucrats who would maximize public benefits. In establishing an Inland Waterways Commission in 1907 to promote this vision, Roosevelt admonished, "The time has come for merging local projects and uses of the inland waters into a comprehensive plan designed for the benefit of the entire country. Such a plan should consider and include all the uses to which streams may be put, and should bring together and coordinate the points of view of all users of water."[34]

Private monopolization of the best dam sites by companies interested only in irrigation or hydropower threatened this vision. With two dam-building agencies in the federal government now, Roosevelt decided it was time to push federal policy in a bold new direction. Taking a risk, and angering a lot of congressmen and well-connected businessmen, Roosevelt vetoed a series of bills authorizing private dams on several navigable rivers, including a dam at Muscle Shoals (Alabama) in 1903, Rainy River (Minnesota) in 1908, and James River (Missouri) in 1909. He objected to the bills granting perpetual rights to the sites; objected to granting development rights without receiving any lease payment or royalty; and, in the case of Muscle Shoals, felt it would be better for the federal government to develop the site under coordinated multipurpose objectives.[35]

Roosevelt's position on this derived from Pinchot and his efforts to establish a permit policy applying to dam sites on the national forests. Because so many of the West's rivers flowed through national forests, new proposals for

Figure 5.1. Hydroelectric Power Trust. Source: U.S. Forest Service website for film *The Greatest Good*, image gallery, http://www.fs.fed.us/greatestgood/images/gallery/cartoons/Cartoons/index.shtml.

developing dams poured into the Forest Service during the first decade of the century. In a 1911 congressional hearing, James Adams of the Forest Service said his agency had processed 60 hydropower applications since 1901 and received 100 additional inquiries. The most dam proposals, in top-five order, came from California, Washington, Oregon, Idaho, and Montana.[36] Clearly the Northwest was hopping with dam proposals. Pinchot adopted two tactics in response. First, in 1907 he established precedent-setting regulations for applicants wishing to develop dams on the national forests: he charged an annual fee for use of the site and, rather than granting a perpetual right to the site, he instead granted a limited time lease, usually fifty years and subject to renewal based on specific criteria.[37] Second, he used every regulatory power available to him to reserve certain dam sites on the national forests for federal

Figure 5.2. Czar Pinchot and His Cossack Rangers. Source: U.S. Forest Service website for film *The Greatest Good*, image gallery, http://www.fs.fed.us/greatestgood/images/gallery/cartoons/Cartoons/index.shtml.

rather than private development. Opponents of Pinchot's policy would accuse him of withholding resources from development, but in fact he was proposing public rather than private development. Pinchot was certainly "breaking new ground"—as he later titled his autobiography. Much of this new assertive policy was implemented in the Northwest and generated a great deal of controversy. The two cartoons in Figures 5.1 and 5.2 show the range of public opinion about these new ambitions and powers of the federal government regarding land and water management.

Roosevelt approved of this policy and decided to apply it to the whole nation. In his 1908 veto of the bill authorizing the Rainy River development, Roosevelt stated that he would no longer sign any bill authorizing a dam unless it contained a fee payable to the federal government and a time limit on the permit, repeating this demand on the veto of the James River dam in 1909 just before leaving office.[38] Many bureaucrats and legislators and CoE administrators felt the government had no right or authority to limit the permit or charge a fee if the facility was not on federal land (even though it was on a federally regulated navigable river), so Roosevelt's policy met resistance. Nevertheless, supporters got a bill through Congress in 1910, amending the 1906 General Dam Act to establish fifty-year renewable licenses for dams on navigable rivers and to authorize the Secretary of War (on behalf of the CoE) to charge an annual fee for the privilege of using a public waterway.[39] Opposed to the fee, the Secretary of War never charged for the hydropower permit. Roosevelt's successor, William H. Taft, and Taft's Secretary of Interior, Richard Ballinger, also opposed the fee, so the law remained unimplemented. A stalemate persisted until Congress passed a more comprehensive measure in 1920: the Federal Water Power Act, discussed in the next chapter.[40]

In addition to these regulations on private development, government plans for utilizing the power generated at federal dams angered private-sector business interests. The 1906 Town Sites and Power Development Act, mentioned earlier, granted public agencies (such as municipal utilities and irrigation districts) preferential access to low-cost power generated at federally owned dams, such as those built by the BoR and the CoE. This "preference" principle, initiated during the Progressive Era, remained fundamental to federal power sales policy throughout the twentieth century and effectively boosted the public power movement, especially in the 1930s.[41] Private utilities disdained this policy and the "unfair" benefits it gave to publicly owned utilities, but industry lobbyists were unable to block these bills or stymie their implementation.[42]

Despite some support from Congress for the activist policies pushed by Roosevelt and his advisors, the legislative body remained relatively conservative. Moreover, the Taft administration that succeeded Roosevelt took a retrenched stand on numerous conservation initiatives and on federal government activism. In fact, Taft's Secretary of Interior, Richard Ballinger, went so far as to suggest that the federal government cede all water-power sites to the states to dispose of as they saw fit.[43] Shortly after leaving office, Ballinger spoke to a gathering of rapt private utility executives at a Stone & Webster Club meeting in Seattle in January 1912, arguing that the recent tendency toward regulation was retarding western development and that water-power development by private companies should not be restricted. Displaying contempt for the very mission of the government department he administered (an odd but not uncommon tradition in American politics), Ballinger railed against federal intervention:

It is bureaucracy that says that the national government shall control our forests, our mines, our water power resources. It is bureaucracy that is undertaking to nationalize all functions of a quasi-public nature and to stifle private enterprise and individual initiative. . . . Were it not for bureaucracy, the states might secure from the Federal government the public lands and forests containing water power resources and regulate them in the interest of the people without the expense, annoyance, and delay of bureau domination centered in the national capitol.[44]

This ideological stance against federal involvement in regional development played well to the Stone & Webster Club audience, but the position and its warm reception by the businessmen in attendance were nevertheless disingenuous.[45] During the Progressive Era, businesses were among the primary agents calling for government involvement in markets, financing, infrastructure development, rights adjudications, and other forms of economic interventionism. While the federal government occasionally stymied development, its primary objective in establishing rules and regulations regarding rivers and hydropower at this time was to facilitate development. It inventoried natural resources, advertised opportunities, offered financial and technical assistance, provided land grants, rationalized property rights and privileges, protected markets, dredged and diked rivers and harbors, and sold cheap power from federal dams. The government virtually laid out a red carpet to businesses eager to engage in resource development for profit and subsidized that development when necessary. Progressives simply wanted to attach some strings to that assistance to see that it served larger national interests other than capital accumulation and to engage occasionally in government-sponsored development rather than always relying on the private sector.

Electrical industry statistics of the time reveal Ballinger's complaint about government regulation "stifling private enterprise" to be greatly overblown. As noted above, in the same year that Ballinger spoke to the Stone & Webster Club in Seattle, the U.S. Bureau of Corporations reported that ten large private concerns controlled 60 percent of the developed water power in the United States; two years later, as Pinchot reported to Congress, those same ten corporations had increased their control to 80 percent. The complete picture of private-sector control of electric power at that time is even more remarkable. U.S. Bureau of Census records show that private utility companies owned 92 percent of the electrical generation capacity in the United States in 1912 and increased their control to 94 percent by the next electrical census of 1917. In 1920 at the end of a period of intensive government intervention in the economy for the war effort, private companies owned 95 percent. All during this period (1912–1920), the electrical industry experienced one of its greatest booms in history with installed generating capacity increasing two-and-a-half-fold in a mere eight years.[46] There is scant evidence of stifled enterprise in these data.

What Ballinger and the Stone & Webster crowd resented was not federal intervention but federal restraint on their acquisitiveness and federal rules regarding the development of federal lands and public waters. In the 1940s, the famous American essayist Bernard DeVoto, in an often-quoted passage, immortalized this contradiction in the attitudes of many western businessmen:

It shakes down to a platform: get out and give us more money. Much of the dream of economic liberation is dependent upon continuous, continually increasing federal subsidies—subsidies which [the West] also insists shall be made without safeguard or regulation. . . . It is the forever recurrent lust to liquidate the West that is so large a part of western history.[47]

Ballinger's perspective reflected a political philosophy of decentralized regulation that served the interests of corporations attempting to maximize their advantages at minimum cost. Since states were more beholden to local corporations than the federal government was, they were usually more pliant toward those businesses than the national government was—at least in regard to access to natural resources. This same contradictory ideology is seen in states' rights and Sagebrush Rebellion political movements in the West later in the twentieth century—a philosophy that has always served as a political counterpoint to the centralizing regulatory tendencies of the past 100 years.[48]

While one would expect business leaders to seek control of the resources and institutions necessary for accumulating wealth and oppose conservation regulations that impinged upon their aims and autonomy, what explains the hostility toward progressive agendas in the wider political and public arenas? Samuel Hays suggests that river regulation and other battles between the conservationists and their opponents should be understood as a larger political contest between those who believed in elite, professional, centralized management of the nation's resources and those who believed in the messy but more "democratic" process of political negotiation among interest groups and their elected representatives. His summary of this bipolar political tension still rings true:

Examination of the evolution of conservation political struggles, therefore, brings into sharp focus two competing political systems in modern America. On the one hand the spirit of science and technology, of rational system and organization, shifted the location of decision-making continually upward so as to narrow the range of influences impinging upon it and to guide that decision-making with large, cosmopolitan considerations, technical expertness, and the objectives of those involved in the wider networks of modern society. . . . On the other, however, were a host of political impulses, often separate and conflicting, diffuse and struggling against each other within the larger political order. Their political activities sustained a more open political system . . . in which complex and esoteric facts possessed by only a few were not permitted to dominate the process of decision-making, and the satisfaction of grass-roots impulses remained a constantly viable element of the political order.[49]

Despite considerable opposition, progressives kept pushing the envelope. Leading progressive policy makers saw visions of efficient, sustainable, and equitable development generated by big federal multipurpose dams selling hydropower to finance irrigation, navigation, flood control, and even recreation. But this vision was about two decades ahead of its time. Not until the election of the second Roosevelt in 1932 did such projects come about on a large scale.

HYDROPOWER AND SALMON: TWO INDUSTRIES IN CONFLICT, TWO FACES OF CONSERVATION

Strange as it may seem to those in the river conservation movement today, conservationists in the early twentieth century enthusiastically advocated dams and hydropower development—all the conservationists, that is, except those involved in fisheries protection. The conservation movement bifurcated when it came to rivers. One side wanted full, efficient technological development, the other wanted unimpaired, free-flowing streams. The hydropower conservationists promoted what they called "conservation reservoirs"[50] where river flow could be stored behind large dams for later use in power generation, while the river and fisheries conservationists vigorously fought these salmon-blocking dams and habitat-diminishing reservoirs. Thus conservationists sometimes fought internally over exactly what was the proper "duty" of water. True to the optimistic, technocratic spirit of the era, however, most progressives clung to a somewhat naive but politically convenient faith that science and enlightened administration would permit full use of every inch of water for all purposes in a balanced, equitable manner. In practice, this optimism often merely provided ideological cover for those unwilling to make difficult political choices or acknowledge real resource trade-offs.

The authors of the 1919 report *Water Powers of British Columbia* exemplify this technological optimism. After advocating for full development of the province's hydroelectric potential, Arthur White and Charles Vick explain the importance of the salmon fishery in the provincial economy, discuss the deleterious effects of dams on salmon runs, and insist that the fishery must be not only sustained but enhanced. How to reconcile these conflicting objectives? With "intelligent regulation and forceful administration."[51] The report called for development of three million more horsepower from the province's rivers—a tenfold increase over then-current levels—while at the same time calling for the enhancement of the salmon fishery through hatcheries and fish passageways at dams. Despite well-meaning rhetoric about coordinated development and managing rivers for all their uses, investors, boosters, and government agencies usually had a particular dominant use in mind. Increasingly in regard to the waters of the Pacific Northwest, that dominant use would be electric power production.

Though similar geographically, the northwestern states and British Columbia eventually followed different paths in managing the conflict between fish and dams. Rivers on both sides of the international border were prodigious salmon incubators. Salmon are born in rivers, migrate to the sea during their first year, mature for several years as ocean fish, then return to their natal streams to spawn and die. The returning adult salmon "run" up rivers in great masses in the spring, summer, and fall. These salmon runs are the focus of the commercial fishing industry, an important sports fishing industry, and a culturally and economically vital Native American fishery. Five species of salmon populate the North Pacific waters and the rivers that empty into them. Prior to the twentieth century, the Columbia River and its tributaries, which cover an area the size of France (see Figure 5.3), had the largest runs of Chinook salmon, Coho salmon, and steelhead trout in the world.

From 1889 to 1920, commercial fishing interests harvested an average of 25 million pounds of salmon per year from the Columbia River. The all-time peak year of harvest was 1883, when the catch reached 43 million pounds.[52] Today, the Columbia salmon runs, a mere shadow of their former glory, are mostly extinct or officially listed as endangered.

The Fraser River in British Columbia is similarly impressive and its alternative history instructive. It is the largest river system in the province, 850 miles long, draining 84,000 square miles—an area the size of the state of Idaho (see Figure 5.4). Before 1914, when a human-caused landslide in the lower canyon blocked most fish passage for several decades until it was partially cleared, the Fraser supported the largest sockeye salmon runs on the planet. Twenty million Fraser River sockeye were canned in 1905 and 1909. In the record year of 1913, thirty-two million salmon were canned.[53] Unlike the Columbia River in the United States, British Columbians have kept dams off the main stem of the Fraser River so that the Fraser still supports the largest sockeye salmon runs in the world, undergirding an important component of the B.C. economy.[54]

The effects of dams on salmon and other anadromous fisheries were well

Figure 5.3. Columbia River Watershed. Source: Bonneville Power Administration Research Library.

understood long before the advent of the hydropower age. The Northwest embarked on its twentieth-century dam-building binge with eyes wide shut. The salmon fisheries on the Atlantic coast had gone into a devastating decline in the mid–nineteenth century with the rise of the commercial logging industry's use of rivers to transport logs and the increase in industrial textile mills damming rivers for mechanical power. Widespread and vociferous battles erupted during the 1830s–1860s, especially between mill owners and the fishermen and farmers affected by mill dams. New England's town councils and state legislatures sometimes sided with the fishers and farmers but more often with the industrialists who heralded the new economy of wage labor, mass-produced goods, and wealth accumulation.[55] By the 1870s, the Atlantic salmon harvest (and many other anadromous fisheries) had collapsed.

Figure 5.4. Fraser River Watershed. Source: Wikimedia Commons, http://en.wikipedia.org/wiki/File:Fraser_River_Watershed.png.

Because this was an interstate problem, the U.S. Congress responded by creating a U.S. Fish Commission in 1871—too little, too late. The first director, Spencer Baird of the Smithsonian Institution, identified three main threats to the fisheries: overfishing, dams, and habitat destruction—the same three key threats to salmon fisheries today. Unable to stop the industrialization of East Coast rivers or effectively regulate fishermen, Baird and other fish advocates in both the United States and Canada turned to "fish culture," the artificial propagation of fish in hatcheries to try to compensate for the losses. Atlantic coast Canadians were on the cutting edge of this technological fix in the 1880s, pioneering hatchery design and egg-collection techniques. Those involved in hatchery science and the policy makers supporting it constructed the perfect ideology for masking conflict and avoiding tough choices. They insisted that hatcheries could sustain fisheries without the need to restrict river development or significantly curtail what was left of the commercial fishing industry. Science would maintain abundance in an unregulated environment. In historian Jay Taylor's words, they "invented a panacea."[56]

Fish culture and its associated technological optimism immediately moved to the West Coast. In 1872 Spencer Baird hired Livingston Stone to try to save the California salmon fishery, which had been devastated in the previous two decades by hydraulic mining in the Sierras and overfishing in the lowlands. He started a hatchery on the McCloud River to try to restore the once fecund Sacramento salmon fishery, with little success. Stone journeyed to the Columbia River in 1877 to establish a hatchery there. He had been hired by a group of salmon cannery owners who put up $20,000 to build the hatchery in hopes of thwarting efforts by the state and federal governments to establish fishing regulations. Over the next several decades, the legislatures of Washington and Oregon and their fish and game commissions established regulations designed to protect the salmon fishery in the Northwest, such as requiring fish passageways in dams and prohibiting the dumping of sawdust and effluent from lumber mills into Puget Sound. But regulations were difficult to enforce and resistance from industries hard to overcome, so fish commissioners turned to the hatchery fix. A typical lament is revealed in the 1901 Washington Fish Commissioner's report detailing the continued widespread violations of fish-protection regulations and the lack of funds for the commission's work, which made it "impossible for us to even attempt anything like a systematic prosecution of these violations of the law." After this complaint, the commissioner more optimistically notes that "the most important work in connection with the Fish Department was the establishment and proper maintenance of fish hatcheries."[57] No money for enforcement, but plenty of money for hatcheries.

Despite promises by hatchery advocates to engineer abundance, the salmon fisheries on the West Coast steadily declined in the early decades of the twentieth century, just as the Atlantic salmon fishery had declined decades before. Salmon harvests on the Columbia River declined fairly steadily from the 1890s, while the decline on Puget Sound and on the Fraser River started in the 1910s. Overfishing caused a good deal of this early decline, but dams increasingly became the bigger problem as they obstructed salmon passage and drowned spawning habitat beneath reservoirs. While hatcheries combined with effective fishing regulations (if enforced) could mitigate the deleterious effects of overfishing, nothing could adequately compensate for the losses caused by big dams. In the 1910s, the U.S. Bureau of Fisheries stated that fishways for salmon passage were not very effective on dams over 20 feet in height. British Columbia experts agreed. Fishway technology has improved since then, but at that time, dams taller than 30–40 feet blocked salmon migrations.[58] For hydropower, the taller the dam, the more productive the facility. Thus, builders sought the greatest feasible height for their dams, putting them on a collision course with fisheries interests.

By the 1920s, electric power companies were responsible for most of the tall, salmon-blocking dams in the Northwest. From that time onward, the fate of the salmon fisheries was tied to the fortunes of the power industry. For example, there were seventy-four hydroelectric dams in Washington state at the start of that decade, producing just under half a million horsepower. Then the Washington State Hydraulics Division excitedly announced in its 1922 report that it had received water rights applications for the development of another 3.8 million horsepower, including four tall dams on the Skagit and Columbia rivers, which would each produce hundreds of thousands of horsepower.[59] All four of them were eventually built, and all blocked salmon runs mostly or completely. After years of study in the late twentieth century, the U.S. government concluded that the hydropower system is responsible for approximately 80 percent of the historic decline of the salmon fishery.[60] Hatcheries failed to compensate for these losses—though not for lack of effort. Fish culturists produced hundreds of millions of salmon fingerlings in the Northwest annually, but they could not replace the productivity of the region's natural rivers or stop the decline of the fishery in the United States.

The story of the Elwha River Dam is a paradigmatic tale of this period in Northwest history. The Elwha River drains out of Olympic National Park on the north side of Washington's Olympic Peninsula. It is near the town of Port Angeles, which looks across the Strait of Juan de Fuca to Canada's Vancouver Island and British Columbia's capital, Victoria. A short but powerful river draining the Olympic mountain rainforest, the Elwha supported runs of all five Pacific salmon species and was legendary for its extraordinarily large Chinook salmon, many weighing 75–100 pounds each. It was one of the most productive salmon streams on the Pacific coast, contributing to the Port Angeles economy and providing sustenance and crucial income for the Lower Elwha Klallam Tribe (often called at that time the S'klallam tribe). In the 1890s, Port Angeles seemed ripe for a boom, with the lumber industry taking off, mills opening, and the port expanding. A Canadian named Thomas Aldwell, dreaming of development, bought himself a 300-acre homestead along the lower river. What Port Angeles needed to fuel its growth, he thought, was a cheap, reliable source of electric power. By the early 1900s he had cut a deal with the main lumber mill in Port Angeles to supply its electricity and patched together investment capital from Canada and Chicago to build a large hydropower dam on the site of his homestead. Designing the dam to optimize hydropower production at the site with no accommodations for salmon, Aldwell soon ran afoul of Washington state laws, the local fish and game warden, and the state fish commissioner.[61]

Construction plans for the Elwha Dam called for a concrete edifice rising 100 feet above the river and anchored into the rocky cliffs that channeled the Elwha's torrent. This was far too high to allow fish passageways (also called fish ladders today), especially in such terrain. Building this dam would completely block all salmon migration above the lower river and devastate the fishery. The local game and fish warden, J. W. Pike, protested the construction plans to Aldwell and reported the problem to the state fish commissioner, John Riseland, who demanded that Aldwell redesign the dam to allow fish passage. Aldwell ignored Pike and Riseland, since that would require building a lower dam with less hydropower production capacity and with costly fish ladders. Without making any adjustments to the design, a Chicago-based construction firm started work on the site in 1911. Washington state had passed a law in 1893 forbidding the construction of dams without fishways on any streams with salmon, so Aldwell's obstinacy was particularly brazen. After construction had begun, Warden Pike visited the site during the seasonal salmon run and found no salmon making it above the dam. He saw thousands of salmon, however, knocking themselves senseless trying to leap the unfamiliar concrete obstruction rising in the riverbed. He wrote an emotional note to his supervisor decrying the scene and begging for assistance. Pike's boss, Riseland, protested again to Aldwell, who continued to ignore the demands.[62]

Fate temporarily intervened in 1912, when the dam was well along toward completion. Aldwell's haste, the investors' cost-cutting measures, and the Chicago construction company's inept design resulted in disaster. A large flood blew out the foundation of the dam, requiring Aldwell and his partners to go back to the drawing board. Riseland took this opportunity to again remind Aldwell of the state law forbidding dams without fishways and instructed him to redesign the dam with passage facilities. Aldwell wanted nothing of that and proceeded to plan for a better-anchored structure of virtually the same height—without fish ladders. Nineteen twelve was an election year, and Aldwell pinned his hopes on a belief that a new governor would be elected and a new fish commissioner would be appointed and all these annoyances would cease. Seeing that Aldwell had no intention of modifying the dam to permit fish passage, Riseland, in frustration, ordered Aldwell to build an electric elevator to convey spawning salmon above the dam. It was a curious and ultimately unworkable scheme, but it was all Riseland had left in his repertoire to try to make the dam legal and to save the Elwha salmon runs. Aldwell ignored the order and fast-tracked reconstruction of the Elwha Dam so that it was completed by December 1913. The dam completely closed the entire upper river, rendering more than 90 percent of spawning and rearing habitat on the Elwha inaccessible to returning salmon.[63]

A new governor did take office in 1913 and did appoint a new fish commissioner: the auspiciously named Leslie Darwin. Darwin was a classic crusading progressive, as evidenced in the rhetorical tone of his annual reports to the governor. One report expressed his commitment to the general principle of the primacy of the public interest over individual self-interest in a peroration that might have been leveled directly at Aldwell and others like him: "The people of this state have an interest in perpetuating and maintaining our food and shellfishery, compared with which the right of any individual, no matter how great his investment therein, sinks into insignificance."[64] He was also a crusading believer in hatcheries. Sometime in 1913 Darwin was reviewing the business of his predecessor and noticed that the Elwha issue remained unresolved. He picked up where Riseland left off and initiated a series of at first polite but then increasingly stern letters to Aldwell. Of course, fish ladders were a moot point by then, and Riseland's directive to build an elevator would probably just embarrass the department if enforced, so Darwin hatched a new plan that would save face for everyone except the salmon. State law allowed obstructions on rivers for the purpose of collecting salmon eggs for hatcheries. Darwin proposed to Aldwell that he compensate for flaunting the law by allowing the state to build a fish hatchery at the base of the Elwha Dam and help to pay for it. Essentially, Darwin offered to legalize the dam and rescind the order for fish passage by designating it as a fish-collection facility. His faith in hatchery solutions made this seem a reasonable resolution. True to form, Aldwell balked and stalled until 1914, when he finally capitulated. The state completed the hatchery at the base of the Elwha Dam in 1915 and waited expectantly for the following year's salmon runs. But most salmon are on four-year life cycles and since the dam ruined salmon spawning starting in 1911 and subsequently blocked the young salmon smolts' migrations to the ocean, the 1916 run was reduced to a trickle. Fewer and fewer salmon returned each subsequent year. By 1922, the year after Darwin retired as fish commissioner, the hatchery was worthless, and the state closed it.[65]

In his 1999 book *Salmon Without Rivers*, fisheries biologist Jim Lichatowich profiled the Elwha case, claiming its significance lay in the fact that it marked Washington state's turn away from enforcing the fish-passage requirement for dams and substituting that policy with an unfounded faith in hatcheries—a decision with bleak consequences for salmon runs. In 1915, the state legislature approved the Elwha compromise as a general principle, allowing the state to permit dams without passageways if hatcheries were built at their base. The Elwha was just one of many tragic consequences of that policy in the following decades and an example of how hatchery policy had evolved by 1915 into an ideological commitment with little foundation in science, a political expedient that provided cover for those willing to sacrifice one use of rivers for another.

In a fascinating postscript to this tale, eighty years later the Elwha Dam was decommissioned and authorized for removal by the federal government in the Elwha River Restoration Act of 1992. By the 1990s, salmon advocates had grown more powerful, and substitutes were available for the electricity produced by the dam, so a broad coalition of private, state, tribal, and federal interests cooperated in an effort to remove the Elwha Dam, along with the Glines Canyon Dam, built several miles upstream from the Elwha Dam in 1927. Predictably, financial and political obstacles stalled progress on the restoration effort, but the federal government finally acquired the two dams in 2000, and removal began about a decade later. The Glines Canyon Dam, at 210 feet in height, will be the tallest ever removed in the United States for fisheries restoration.[66]

Fish managers and policy makers in British Columbia were on a track similar to that of their American counterparts through the 1910s but eventually took a different path. A cautionary and precedent-setting experience occurred between 1898 and 1905 on a major tributary of the Fraser River called the Quesnel-Horsefly basin, exemplifying the conflict between dams and fisheries. The Horsefly and Quesnel rivers respectively flow into and out of Quesnel Lake, a highly productive sockeye salmon rearing ground. The lake and the two rivers together provided about 25 percent of the total Fraser River salmon spawning habitat, supporting sockeye runs numbering 10 million fish during big run years. In the 1890s, a gold-mining company with mining rights along the Quesnel River decided to build a diversion dam at the outlet of Quesnel Lake to make it easier to seek placer gold in the riverbed. The province of British Columbia granted the company a permit to do so in 1895, and the work commenced a couple of years later. Representatives of the powerful salmon-canning industry in Vancouver at the mouth of the Fraser River protested the mining company's plans, to no avail. There was no provincial fish commission at the time, and the dominion's Department of Marine and Fisheries was distant and largely unresponsive. The mining company built a simple rock dam 18 feet high and more than 700 feet long at the outlet of Quesnel Lake. At that height, it could easily have supported an effective fishway, but none was built. Finished in 1898, it blocked spawning salmon almost completely from the upper watershed. Four years later, the millions-strong Quesnel run had trickled to a few thousand. Salmon canners in the lower Fraser River felt the effects. Big runs occur every four years on the Fraser, and the "big year" of 1905 yielded 500,000 fewer cases of canned salmon than the big year of 1901.[67]

Cannery owners continued to press for remedies, and by 1902 they had a competent and forceful provincial ally in the newly established commissioner of fisheries for British Columbia, John Pease Babcock. Babcock protested the lack of an adequate fishway in Quesnel Dam in his earliest reports, while the canners lobbied the B.C. premier to intervene. Ironically, the mining company found the dam unusable for its intended purpose and abandoned it after only two years, yet the obstruction remained in place. Finally, in 1903 the provincial government secured permission from the owners to build a proper fish ladder, which the B.C. Fisheries Commission completed in 1904, allowing salmon to once again enter Quesnel Lake and its tributaries. Four years later the adult sockeye spawning run up the Quesnel-Horsefly drainage returned in force. In the big year of 1909, Commissioner Babcock estimated that four million salmon entered Quesnel Lake through the new fishway.[68] This story had a happy ending, but only because the dam was small and relatively insignificant, the fishery interests were united, and the government was willing to intervene. In the coming years, the battles would grow more fierce as fish interests fought hydropower dams that generated significant revenues and supported the urban-industrial economy.

Another even larger salmon disaster on the Fraser River in the next decade served to further solidify Canadian fishery interests' commitment to maintaining salmon passage on rivers and protecting spawning grounds. As mentioned earlier, massive rockslides in the steep Hell's Gate portion of the lower Fraser River from 1911 to 1914 created a very narrow channel of raging torrents that blocked the great majority of salmon from passing upstream. The landslides were caused by construction activities for the Canadian Northern Railway, which was blasting a railbed through the loose rocky talus in the steep-walled canyon. The debris blocking the channel was so extensive and the river itself so powerful that it took decades just to clear a few navigable pathways for salmon through the boulders and rushing currents. Upstream of Hell's Gate, the unfortunate Quesnel-Horsefly basin saw its 1913 "big year" drop to 550,000 fish, down from 4 million just four years earlier. Four years after the initial landslides, the whole Fraser River sockeye fishery predictably collapsed, not to be resurrected until the obstructions were finally ameliorated in the 1940s.[69]

This Hell's Gate disaster and the memory of the Quesnel Dam fiasco spurred quite a bit of policy innovation and government action. The B.C. government passed the Water Act of 1914, with a section (35) that required all dam licensees to build "durable and efficient fishways." In a perhaps related action the same year, Canada's Parliament passed the 1914 Dominion Fisheries Act, giving clear direction and authority to the minister of marine and fisheries to require fishways in dams permitted by the Crown.[70] Of course, as we have seen, having a fish commissioner and a law requiring fish passageways in dams does not guarantee enforcement of those laws or preservation of the fishery. But it was a step forward.

In other government action typical of the racial politics of the era, B.C. policy makers tried to shut down aboriginal fishing on the Fraser River after the landslides at the request of the canneries in Vancouver. The canneries continued to buy and process fish, and they wanted the Indian fishery curtailed, at least until the slide could be cleared and the runs recovered. The Indians refused to comply with the provincial government's order not to fish, arguing that their catch was miniscule in comparison to the harvest taken by the canners, that the order unfairly targeted them when they had done nothing to cause the slide and its decimation of the sockeye run, and that in any case the desist order violated their long-established rights to fish. First Nations groups organized successful protests and civil disobedience and made some headway in advancing their claims during the 1910s.[71]

The political potency of the canning industry and its contributions to the Vancouver economy helped it to successfully block proposals for dams on the main stem of the Fraser River in the coming decades. Its efforts were aided by the B.C. Fisheries Commission and by the fact that the electricity market was insufficient to justify damming the mighty Fraser at that time. Instead, the hydropower interests focused on the Fraser's tributaries and other nearby rivers. Aiding the fishery interests, a tributary of the Fraser River near Vancouver with excellent hydropower potential—the Stave River—had no salmon runs. Blocked by a waterfall, salmon could not enter the upper Stave River watershed, so protest against high dams there was muted. One large dam was built in 1912 at Stave Falls and another built downstream of the falls in 1927. One by one over the next several decades, both near and far from Vancouver and Victoria, power companies dammed B.C. rivers, causing fisheries to decline. This kept fisheries interests on alert and underscored the increasing importance of keeping dams off the main stem of the Fraser.[72]

While fighting all these river obstructions, most fish advocates in British Columbia embraced hatchery technology—at least until the 1920s. Technological optimism still reigned, but weakened in the face of continued salmon decline. By the 1920s, B.C. fisheries commissioner Babcock had turned decidedly against dams and the mitigation potential of ladders and hatcheries. While he believed effective fish ladders could be built on many dams to get salmon upriver to spawn, he recognized that dams posed a far more problematic obstacle to the downstream migration of salmon smolts to the sea. The young fish do not go down the fish ladders and therefore must travel through the deadly hydropower turbines. In 1924 Babcock made clear his view that hatcheries and fish ladders were futile stopgap measures, and that dams and salmon were fundamentally incompatible. "In my judgment the establishment of a large power plant in any salmon stream will be fatal to the salmon."[73] The only way to ensure a healthy salmon fishery was to maintain a natural river.

A second blow to the technological optimists came in the 1920s. Fish culturists considered themselves scientists and considered hatcheries a science-based means for enhancing fish populations, so the continued decline of salmon even on rivers with hatcheries was baffling. It was embarrassing, too, since hatcheries promised to compensate for salmon-blocking dams by adding millions of fish to the rivers below the dams. With leaders like Babcock questioning the compatibility of dams and salmon, fisheries scientists in the Northwest finally agreed to conduct a carefully controlled experiment on Cultus Lake in the Fraser basin to test the efficacy of hatcheries. From the beginning, fish culturists counted their success in terms of the number of eggs collected and smolts released to the river. The numbers were staggering and assured hatchery managers of continued funding. But were hatcheries really any more effective in producing salmon than natural reproduction on an unobstructed river? No one had ever rigorously tested the question. So in 1925, spurred by a request by the British Columbia Fisheries Commission, biologist Russell Foerster launched a ten-year study to settle the matter. In 1936 he reported conclusively that artificial propagation was no more effective than natural propagation. If a stream could support salmon, then nature was as efficient and successful an incubator as a hatchery. With the scientific justification for hatcheries removed, Canadians quickly abandoned their hatchery programs, closing them all by 1936. Instead, fishery advocates in British Columbia increasingly emphasized the importance of salmon refuges: salmon-friendly free-flowing rivers.[74]

Americans took a different path. Not wanting to abandon artificial propagation or weaken support for continued dam-building in the Northwest states, fishery agencies and policy makers in the United States held much more tenaciously to a belief in the efficacy of hatcheries. In his book *Salmon Without Rivers*, Northwest fish biologist Jim Lichatowich claimed that the ideological and political commitment to hatcheries as a panacea for salmon decline prevented American decision-makers from acknowledging the policy implications of the Canadian study. Instead fish managers and policy makers in the U.S. Northwest "invested in a 'conspiracy of optimism,' clinging to an unfounded hope that hatcheries could restore the salmon. As a result, wild coho [salmon] in the lower Columbia River have disappeared, and populations of salmon and steelhead in other parts of the basin have become severely depressed."[75] Canada's solution seemed the more successful path to simultaneously developing hydropower and conserving fish.

CONCLUSION

As U.S. states and Canadian provinces groped their way toward public regulation of electric power systems, the provincial and national governments of the United States and Canada similarly groped their way toward public regulation of rivers and hydroelectric dams. A synergistic relationship between governments and private corporations developed in stages. The chaos that often accompanied unregulated competitive development first led to efforts by governments to clarify property rights and privileges (franchises and licenses). The business community generally welcomed, even lobbied for, the clarity and security brought by state-endorsed rights and privileges. But after some time, the increasing scale of development, scope of environmental impacts, and complexity of social repercussions forced governing institutions to intervene more assertively. They carefully sought legal rationales for expanding their authority and promulgated increasingly detailed rules and requirements for private interests seeking to develop public resources such as the nations' waterways.

A little further down the road, Progressive Era conservationists pushed for more rationalized, comprehensive development, seeking a closer partnership between government and the private sector to enhance economic expansion, efficiency, and national welfare. National governments more actively set goals and prohibitions, provided incentives, and channeled investments. Finally, at the municipal, regional, and national levels, governments began actively engaging in direct resource development, such as the city of Seattle's hydropower dams, Ontario's public power system, and U.S. Bureau of Reclamation dams in southern Idaho. Often in these cases, private business interests supported public development, although just as often businesses opposed it in practice and principle, especially if they considered public development unwelcome competition.

This evolution of an activist government occurred because citizens demanded it. Far from an imposition on an unwilling populace, the regulatory state evolved democratically as a means for solving real problems and achieving popular social goals. In the United States, conventional wisdom asserts that the activist state channeling development toward broad social welfare goals was born during the Depression in the policies of Franklin Roosevelt's New Deal. But ample precedent existed for this political turn in the three decades preceding the Great Depression. There was much more of a continuum of ever increasing state intervention in the early twentieth century than the sudden unprecedented surge of federal government activism in the 1930s presumed by popular imagination. Historian Jay Brigham effectively demonstrates this point in his book *Empowering the West*.[76] All during the 1910s and 1920s, he says, "people showed a willingness, if not a preference, for government involvement to achieve their goals." A significant number of Americans "thought government could and should be a positive social and political influence on the local, state, and federal levels. . . . When Americans looked to the government for assistance in the 1930s, it was not without precedent."[77]

The philosophy of conservation embodied this era's emerging style of state-sponsored capitalism. It combined the goals of business with the vision of moral reformers seeking to perfect society through an engineered prosperity widely disseminated. In hindsight, this optimistic vision was naive. It placed too much faith in technological solutions to resource scarcity. Its extreme utilitarianism marginalized noncommodity values. It privileged dams over free-flowing rivers. Its technological optimism provided cover for those seeking to avoid making hard choices between incompatible uses (fish and hydropower). And in practice, the benefits of development were not as widely disseminated as promised. Still, compared to what came before, Progressive Era conservation represented a significant step toward setting up an institutional framework capable of striving for what we might now call responsible, sustainable development. It combined an activist government with entrepreneurial capitalism to achieve socially desired development responsive to public opinion.

CHAPTER 6

War-Accelerated Transitions I

When at war every important force in our nation must be enlisted to the fullest extent and in the most efficient manner. . . . We are now to forget our education in competition, and think of nothing but cooperation . . . for that is vital in winning the war.[1]

—E. W. Rice, Jr., President, General Electric Company, 1918

Wars are transformative. They cause far-reaching economic, political, and cultural repercussions in affected societies. This is as true today as it was a hundred years ago or a thousand years ago. Because of this, historians from ancient times to the present have written frequently and prominently about military conflicts and their consequences. A classic study of the transformative power of war on the American West is Gerald Nash's *The American West Transformed: The Impact of the Second World War* (1990).[2] Nash shows in compelling detail how production; labor; markets; demographics; urbanism; popular culture; and local, state, and national policies all underwent dramatic transformations during World War II, with many of those changes continuing after the war.

World War I had less of an impact on the United States and Canada, but the impact was nevertheless significant and included changes to the rapidly evolving electric power system, which had become a crucial source of energy in manufacturing and transportation. In his 1983 monograph *Networks of Power*, technology historian Thomas P. Hughes devotes an entire chapter to the effects of World War I on the electric power systems of the United States, Britain, and Germany.[3] Hughes notes that under serious perceived threats, governments, business leaders, and popular opinion all acquiesce to "emergency" measures that otherwise would not be acceptable, often leading to dramatic transformations in socio-technological systems. Rather than asserting deterministic cause-and-effect relationships, Hughes advances a more nuanced historical-contingency argument. At every moment in time, he claims, technological systems move along on the momentum of past decisions, designs, and infrastructure—what scholars now often call path-dependence.[4] But these systems are also contested and constantly evolving in incremental ways. Wars and other crises can trigger rapid and significant transitions from path-dependent inertia. In the three nations Hughes studies, he finds that World War I "accelerated development of certain technological characteristics," including the construction of very large central power plants and a more extensive interconnection of formerly separate power grids. The war emergency increased "the rate and direction of technological change" and "broke a conservative crust" that had favored certain developmental options while restraining others.[5] In other words, many potentialities lie latent or inhibited in large sociotechnological systems, and crises can release those potentialities, reconfiguring system design, infrastructure, costs, and benefits. The war indeed had this effect in the Pacific Northwest.

Wars accelerate government intervention in domestic affairs. During war mobilization, national governments harness manufacturing, transportation, and energy systems to support the war effort. In the United States, the war accelerated the electrification of industrial production, greatly increased electric power generation, and advanced the interconnection of utility systems to ensure power supplies for manufacturing.[6] It also clinched Progressive Era trends toward an active, regulatory government pursuing state and national interests in partnership with the private sector. U.S. historians often mark World War I as the end of the Progressive Era and the start of a conservative resurgence of laissez-faire economic policy but, ironically, the war boosted Progressive Era government activism in promoting and shaping electric power systems.[7]

Canada's story diverges a bit from that of the United States at this time. World War I did not cause a dramatic boom in electrical production and consumption in Canada as it did in the United States, though it did accelerate government intervention in economic planning. That intervention, however, was narrowly constrained to support the war effort. The assertive regulatory institutions overseeing utilities in the United States did not proliferate as much in Canada at this time. As mentioned in an earlier chapter, British Columbia did not establish a public utility commission until 1938. Nevertheless, for both nations and for the bi-national Pacific Northwest as a whole, World War I marked a turning point, foreshadowing the power system trends of the coming decade.

ELECTRICAL POLITICS DURING WORLD WAR I

Historian Martin Melosi observes that World War I clinched several trends begun during the Progressive Era: rapidly rising per capita energy consumption that paralleled a rising gross national product, enhanced centralization of both government and corporate power, and more coordinated economic planning between governments and private industries. The war "emergency" provided a rationale for many conservatives to loosen their resistance to the type of government activism promoted by progressives. The surge in government-industry coordination bolstered by the war was just the kind of "New Nationalism" Roosevelt promoted in his failed bid for election on the Progressive Party ticket in 1912. While some retrenchment occurred after the war, the long-term trend was clearly in the direction of an activist federal government and coordinated economic planning both by government and by the private utility industry. As Melosi concludes, if there ever was a laissez-faire energy economy in the United States, World War I ended it.[8]

Interestingly, Canada did not have the same experience, at least not to the same degree. The war stimulated industrial production as Canadians assisted Great Britain's war effort, and electrical generation consequently increased, but only incrementally and unevenly. While electrical generation in the United States grew by 42 percent from 1917 to 1920, electrical generation in Canada grew by only 3 percent.[9] Areas with war industries boomed—Ontario's Hydro-Electric Power Commission tripled its power production during the war years[10]—while other areas suffered economic doldrums because of the diversion of workers into the military and the diversion of capital into concentrated zones of war material production. British Columbia provides an excellent example of this pattern. Jeremy Mouat's history of West Kootenay Power and Light (serving the area east of the Cascade Mountains) devotes an entire chapter to the boom associated with the rapid expansion of zinc production for the war effort in Trail, British Columbia.[11] In contrast, Cecil Maiden's history of BC Electric (serving the urban areas on the west side of the Cascades) hardly mentions the war at all, except to note that while freight tonnage on the electric railway was up by 100,000 tons in 1917, ridership on the city trolleys and new customers for light and power service stagnated.[12] The war seemed to strain the economy of British Columbia's major cities rather than bolster it. As historian Patricia Roy observes, the war exacerbated an already "unhealthy" financial situation in British Columbia, causing the provincial government to retrench. B.C. attorney-general W. J. Bowser acknowledged in 1914 that, because of financial strains caused by the war, the province would "do nothing other than carry on works that are absolutely necessary for the good of the Province."[13]

The boom in the Boundary Mining District served by West Kootenay Power and Light came as a result of government intervention—an intervention that delighted some and worried others. To assist the British Empire, Canada ramped up its production of weapons and ammunition. Bullet shell casings, needed by the millions, were usually made of brass, which is a blend of copper and zinc. In 1914, when the war broke out in Europe, Canada could mine both metals but had no processing facilities to refine the raw ore into manufacturable metals. Consequently, Canada bought copper and zinc from the United States, which at the time was a neutral country full of anti-war sentiment. To secure its own supply, the Canadian government established a zinc commission to determine how best to jump-start domestic zinc production. In 1915 the commission chose Trail, British Columbia, as a prime location for a zinc concentrator, since the lead mines in that district contained ample amounts of low-grade zinc ore. New technologies for electrolytic processing of zinc ore were just being developed, and generous government subsidies helped establish one of these experimental plants in Trail. Electrolytic processing requires extraordinary amounts of electricity, so West Kootenay's abundant hydropower from its Bonnington Falls facility provided the other key reason the Zinc Commission chose Trail. As the mining industry expanded, so did the market for electric power.[14]

The Consolidated Mining and Smelting Company of Canada (COMINCO), the largest mining company in the boundary region and a subsidiary of the Canadian Pacific Railway, got the contract and the government subsidies for the zinc plant in Trail. To ensure a continuing abundant and cheap power supply, COMINCO bought a controlling interest in West Kootenay Power and Light in 1916. The two firms' destinies were then permanently interlinked. Their interdependence was abundantly obvious to both companies: by 1925, the zinc plant consumed 70 percent of West Kootenay Power and Light's total electric load. Interestingly, the power company was also a subsidiary of the Canadian Pacific Railway, an example of the degree to which economic power was concentrated in British Columbia. COMINCO became a gargantuan force in the province, achieving distinction in 1941 as British Columbia's most profitable enterprise ever. By 1966 it had produced half of all the mineral wealth of the province. Technological innovation in electrolytic processing of metals, hydropower from the Kootenay River, and government contracts and subsidies made it all possible.[15]

Controversy accompanied COMINCO's growth. Smaller mining firms in the region resented COMINCO's dominance of the market and the power supply, and they banded together to lobby for government assistance to compete against the privileged giant—to no avail. As the expansion of smelter operations continued after the war, West Kootenay needed more dams for more power, and since the Kootenay was an international river, damming it spawned international concern. Even more controversial, COMINCO's smelters spewed prodigious quantities of sulfur dioxide (the main pollutant in acid rain) into the airshed of the border region, ruining crops and forests in a wide swath aligned with prevailing winds on both sides of the international border. For the rest of the twentieth century this pollution dogged international relations.[16]

Despite the boost in power production to feed the mining industry near Trail, British Columbia's electrical generation capacity as a whole shrank 8 percent from 1917 to 1920.[17] The postwar economic slump partly explains this trend, but electric system development in British Columbia continued to lag behind that of the United States throughout the early 1920s. Even company booster and chronicler Cecil Maiden acknowledges that recovery after World War I was "very slow," although with hindsight he adds that it was also "very sure."[18] In later years this lag translated into dissatisfaction with the performance of private electric utilities and increasing—and eventually successful—agitation for a regional public-power system in the province. But that was decades away. In the meantime, private-sector consolidation of the electric industry in British Columbia proceeded. At the start of the war, three large companies in the province controlled 60 percent of the business. By 1920 there were only two. BC Electric bought out its main competitor, the Western Canada Power Company, merging the two systems into one conglomerate that dominated the west slope of the Cascades, with West Kootenay Power and Light dominating the east slope.[19]

In a pattern extending throughout the twentieth century, government initiatives promoting economic development helped shape the structure and growth of the power system. In the Kootenay, the national effort to boost munitions production and later simply to boost Canadian self-reliance profoundly shaped the regional economy and power supply. When West Kootenay Power and Light went in search of new markets after the war, it extended its lines into the Okanagan Valley to the west, hoping to tap the power needs of the irrigation district developing there with government assistance.[20] Returning veterans needed homes and jobs, and the agrarian life still seemed an attractive option for many. Canadian legislators adopted a nationally sponsored irrigation crusade similar to the one that watered the arid regions south of the border in the American West.[21]

The war boom in the United States was larger, more widespread, and more sustained. While the energy sector in general—including electricity, coal, petroleum, and natural gas—grew rapidly during the war years, electricity outpaced all the other energy sources. From 1916 to 1917 alone, electricity production in the United States increased by 40 percent. Government orders for war materials gave a tremendous boost to industrial production, and wartime industries greatly accelerated their conversion from steam power to electric power.[22] In contrast to the contraction experienced in British Columbia, the electric utilities in the U.S. Northwest generally flourished, especially those serving larger cities with manufacturing facilities. Portland General Electric's historian Craig Wollner notes, "The need for power in the shipbuilding, lumber, machinery, and woolens industries in the Northwest greatly strained Portland Railway Light and Power to meet its load demand."[23] The Oregon Public Service Commission reported that electricity output for Portland Railway, Light and Power more than doubled from 1916 through 1917. In this same two-year period, the Oregon Power Company more than doubled its production of electricity, and Eastern Oregon Light & Power nearly doubled its electrical generation. Among the big players in Oregon, only Pacific Power & Light saw a negligible increase in its output in these two war years. The Idaho Power Company's output skyrocketed from 46 million kwh to 112.7 million kwh, with its power sales to the state of Oregon tripling.[24]

Likewise, electric demand for the municipal-owned Seattle City Light on economically prosperous Puget Sound grew "at an unprecedented rate," according to the utility, so rapidly that it, too, had trouble keeping pace. Growth in customers and growth in manufacturing fueled the boom. From 1915 to 1919, the number of City Light customers increased more than 50 percent from approximately 40,000 to over 65,000. City Light also provided electricity to many factories, including four local shipyards, two of which built and launched 100 ships during the war years. To meet all this new load demand, City Light authorized a second steam-powered plant at its Lake Union Steam Station in 1916, and when it came online in 1918, load demand had already reached the full capacity of the new plant, so the utility had to immediately start construction of a third steam unit at Lake Union.[25]

Seeking additional power to meet demand put the municipal utility into conflict with for-profit electric companies in the Puget Sound region. City Light had an inefficient hydropower plant with a leaky reservoir at Cedar Falls on the Cedar River; it was trying to improve the reservoir during the war years by lining it with clay to inhibit water loss. But even with the improvements, the utility would need more power, so it sought to acquire another, better dam site. That is when it ran into a series of roadblocks put up by the giant utility holding company Stone & Webster. City Light preferred a site on the Skagit River in the North Cascades, but a group of private investors organized as the

Skagit Power Company held the permit to develop the river and refused to sell it to Seattle City Light. Coursing out of a massive field of glaciers in the North Cascade Mountains on the U.S.-Canada border, the Skagit River had lots of water and lots of vertical drop. An excellent dam site existed on national forest land (then called the Washington National Forest but now the Mount Baker–Snoqualmie National Forest). Seeking to control valuable future hydropower sites for its subsidiary companies in the region, the Stone & Webster holding company secretly acquired the Skagit Power Company, which had received a permit to develop the site from the U.S. Forest Service in 1910, and endeavored to keep it out of the hands of the municipal utility. Closed out of this opportunity, City Light began investigating alternative sites in 1917 and settled on three: the Hebb site on the White River; Sunset Falls near Index, Washington; and the Cushman site in the Olympic Mountains. While Seattle City Light investigated these options, Stone & Webster quickly and quietly purchased the privately owned Hebb and Sunset Falls sites to keep City Light from acquiring them. City Light was legally barred from developing the third Olympic Mountains site due to jurisdiction issues. Thus, Stone & Webster successfully blocked Seattle's municipal utility from expanding its hydropower facilities in the middle of a world war, at a time when electricity demand exceeded the utility's supply capacity.[26]

In frustration, J. D. Ross, the director of City Light, filed a competing request with the Forest Service in 1917 to develop the Skagit River, its original preferred site. As noted in chapter 5, the General Dam Act of 1906 discouraged the monopolization of dam sites by private speculators by requiring permit holders to begin construction within a year of permit approval and to complete projects in a reasonable time frame. This provided City Light an extraordinary opening. Stone & Webster's Skagit Power Company had not started construction and had also fallen behind on their lease payments to the government for the rights to the site. The U.S. Forest Service, founded by public-power advocate Gifford Pinchot, was sympathetic to Seattle City Light in this instance. As this was 1917, President Woodrow Wilson was keen to ensure adequate power would be available to support war material production. The Skagit problem looked exactly like the kind of situation the architects of the General Dam Act of 1906 hoped to eliminate. After effective lobbying of federal officials in the Northwest and in Washington, D.C., City Light leaders convinced the federal government to revoke the Skagit Power Company's permit and reassign it to Seattle City Light in 1918. Looking back triumphantly on the episode, the City Light Annual Report concludes that Stone & Webster "put themselves in the predicament of the boy in Aesop's fable who put his hand in the jar of nuts and grabbed them all but could not get his hand out without dropping everything." At the end of 1922, the first unit of the Skagit Project, Gorge Dam, was nearly complete, while Stone & Webster still had not developed either the Hebb or Sunset Falls site that they purchased to keep from City Light control.[27] City Light eventually built three dams on the Skagit River under their federal permit: Gorge Dam completed in 1923, Diablo Dam completed in 1930, and Ruby (aka Ross) Dam, the largest of the lot, completed in 1940. Today, the three dams supply about a quarter of Seattle City Light's power needs.

In many other locations besides Seattle, the war similarly boosted the public-power sector of the electrical industry, although there were regional differences. While total electrical horsepower capacity remained virtually unchanged in Canada from 1917 to 1920, the horsepower capacity in municipally owned public-power systems increased by 20 percent. This 20 percent increase in public-power generation in Canada was tied to the rise of provincially owned power-supply systems, which sold low-cost electricity to municipalities, thereby encouraging small towns to develop publicly owned electric utilities. In 1919 and 1920, three other provinces followed Ontario's lead and established public-power commissions.[28] These commissions usually took over a portion of the privately owned power systems in their provinces, thus increasing the percentage of publicly owned power without increasing the total supply of power. British Columbia did not establish a Public Power Commission until 1945, and in comparison to provinces that did so earlier, its electricity industry, both public and private, languished. Total horsepower for privately owned central electric stations in British Columbia declined 9 percent from 1917 to 1920. In the same years, horsepower capacity for municipally owned utilities in British Columbia increased 8 percent, although municipally owned power represented less than 5 percent of the total, so the overall trend in electricity production was significantly down.[29] Most municipally owned electric systems in British Columbia actually purchased their power at bulk rates from the private utilities since small towns lacked the capital and expertise to construct major dams and distribution systems and no public-power commission was yet available to do that work for the municipal utilities. So even though British Columbia had a large number of public-power systems, it produced very little power of its own. Private companies totally dominated electricity generation until 1945, when the province created its public-power authority.

It is difficult to identify the reasons for the decline in horsepower in the private sector in British Columbia and the increase in horsepower in the public sector during World War I, but anecdotal and statistical evidence from the electrical censuses suggests that the economic slump led private companies to scale back investment, leaving small and remote towns little choice but to develop their own local power supplies—usually small-scale hydropower on lesser streams or small coal- or oil-fired generators in towns.[30] This trend in British Columbia reversed after the war as the private-sector utilities expanded during the 1920s (more on this in the next chapter).

Unlike the experience in British Columbia, the for-profit commercial electric utilities in the U.S. Northwest expanded during World War I at an even more rapid pace than the publicly owned utilities, increasing their share of the electricity market. The war also greatly boosted the federal government's role in the energy sector, both as a regulator and as a supplier. The U.S. government aggressively intervened to ensure an adequate power supply to meet manufacturing needs during war mobilization, and soon after the war, it finally created a Federal Power Commission to regulate water-power development at the national level. On the supply side, because of the need for nitrates for explosives, Congress passed the National Defense Act of 1916 authorizing construction of a federal hydropower dam at Muscle Shoals, Alabama, to produce nitrates for a munitions plant the government was also building nearby. The nitrate plant later doubled as a fertilizer plant in peacetime. President Roosevelt's 1903 veto of a law authorizing private development of Muscle Shoals (discussed in chapter 5) preceded federal government development of the site. Unlike previous federal dams built by the CoE and the BoR, this one's purpose was primarily hydropower generation rather than navigation or irrigation. At Muscle Shoals in 1916, the national government finally entered the electric power-generation business through the front door rather than the back door.

Ironically, the war ended before the Muscle Shoals hydroelectric facility was completed. With the war emergency over, conservatives renewed their opposition to the project's being owned and managed by the government. After conservative presidents Warren Harding and Calvin Coolidge gained the White House in the 1920s, the U.S. government ended its investments in the Muscle Shoals facility and put it up for sale to the private sector. But lack of interest by the utility industry and political opposition from remnant progressives ultimately kept Muscle Shoals in federal ownership. The very vigorous debate over the fate of Muscle Shoals, however, perfectly encapsulates the larger debate over the future of electric power in the United States: an ideological debate over public versus private power and nationalism versus localism. The resurgence of federal government activism during the Great Depression finally ended this debate over Muscle Shoals, when it became a permanent federal facility.[31]

In the U.S. Northwest, especially Washington state, public power at the municipal level expanded, but not as a result of state action. Unlike Canadian provinces, U.S. states at that time did not form power commissions to develop and distribute electricity. States instead committed to their public utility regulatory agencies and left public-power development to cities and the federal government. The federal government built only a relatively few dams in the region before the New Deal era, so towns and cities were generally on their own competing with the large private power utilities. While in British Columbia most municipally owned utilities bought bulk power from investor-owned utilities, this practice was less common in the United States because of the hostility between public and private power systems. Mostly, the private sector in the United States sought to buy out the publicly owned utilities and sold bulk power to the public systems only when private owners could not purchase those systems outright or when public and private owners were clearly not competing with each other for customers. Nevertheless, municipal systems continued to expand in the Northwest in the 1920s long after they started declining nationwide.

THE FEDERAL WATER-POWER ACT OF 1920

As noted earlier, the war provided the impetus to resolve congressional differences over a comprehensive water-power regulation bill that progressives had been pushing since passage of the General Dam Act in 1910. During the war, while the federal government sought ways to better coordinate war production to avoid material and energy bottlenecks, the chief engineer for the U.S. Forest Service, O. C. Merrill, drafted a proposal in 1917 to create a Water-Power Commission to better coordinate the then-fragmented oversight of hydropower projects on federal lands and navigable rivers. Three federal departments were independently involved in evaluating and approving permits to develop hydropower: the Army Corps of Engineers, the U.S. Forest Service, and the Department of Interior. The war emergency highlighted the inefficiency of this ad hoc accretion of oversight authorities. As Milton Conover explains in his 1923 legislative history of the Federal Power Commission, the war greatly increased the demand for electricity and raised construction and fuel costs for steam-driven power plants, making hydropower especially attractive to government and industry. Hydropower was not only cheaper over the long run, but also conserved coal and fuel oil for direct use in industrial production. It took three more years, but progressive proponents of a single federal regulatory agency to coordinate water power finally triumphed over the lobbyists and supporters of private power producers who preferred privatization of federal resources and less regulation.[32]

The Federal Water-Power Act of 1920 created the Federal Power Commission (now known as the Federal Energy Regulatory Commission—FERC) composed of the heads of the three departments—War, Agriculture, and Interior—plus a secretary and an engineer. The commission drew on the staff and expertise of the three departments to accomplish its work. The commission's main purpose was to coordinate policies and procedures to facilitate rationally planned hydropower development. Like the state public utility commissions, the FPC had authority to promulgate rules and regulations, direct staff to collect and report information, require applicants to submit engineering plans and open their financial records to inspection, hold hearings, make valuation determinations, and issue orders and licenses. Reflecting the triumph of the progressives, the act authorized fifty-year renewable leases, not perpetual rights, for dam sites and authorized the FPC to charge fees to cover the cost of oversight and to partially compensate the United States for allowing the private use of public property.[33]

Interestingly, section 10(e) of the act that authorized these fees also authorized the FPC to expropriate "excessive profits" earned by licensees in states that did not already have regulatory agencies designed to curtail unreasonable profiteering.[34] In other words, if the FPC determined that a company using public property was earning excessive profits at the public's expense, and if the state where that project resided had no legal prohibitions against excess profits, the fee for use of the hydropower site could be raised high enough to capture the unjust earnings. This clause stimulated several states to adopt legislation designed to keep such authority at the state level. The law also required extensive financial evaluations to determine "fair return" on investment. In its early years, the FPC lacked the staff and resources to make such determinations, so this clause was a paper tiger that simply served notice to corporations that fair electricity rates were now a federal policy and that profiteering could be punished. Once again, America's political institutions sought to facilitate economic development through a partnership between government and capitalists that allowed the private sector to organize businesses and accumulate profits within a moderately regulated environment that placed some limits on greed and protected some of the interests of consumers and the general public. It was a difficult balancing act, with the FPC sometimes favoring corporations and other times favoring consumers, depending on administrative policy and the ebb and flow of political tides.

The first few reports of the FPC are revealing. Aware that it was a controversial new federal regulatory body whose success would be judged by both the amount of hydropower development it approved and the protection it afforded to the public interest when approving those projects, the early FPC reports engaged in a one-sided conversation with imagined critics and supporters. In particular, the third annual report (1923) took extra measures to review the commission's performance and authorities. The introductory narrative argued that the commission had already accomplished an extraordinary amount of work and that it aimed to cooperate with rather than supersede state public utility commissions (PUCs). Despite the optimism, the report also carefully explained that it did not have the resources or the authority needed to fully accomplish its statutory mandates and recommended amendments to the law to resolve the problems.[35] Like the state utility commissions, the FPC started out with extensive responsibilities handled by a small staff with a meager budget. To service hundreds of applications across all fifty states, the FPC in 1923 had only two accountants checking valuations. Clearly, with such a limited staff it could not "safeguard the public interests [and] carry out in the manner the law directs the duties which it imposes."[36] So the FPC asked Congress to increase its budget and extend its authority. Over time, the national legislature complied.

To blunt anti-regulation critics and those who felt government bureaucracies were inefficient, the commission prioritized "getting water-power development under way" in its first few years. The task before it was overwhelming. When the act passed in 1920, the United States had 9 million horsepower of installed hydroelectric capacity. In the next three years alone the FPC received applications for projects equaling 21.5 million more horsepower. The 1923 report proudly stated, "the commission has dealt with applications involving six times as much horsepower, has issued permits and licenses involving three times as much horsepower, and has twice as much horsepower built or building as the individual departments working independently in the preceding 20 years." The FPC accomplished all this "with practically no increase in personnel over what had previously been employed by the individual departments in the conduct of their independent water-power activities." The Pacific Northwest played a significant role in this permitting business. The very first project the commission analyzed was for a dam on the Deschutes River in Oregon, and applications from Washington, Oregon, and Idaho constituted 12 percent of the horsepower capacity of permit and license requests to the FPC during the years 1920 to 1923.[37]

Many critics of federal regulation argued that the states were the proper venue for utility oversight. Of course, utilities generally favored this position partly because states had more limited resources and jurisdictions than the federal government. But as utilities grew into regional and national corporations, the states were less adequate to the task of promoting development and protecting the public interest. The FPC entered the field at first as a partner to the states, claiming it had no intention of imposing policies on states. One of its tasks was to fill gaps where state regulation of electric utilities had not yet been adopted. It also promoted uniform accounting standards across the nation to make valuations more consistent, reliable, and fair to utility customers. The 1923 FPC report notes that some accounting procedures used by state commissions to determine property values and electricity rates allowed utilities to inflate their property values beyond what the company actually invested in facilities. Though technically legal, those higher property values meant the utility could recover a higher return from ratepayers since regulators set electric rates to allow a fair return on investment. The FPC in 1923 took the position that valuations "shall not be capitalized in excess of amounts actually expended." The commission hoped its positions on accounting, valuation, and excess profits would serve as a yardstick for state commissions, strengthening protections for ratepayers and enhancing regulatory efficiency while still promoting the rapid development of hydropower.[38]

In the 1920s, neither the new Federal Power Commission nor other state and federal policy makers considered that the public interest might be well-served by not licensing dams; that is, by leaving some rivers free-flowing. The FPC had a singular mission: to promote, license, and oversee water-power development across the nation. Later in the twentieth century, the agency's mission and regulatory authority expanded well beyond hydropower (acknowledged in its new name: Federal Energy Regulatory Commission), to eventually include a mandate to judge whether a dam up for relicensing was still in the public interest and, if not, whether it should be removed and the river restored. But a lot of water would pass through the dams before Congress extended this mandate to FERC.

INTERCONNECTING ELECTRIC UTILITIES INTO REGIONAL GRIDS

World War I facilitated another progressive initiative: increased interconnection and coordination of electric utilities on a regional scale, a trend that greatly accelerated in the 1920s. The government initially encouraged interconnections to address concerns about electricity shortages during the war. Interconnections offered a quick and relatively inexpensive means of increasing the energy supply. By tying various electric systems together, utilities with surpluses could transfer power to those with deficits without anyone's having to build a new power plant. Progressives appreciated the improved efficiency that resulted from these interconnections and the enhanced coordination that interconnections required. The U.S. government encouraged interconnections during the early war effort in order to quickly meet the growing demand for electricity to support war production. But mere encouragement proved inadequate; a power shortage loomed in the summer of 1918. In response, the federal government seriously considered nationalizing the electric utility industry just as it had done for the railway industry earlier in the war. But staunch opposition resulted in a compromise in which the government simply ordered regional utility company interconnections to help avoid additional power shortages that fall and winter of 1918–1919.[39] Again, the war served as a rationale to elicit support from politicians who would otherwise oppose government involvement in the planning and coordination of economic affairs.

Interconnections mainly occurred between investor-owned utilities because those companies dominated the market. Municipally owned systems generally did not have many transmission facilities outside their urban service area, so they were less involved in large regional interconnections than utilities serving more extensive areas. In 1916, Pacific Power & Light (PP&L) reached its capacity to supply power from its own facilities and faced a shortfall, which it forestalled by connecting with Washington Water Power (WWP) in late 1917, which had a large surplus from its newly completed Long Lake hydropower facility, west of Spokane. The two companies signed an agreement to sell each other "waste power . . . from time to time" and thus became one of the first interconnections in the Northwest. Technical problems arose tying the two systems together because they used different voltages in their transmission lines. PP&L had to purchase expensive and specialized equipment to facilitate the tie-in.[40]

Such technical and financial obstacles to easy integration led to increased coordination among companies planning system expansions and helped facilitate the institutional mergers and "power pools" of later years. The power exchanges between PP&L and WWP facilitated by interconnection proved so useful and profitable that, between 1923 and 1940, WWP connected its system through high-voltage transmission lines to other Electric Bond and Share subsidiaries to the east and south, including Idaho Power, Montana Power, and Utah Power & Light. This interconnected system integrating giant regional utilities across six states was at the time one of the largest utility interties in the United States.[41]

Other private utility interties occurred in the Northwest in 1918. Northwestern Electric initiated a high-voltage transmission line to connect with Portland Railway, Light and Power (later Portland General Electric). Then PP&L connected with Northwestern to allow transfer of surplus power among all three private utilities. PP&L's company bulletin proudly proclaimed in 1918 that any surplus power it sold to Northwestern would mingle with power going to Portland and the naval shipyards there, thus allowing PP&L to contribute to the war effort.[42] The two municipal systems in Seattle and Tacoma interconnected five years later in 1923 but did not connect to the surrounding private utilities.[43] Interties between the United States and Canada expanded at this time, too. Puget Sound Power & Light noted in a 1920 report that its Bellingham, Washington, division interconnected with the BC Electric system at the border, bringing an additional 7,500-horsepower capacity into PSP&L's grid.[44]

Before the war, private utilities had discussed interconnection for the purpose of balancing loads, but not until World War I and afterward did this become a standard industry practice. As Thomas Hughes describes, interconnection facilitated initial coordination of formerly independent systems, which led to more formalized coordinating structures, which then led ultimately to what he calls "institutional mergers": a merging of systems not at the corporation level but at a larger multi-state, intergovernmental, interorganizational level, such as the Northwest Power Pool that public and private utilities created with government support during World War II (chapter 11).[45] Transmission line interties and coordination of systems became a twentieth-century trend that continued unabated into the twenty-first century. In a bold proposal for the time and a sign of things to come, a professor at the University of Washington, Carl Magnusson, proposed in 1919 the construction of a 220,000-volt transmission line to run from Vancouver, British Columbia, all the way to Los Angeles, tying the entire Pacific coast together into a single integrated electric grid.[46] In its 1923 report, the Federal Power Commission notes that "only four gaps with a total of about 25 miles require to be closed in order to have an interconnected system along the Pacific coast reaching from British Columbia to Mexico, a distance of some 1,400 miles."[47] Several decades later this West Coast intertie was realized on a monumental scale with the Pacific Northwest–Pacific Southwest Intertie. In pursuit of efficiency, profit, and a more secure power supply, business and government launched an era of technological coordination and system integration that came to define the power grid of the twentieth century.

SUPERPOWER

These first steps toward regional integration of power grids in the U.S. Northwest arose in conjunction with a national debate over how best to improve and expand electrification nationwide. As mentioned above, concerns during World War I over bottlenecks and inefficiencies in the national electric supply launched this debate. Louis Hunter and Lynwood Bryant note in their three-volume *History of Industrial Power* that energy shortages during World War I "made a strong impression on American statesmen, economists, and engineers."[48] Ordering interconnections during the war represented only one immediate response to the problem. More visionary and long-term proposals quickly evolved. To many, the electricity grid seemed an increasingly useful infrastructure for transporting energy from place to place and utilizing it more efficiently. Technological optimists and policy makers dreamed utopian visions of a new industrial revolution spawned by widely available and inexpensive electric power coordinated at the regional level by new public and/or private institutions overseeing both supply and transmission. Engineers and economists provided studies to articulate the possibilities, while government and industry leaders debated various paths forward.

A series of influential studies on energy resources in America completed between 1917 and 1921 by Charles Gilbert and Joseph Pogue for the Smithsonian Institution proposed some novel innovations that set the terms of this debate for the next decade. Gilbert and Pogue suggested that giant coal-fired steam plants should be built at the mouths of coal mines (rather than transporting the coal long distance to points of use) with the energy transmitted on super-high-voltage power lines linking entire regions and tied in with hydropower and other large generation facilities. More importantly, the authors argued that these transmission lines should be treated like public transportation systems, built and managed by a national government agency to ensure the widest and fairest distribution of the benefits of electrification.

In short order, competing proposals emerged from advocates of private power and their public-power adversaries. In the northeastern states the debate crystallized in the Murray "Superpower" plan of 1921 and the Cooke/Pinchot "Giant Power" plan of 1925. William S. Murray's proposal, originally published by the U.S. Geological Survey, was actually launched and partially funded by private power interests. It proposed to link existing utilities from Maryland to Maine to make one interconnected grid. The federal government would provide funding to support grid integration and construction of a base-load power supply for the grid from giant central stations. The private utilities, however, would maintain ownership of the for-profit system. Predictably, the report urged federal and state governments to lift regulatory restraints and provide financial incentives to the private sector to achieve the Superpower goals. Though it was published by the USGS, Murray was an electrical engineer in a successful private practice involved in designing large railway electrification and hydropower projects on the East Coast. In 1925, McGraw Hill publishing company in New York issued a revised version of Murray's report, titled *Superpower: Its Genesis and Future.*[49]

Gifford Pinchot, while in his first term as governor of Pennsylvania, sponsored an alternative program called Giant Power, launched in 1923 with assistance from a Giant Power Survey Board that included engineer Morris L. Cooke, who became deeply involved in rural electrification programs. Pinchot and the Giant Power Board delivered to the Pennsylvania legislature in 1925 a formal proposal for a statewide, integrated power grid to be built and managed by the state government as a nonprofit public service. The state would also build mine-mouth coal-fired power plants to provide a steady base load at low cost, and it would systematically electrify all remaining settled regions of the state, including rural areas that had been largely neglected by the for-profit utility companies. Aimed at maximizing public benefits as opposed to private profits, Giant Power sought to lower electric rates and universally distribute the benefits of electricity. Giant Power envisioned an ultimate goal of numerous regional publicly owned power systems coordinated by interstate agreements and/or federal authority.[50]

Clearly, the two proposals were similar except for the degree of public ownership and the commitment to universal service. But as staunchly pro-business presidents held the U.S. White House from 1921 to 1932, Superpower received more federal government endorsement. In fact, while he served as secretary of commerce, future president Herbert Hoover gave a speech in 1922 at the annual meeting of the National Electric Light Association (NELA), a private industry trade group, urging business leaders to get on board the Superpower crusade.[51] And while both proposals sounded at the time like utopian dreams, neither in fact was. Superpower principles were already being implemented by industry leaders like Samuel Insull in the Midwest. Likewise, the existing public-power system in Ontario approximated the Giant Power concept and probably inspired Pinchot's and Cooke's proposal. Even the Federal Power Commission report of 1923 dedicated several pages to a discussion of Superpower, saying, "It is with us now and has been for several years." Acknowledging trends toward bigger central power plants, high-voltage transmission lines, and utility interconnections, the FPC claimed, "superpower, in fact, merely means the general employment of the best of present-day practice."[52]

Although political discourse hardened into public-versus-private partisan divisions, in reality the evolving electric system remained a hybrid of the two. While privately owned for-profit companies substantially dominated the supply side in the 1920s, they were subject to significant public regulation and challenged by a small but persistent public-power sector, especially in the Pacific Northwest. Proposals to create integrated regional electric grids highlighted and exacerbated the divisions between public and private power, particularly when considering who would own and manage the regional grid. Early integrations in the Northwest, as noted above, tied private systems to other private systems or public systems to other public systems. But it became increasingly clear that comprehensive grid integration would be the wave of the future. Partisans on each side hoped to banish the other from evolving regional systems, but in most cases this did not happen. Instead, public and private utilities eventually learned to share regional grids (see chapter 12).

In the 1920s, public policy generally favored government incentives to the private sector to implement Superpower principles. But the stock market crash of 1929 and antitrust proceedings against the utility industry slowed down the implementation of private-sector-dominated Superpower. With the collapse of capital markets in the Depression of the 1930s, the federal government took a much more vigorous role and favored publicly owned power systems, advancing their presence and influence in the electric industry. New Deal programs provided a springboard for limited implementation of government-sponsored Giant Power principles, with the Tennessee Valley Authority a prime example. In the Northwest, congressional opponents blocked similar proposals for a Columbia Valley Authority in the 1930s and 1940s, but the construction of the Bonneville and Grand Coulee dams and the establishment of the Bonneville Power Administration in 1937 made public power a major force in the region and created a Giant Power–style regional grid uniting both public and private power utilities.

CONCLUSION

World War I significantly boosted several trends in the evolution of the Pacific Northwest's electric power system. It accelerated a long-term trend toward greater efficiencies in production and distribution of power. It also accelerated nascent cooperation among utilities and integration among formerly independent power transmission grids. Cooperation among utilities was not universal, however. In the U.S. Northwest, the interconnections established in this era bifurcated between the investor-owned sector and the public-owned sector. The war did not much affect the separation and antagonism between public and private power companies. In British Columbia, the integration of systems came primarily through utility consolidation as two private companies came to dominate the province. Most municipal systems in British Columbia bought their power wholesale from one of these two companies, so in that sense there was much more cooperation between public and private entities in British Columbia than in the United States.

The war also accelerated central government involvement in planning and facilitating economic activity, despite a political culture in both Canada and the United States that strongly favored private ownership and control of the economy. Before the war, government intervention mostly took the form of incentives, subsidies, contract stipulations, and the like. The war emergency allowed both nations' governments to become much more aggressive in directing their economies. Canada's decision to subsidize homegrown metal-processing facilities for the war effort exemplified this trend. Similar subsidies were granted in the United States, but progressives in the government took some remarkable steps beyond mere incentives to private industry during the war by temporarily nationalizing the railway, telegraph, and telephone systems; creating a board to direct industrial production for the war effort; ordering utility interties and oil, coal, and gas development to strengthen the energy supply; and approving government construction and ownership of a nitrogen-fixing plant for munitions at Muscle Shoals. Beyond economic planning and subsidies, the war accelerated the U.S. government's involvement in the regulation of business. Electricity shortages and muddled authorities overseeing river development during World War I led to congressional approval of the Federal Water-Power Act in 1920 to facilitate hydropower development while protecting the public against predatory profiteering, a clear extension of the policies embedded in the previous decade's trend toward state regulation of utilities.

While economies of scale and system integration trends in the electric utility industry continued after the war, national policy makers reversed or blunted many of the government interventions established during World War I. The U.S. government returned the railroads to private control, dissolved the War Industries Board, cancelled its war production contracts, and generally stepped back from its role as director of national economic activity. Comparing the United States, Britain, and Germany, Thomas Hughes notes in *Net-*

works of Power that in many cases, "efforts made by the three governments to maintain in peacetime the momentum of their wartime planning and nurturing of technology . . . failed . . . [due to] a conservative reaction."[53] With the war emergency over, traditional corporate resistance to active government intervention in economic affairs resurged. Despite the persistence of state and new federal regulatory institutions overseeing public utilities, a laissez-faire policy predominated at the national level in the 1920s, and the private-sector power industry flourished with few restraints or competition. As the next two chapters reveal, the good times for the private sector did not last long.

CHAPTER 7

Roaring Twenties

Electricity Is Cheap—Use More of It.

—*Electrical West*, 1928[1]

During the 1920s, the electric power industry experienced a remarkable boom. Every economic indicator for the industry was extravagantly positive. While some sectors of the economy did not prosper during the decade, businesses associated with electric power certainly did. In particular, the commercial (privately owned) electric power industry in the United States greatly expanded, accumulated unprecedented amounts of capital, merged into ever larger regional systems, and invented myriad financial instruments to enhance revenues and profits. It cultivated public support through improved service, lower electric rates, and extensive public relations campaigns, while cultivating political support through campaign contributions and personal influence. In Canada the electrical industry experienced an equivalent boom but with less extravagant accumulations of wealth. In fact, the Canadian commercial utilities lost some ground to provincial-owned public-power systems during the decade. British Columbia was an exception to that rule. In both the U.S. Northwest and British Columbia, the movement to expand publicly owned nonprofit electric utilities remained a minor force and failed to make any significant gains until the Depression years. Nevertheless, private industry in the United States generally considered public power a mortal threat, and the two contenders engaged in vituperative verbal warfare throughout the 1920s.

In the United States, the combined kilowatt-hour output of privately and publicly owned electric utilities more than doubled between 1920 and 1928. In comparison, the U.S. population increased by only 13 percent during that period. Remarkably, this dramatic increase in power generation occurred at a time when electric railways and electric mass transit were declining. Formerly among the largest consumers of electric power, the urban transit systems were converting to automobile and bus transportation. Despite the losses in this huge sector of the power market, total power consumption climbed rapidly, underscoring how spectacular the increases in manufacturing, agriculture, and domestic use were. Even more impressively, Canada's kilowatt-hour output nearly tripled in those same years while its population grew by just 17 percent. Making up for the sluggish growth during World War I, Canada experienced 20 percent increases in electrical generation per year in 1922, 1923, 1926, and 1927. Electricity production accelerated at around twenty times the rate of population growth in both Canada and the United States in the 1920s.[2] The Northwest mirrored this trend as electrical generation, the number of customers, and per capita consumption increased at a far faster rate than population growth.[3] This decade was indeed the "roaring twenties" for the electrical industry.

Consumers benefited from the boom as much as did employees and owners of the utilities and electrical manufacturing firms. The rates people paid for power declined quickly and often dramatically. Per capita consumption skyrocketed as a result, as did the number of households served by utility companies. Job growth in the industry was impressive, too: U.S. employment in the electrical industry more than doubled between 1917 and 1927. But even more impressive was the increase in salaries and wages in the electrical industry, which rose at more than twice the rate of employment gains.[4] Few expected the bubble to burst. Enough people benefited from the growth that warnings by folks such as Democratic senator Thomas Walsh of Montana and Republican senator George Norris of Nebraska about the utility industry's risky and irresponsible financial practices went largely unheeded.

INNOVATION AND REGULATION

The boom in electrification and the good fortunes of the electrical industry were driven to a significant degree by technological progress. The trends of previous decades continued as the size and efficiency of electric generators, the voltages and distances of transmission lines, and the scale and organization of power systems greatly increased. Between 1912 and 1927 the maximum kilowatt rating for steam turbines jumped more than eightfold, from 20,000 kilowatts to 165,000 kilowatts. The upper limit of working pressure in boilers for steam turbines increased sixfold in these same years, from 200 psi to 1,200 psi. In 1912 the standard voltage for long-distance transmission was 66,000 volts. By 1927 there were lines transmitting at 220,000 volts.[5] Technological progress occurred in the water-power sector too. In British Columbia in 1925, West Kootenay Power and Light replaced its original hydropower facility at Lower Bonnington Falls with new turbines that produced fifteen times as much power as those replaced.[6] All these technological developments lowered the cost of generating electricity, making it possible to lower the price for consumers while simultaneously increasing profits for producers. Lower prices stimulated greater consumption, which created a need for more investment in production facilities, which led to more revenue and more profits. This is the sweet spot for any consumer technology—an extraordinary situation in which for-profit companies, customers, investors, regulators, and political leaders all sing the praises of the new technology and dream of the social transformations it will bring.[7]

In the Pacific Northwest, northern Rocky Mountains, and British Columbia, most electrical generation at this time still derived from hydroelectric power rather than fuel-burning steam power, unlike in the United States as a whole, where hydropower constituted only a quarter of the nation's rated kilowatt capacity in 1927. The 1927 Electrical Census noted that the eleven western states held 11 percent of the U.S. population but produced 38 percent of the nation's hydropower. In the Northwest, hydropower constituted 98 percent of Idaho's kilowatt capacity, 96 percent of Montana's, 83 percent of Washington's, and 53 percent of Oregon's. Like Idaho and Montana, over 99 percent of British Columbia's electricity derived from hydropower in 1927. For Canada as a whole, the figure was 98 percent.[8] Clearly, rivers were central to electrical development in the region, and some of the technological enthusiasm for electricity spilled over into enthusiasm for the region's water resources. In a biennial report from this decade, the Washington State Department of Conservation and Development crowed that seven to ten million horsepower of latent, undeveloped hydropower existed in the state. "This is by far the most valuable single natural resource of our State. Our timber is fast disappearing; our mines will one day be worked out; but our rivers will rush ever onward to the sea and when harnessed will furnish energy equivalent to the labor of millions of men."[9] A 1930 report by the British Columbia Department of Lands proclaimed, "Among the great agencies that contribute to the development of the modern state, cheap power stands in the front rank. Abundant resources of water-power that may be cheaply developed add largely to the natural wealth and to the industrial opportunities of a new country. British Columbia is possessed in large measure of this important factor of growth."[10]

As explained in the two previous chapters, the U.S. government and Canadian provincial governments took an increasing interest in regulating the development of dams and other river infrastructure, including hydropower, in the 1910s and 1920s. British Columbia came late to the table, not establishing a provincial regulatory authority for public utilities until 1938, almost a decade into the Great Depression, although its Department of Lands did serve as a promoter of hydroelectric development. Businesses and their allies in the 1920s often promoted the view that government regulation was obstructionist, but every economic indicator shows otherwise. In fact, the state utility commissions and the Federal Power Commission demonstrably enhanced electrical development and profitability. Believing that its primary mission was to facilitate hydropower development, the FPC dramatically accelerated the issuance of permits for dams in the 1920s. The FPC's annual report of 1926 crowed that in the first four years of the agency's operational existence (1921–1925), the United States "witnessed the greatest water-power development in our history." Even so, this remarkable boom paled in comparison to fiscal year 1926, in which the FPC licensed projects whose hydropower capacity equaled 40 percent of the total licensed in the previous four years. The FPC so thoroughly dedicated its first decade to expanding hydropower development that it was largely unable to fulfill its other responsibilities related to protecting the public interest in ensuring fair rates for interstate sales of electricity, accurate facility valuations, and protection against securities fraud. Repeatedly in its reports during this decade the FPC complained that its meager organizational resources required it to postpone most of its other obligated duties except for fast-tracking project applications.[11] Despite this caveat, the FPC was optimistic about its accomplishments, stating, "it can scarcely be doubted that one of the primary factors in this development had been the Federal water power act [sic] under the terms of which there had been offered for the first time conditions safeguarding the large investments required."[12] Likewise, at the state level, by guaranteeing a fair return on investment and protecting utility monopolies, the public utility commissions stabilized the business and lowered investment risk, attracting even more capital. If anything, the electric power industry would have benefited from stronger regulation to avert the abuses that helped usher in the stock market crash of 1929.

FOR-PROFIT VERSUS NONPROFIT

In the competition between for-profit and nonprofit power systems, the for-profit utility sector fared quite well between 1920 and 1928, especially in the United States. The share of the market controlled by investor-owned electrical utilities in the United States in these eight years fluctuated between 97 percent in 1920 and 95 percent in 1928. Clearly, private industry had a lock on the power supply in the United States, its hostile rhetoric about public power mostly hyperbole. The 1920s were a boom period in Canada, too, though somewhat less so for investor-owned utilities than in the United States. Publicly owned municipal power systems flourished in Canada, mostly due to the expansion of several provincial public-power authorities. At the start of the decade, privately owned power systems in Canada controlled 76 percent of the market. By 1928 their control had slipped to 70 percent. Investor-owned power systems in Canada expanded 2.6-fold in those years, while municipal power in Canada expanded 3.4-fold. Despite the slower growth rate for private utilities, that still represents an impressive expansion. Once again, public power fared better in Canada than in the United States, although in both nations the private sector still controlled the overwhelming majority of electrical production.[13]

As usual, disaggregating national statistics reveals some unique trends at the regional level. Interestingly, both British Columbia and the Northwest states bucked their respective national trends in regard to public power versus private power in the 1920s. Municipally generated electricity made up only 3 percent of power supply in British Columbia in 1920, subsequently dropping to 2.3 percent by 1928. Public power in British Columbia, almost irrelevant to start with, lost ground in that decade, contrary to the larger national trend in Canada. In the U.S. Northwest, on the other hand, the number of public-power systems expanded slightly throughout the decade. Historian Gene Tollefson attributes this increase mainly to the availability of surplus power from a few publicly owned hydroelectric dams in the Northwest, which sold cheap electricity to several nearby municipal systems, including the federal dam at Minidoka, Idaho, and the municipal power system in Tacoma, Washington. This slight increase in public power systems in the U.S. Northwest was contrary to the national trend in which fewer municipally owned systems survived the decade. Nationally, the overall number of municipal systems declined primarily because private power companies with more capital and expertise often bought out small, aging, or poorly managed municipal systems and integrated them into larger intercity and sometimes interstate transmission grids. Municipal systems normally had the authority to serve only customers within their city limits, so when a small municipal system was in financial or technological trouble, commercial electric companies were the only alternative available at that time. A municipal utility in one city could not buy and operate a system in another city. Not until the 1930s did Washington and Oregon authorize the creation of public utility districts to extend not-for-profit electrical service beyond municipal boundaries.[14]

A decline or increase in the number of municipal systems provides one perspective on the fortunes of public power, but it can be misleading. Looking at kilowatt-hour output for private versus public systems and comparing

Table 7.1. Number of Power Plants and Kilowatt-Hour Output: U.S. and Northwest Private Power versus Public Power, 1920–1928

		Number of Power Plants		Output in Millions of kwh		Percentage of Market Control	
State	Census Date	Privately Owned	Publicly Owned	Privately Owned	Publicly Owned	Privately Owned	Publicly Owned
WA	1920	60	10	963.2	216.2	82	18
	1924	58	12	1,189.5	290.0	80	20
	1928	58	10	1,817.7	510.3	78	22
OR	1920	59	6	466.2	8.8	98	2
	1924	51	7	637.8	17.9	97	3
	1928	53	9	988.1	30.1	97	3
ID	1920	42	2	531.6	59.4	90	10
	1924	39	4	739.5	51.0	93.5	6.5
	1928	37	7	833.6	112.2	88	12
Total U.S.	1920	2,949	891	37,788.9	1,275.0	97	3
	1924	2,866	899	51,963.4	2,072.3	96	4
	1928	3,018	819	79,179.7	4,011.5	95	5

Data source: Federal Power Commission, *National Power Survey: Statistics of Electric Generating Plants in the United States, 1920–1935* (Washington, DC, 1936), 2–4, 62, 70, 78, 94, 103, 110.

market share percentages for each offer a more revealing view. At the national level in the United States, the number of public-power systems decreased by almost 10 percent from 1924 to 1928, yet public power's market share in those four years crept slightly upward, from 4 percent to 5 percent. While many municipal systems were taken over by investor-owned utilities, many other municipal systems expanded their service territories and productive capacities, Seattle City Light being a prime example as it dramatically boosted its customer base and its power production in the 1920s. Among the Northwest states, only Washington had a significant public-power sector. The number of publicly owned systems in Washington state declined from twelve to ten from 1924 to 1928, yet their market share in the state increased. In 1928, public-power systems supplied 22 percent of Washington's power at a time when municipal systems nationally supplied only 5 percent of the nation's power. In contrast to Washington, the number of municipal systems in Oregon grew from seven to nine from 1924 to 1928, but their control of market share in that state remained the same, mired at an insignificant 3 percent. Idaho's public-power sector fluctuated between 6.5 and 12 percent of the market in these years.

In summary, during the electrification boom the overall ratio of public power to private power did not change significantly in the Northwest states and the United States in general. In Canada, however, public power made strong gains in several provinces and the nation as a whole, though not in British Columbia, where public power remained virtually irrelevant. In fact, the electric power statistics for British Columbia are much closer to the U.S. record than to Canada's overall record.

THE CONSUMER EXPERIENCE

Electricity continued to transform production, work environments, domestic life, and commerce in the 1920s, at a pace that at times seemed frenetic. Historians have noted that the "roaring twenties" did not roar for all sectors of the economy or for all communities. The farm sector, for example, remained in financial doldrums after World War I due to overproduction and consequent low crop prices. In contrast, the electrical industry's galloping growth seemed a beacon of prosperity and opportunity. Those seeking economic advancement looked to electrification as the stimulus. Those seeking increased profits from production or sales looked to electrification for that competitive edge. Those seeking a progressive image surrounded themselves with the accoutrements of modern electrical technology. And the electrical industry reveled in these trends, encouraging them with advertising campaigns, news releases, editorials, and social events. More and more electrical devices came on the market, accompanied by ever more aggressive sales campaigns. But the best incentive for sales was affordability. Throughout the 1920s the cost of both electric appliances and the juice to drive them dropped steadily.[15]

Remarkably, nominal electric rates (not adjusted for inflation) declined throughout the first four decades of the twentieth century. Very few commodities saw nominal price reductions over such a long period of time. Most commodity prices in fact rose during the 1920s, so this countervailing trend is quite significant. The rate reductions occurred most notably in the first two decades of the century but extended into the 1920s and 1930s due to continued economies of scale, efficiency gains from consolidations, the actions of state regulatory commissions, and especially technological progress. Unlike the farmers who couldn't make a decent profit as crop prices plummeted, the electric companies were comfortably and reliably profitable even as the prices they charged for electricity dropped. Figure 7.1 shows the steady decline in average U.S. electric rates (nominal prices) during the first fifty years of the industry's history.

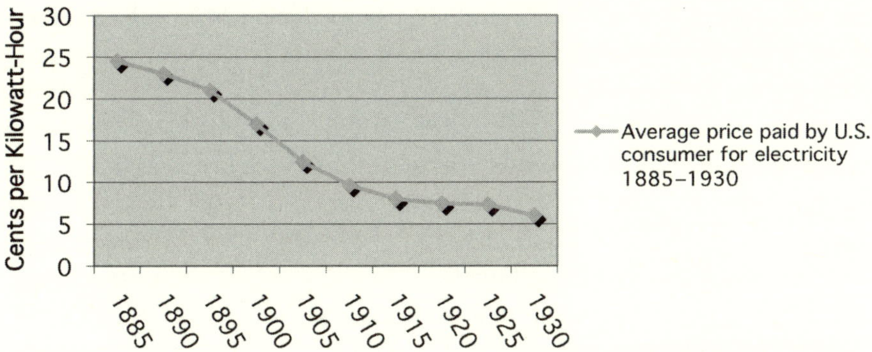

Figure 7.1. Average U.S. Domestic Electric Rates. Data source: U.S. Census, reported in Edison Electric Institute Statistical Bulletin 4, *The Electric Light and Power Industries in the United States* (New York: EEI, January 1937), 17. Graph by Paul Hirt.

The same pattern of falling electric rates prevailed in Canada, where the average domestic retail price of electricity dropped by about a third between 1913 and 1926. British Columbia followed Canada's overall pattern of moderate rate reductions, although this was not uniform across the province. Very large rate reductions occurred in Vancouver between 1913 and 1926 (9 cents/kwh to 4.5 cents/kwh) and Prince Rupert (16.75 cents/kwh to 6 cents/kwh), with smaller but still significant reductions in Victoria, Kelowna, and Prince George. Rates remained stable in fourteen other B.C. towns, with a few going up slightly. These "retail" rates for residential electric service were quite a bit higher than what large commercial, industrial, and transportation customers paid per kwh.[16]

Electric rates in the United States also varied considerably by community, utility company, region, class of use, and consumption rates, so aggregate figures can obscure more than they reveal. Complicating matters further, unlike today, utility companies back then charged dozens of different rates for different types of uses and appliances, and they had elaborate sliding scales charging lower rates per kwh for each additional increment of power consumed. As a result of this mind-numbing complexity, it is virtually impossible to precisely calculate and compare electric rates across utilities and regions over time, so rough averages and general trends must suffice.[17] Because hydropower provided most of the electricity in the Northwest, rates tended to be lower there than in regions where electricity was produced mainly or exclusively by fuel-burning steam plants. Moreover, rates tended to be lower for publicly owned power systems than for for-profit power systems, although there were exceptions to these rules. A striking example of the decline in rates is the for-profit Washington Water Power Company headquartered in Spokane. Its average retail electric rates steadily declined across its service area from about 20 cents/kwh at the start of the twentieth century to the astonishingly low rate of 1.9 cents/kwh in 1925. More typically, in the eight years from 1922 to 1930, the Pacific Power & Light Company's residential electric rates for Walla Walla dropped from almost 12 cents/kwh to just over 4 cents/kwh.[18]

While company reports and public relations materials from the commercial sector of the industry enthusiastically touted these rate reductions, portraying them as proof of their company's commitment to reasonable rates and the public welfare, in fact the reduced rates were mainly a marketing strategy to encourage greater consumption. Moreover, most rate reductions after the 1910s were ordered by the public utility commissions. In my review of the annual PUC reports for Washington, Oregon, and Idaho from 1910 to 1930, rarely did I find utilities petitioning the commissions to approve rate reductions. In nearly all cases, rates were lowered intentionally by the commissions after extensive reviews of utility financial records and independent valuations of company assets, with company attorneys and accountants often challenging the valuations and tariffs set by the commissions.

The utility commissions were not the only force exerting a downward pressure on electric rates, however. Municipal-owned utilities often stimulated rate reductions in neighboring commercial utilities merely by example and competition. In Seattle, the municipal-owned utility Seattle City Light served properties directly adjacent to properties served by the large commercial utility, Puget Sound Power & Light (PSP&L). Whenever Seattle City Light reduced its rates, the commercial utility followed suit. In the old days this kind of competition often led to the sale of power below cost in an effort to drive the competitor out of business, and this too often led to bankruptcies and unreliable service. With the advent of state regulation, this cutthroat competition ended. Despite its rate reductions, PSP&L continued to earn healthy profits—though less so in its service territories adjacent to Seattle City Light. For its part, City Light proudly noted in 1922 that "the municipal lighting plant has returned a handsome surplus of over $5,300,000 above all costs of operation, interest and depreciation, and has cleared a surplus every year of its life except the first two."[19]

Sliding scales for electric rates dropped the price of electricity significantly for increasing increments of power, so customers' average electric rate was often well below the top rate. For example, Seattle City Light's top monthly rate in the 1920s for the first 40 kilowatt-hours was between 5 and 6 cents/kwh, after which the rate fell abruptly to 2 cents/kwh for any additional power that month. Since consumption steadily rose, customers consumed much of their power at significantly less than the top rate. As a result, the average price paid by residential customers of Seattle City Light dropped by almost half between 1918 and 1930, from 5 cents/kwh to 2.8 cents/kwh. A 1931 study cited in a Seattle City Light annual report compared average residential electric rates for commercial utilities with those of publicly owned utilities in the United States and Canada. The study found that the average cost for residential service for all U.S. private power firms in aggregate came to 5.78 cents/kwh in 1931. In contrast, Seattle's municipal utility rates averaged 2.8 cents/kwh, Tacoma's municipal utility rates averaged 1.7 cents/kwh, and for twenty-six towns in Ontario supplied by the nonprofit provincial power authority the average residential rate was 1.45 cents/kwh. This same report also showed that the lower the electric rates, the more energy residential customers consumed. Average monthly consumption per household for the investor-owned utilities was 48.6 kwh. The average monthly consumption per household for Seattle City Light was nearly double that, and for Tacoma's municipal utility, nearly triple.[20] Consumers of all classes were getting more for less in the 1920s, but especially so for those served by publicly owned utilities.

Both municipal and commercial utilities encouraged consumption, especially the latter, since increased consumption was the best way to raise revenue due to utility commission restraints on profit margins. Unlike utilities today, utilities then never talked about conservation, except to argue that damming more rivers for hydropower was an ideal way to "conserve" the nation's rivers. Thus, for the commercial sector of the industry in particular, promotion and sales campaigns dominated public relations. Utilities barraged consumers of all classes with advertisements for new products, exhortations to convert gas appliances to electric appliances, pleas to husbands and wives to "modernize" their households with more of the latest electric gadgets, and boasts of the municipal and industrial prosperity wrought by electrification.

The primary trade publications for the commercial electric power industry in the twentieth century were *Electrical World* and its West Coast cousin, *Electrical West.* Packed with advertisements, short accessible articles, editorials, news briefs, and periodic assessments of the industry, Electrical West offers an illuminating window on the western commercial power sector in the United States and Canada. Manufacturers, farmers, and homeowners were all key targets of sales campaigns, as well as electrical industry employees themselves, who were urged to set proper examples as enthusiastic consumers and to serve as boosters for their industry. As an article in *Electrical West* succinctly put it, "In the final analysis there is only one real problem in industry and that is marketing."[21] Leaders of the electrical industry in the United States adopted many promotional campaigns in the 1920s, including one program aimed at utility company executives to increase retail energy consumption with a very specific target: "one more kilowatt hour per day per domestic customer."[22] With 16 million residential service customers in the United States in 1927 (1.3 million in the Mountain and Pacific states), this would result in substantial additional revenue to the utility companies, not to mention the boost to electrical appliance manufacturing companies like G.E., Westinghouse, and Hotpoint.[23]

With sales goals and revenue-enhancement pressures from above, utility operating companies at the community level adopted a plethora of their own marketing campaigns. The Portland Electric Power Company and Northwestern Electric coordinated with electric appliance dealers in Portland to form the Electrical Appliance Retail Association in early 1928 with the hope that a coordinated marketing campaign would increase overall sales for everyone. Jointly funded print ads launched in May suggested that "Spring Cleaning Time is Electric Range Time." Because some consumers worried that certain brands of ranges might be incompatible with the service provided by their utility company, the association generated a list of approved ranges with guaranteed installation and service. In 1927 and 1928 both Hotpoint and Westinghouse launched their own coordinated sales campaigns for electric ranges, which were significant moneymakers for both the appliance dealers and the utilities. *Electrical West* reported, in an article titled "Range Sales up, Up, UP," a 28 percent increase in sales following the first two campaigns. According to Washington Water Power's historian Steve Blewett, "Due to aggressive sales campaigns, WWP had one of the highest rates of electric range and water heater use in the nation. . . . In addition to ranges and water heaters, WWP was also setting records in sales of the Thor washing machine and promoting other appliances and uses through employee involvement in sales."[24]

The great majority of sales in this era involved personal contact. Utilities and appliance dealers organized office displays, hired small armies of door-to-door sales agents, created travelling exhibits, and sponsored promotions at county and state fairs. Washington Water Power, for example, sponsored an annual Housekeeper's Week at the University of Idaho (Moscow, Idaho) and created an "appliance shop on wheels" for door-to-door sales. An effective verbal sales pitch was therefore crucial. The Pacific Coast Electrical Association sponsored a Three Minute Speech Contest at its annual meetings in the late 1920s, encouraging contestants to deliver their most effective sales pitch in three minutes or less. The quote below comparing electric motors to happy slaves is excerpted from the prize-winning speech from the 1928 convention.

Figure 7.2. Washington Water Power Electric Cooking Advertisement. Source: Steve Blewett, *A History of the Washington Water Power Company, 1889–1989, Building on a Century of Service* (Spokane: Washington Water Power, 1989), 31.

> Electric motors have well been likened to an army of willing slaves, begging the privilege of doing our hardest tasks—for three cents an hour! Each tiny motor, in the role of a smiling, ever-ready slave, comes bowing before us, pleading:
>
> "Use me! For though the general cost of living has increased 65 per cent since 1914, I am even cheaper than I was then! And, as you increase my work, my cost becomes proportionately less! For economy and efficiency, for cleanliness, and for the sake of your health, make use of me!"[25]

The sales pitch quoted above included an exhortation to both husband and wife about the use of labor-saving electric appliances in the home:

> [To the husband:] Would you knowingly allow your wife or your mother to work for 3 cents an hour? Yet that is the ridiculous wage of any woman who does anything a small electric motor could do for her!
>
> [To the wife:] Madam! Electricity will sweep your floors for 1¾ cents an hour! It will wash your clothes for 3 cents an hour, and pedal your sewing machine for ¾ of a cent an hour! It will cook for you, iron for you, warm you, cool you, clean for you and make your ice—at an average rate of only 3 cents an hour![26]

To assist with the effort to sell more appliances in the home, male executives and sales managers established "women's committees" to help with sales campaigns for the domestic market. Many articles on this topic appeared during the 1920s in *Electrical West*, most of them lamenting how wives too often resisted adopting electrical appliances and how men who worked for the industry should best convince their wives to use the new technologies and help promote them. A power company executive in the Northwest suggested that the women's committees should be composed predominantly of the wives of employees, adding that "every wife is a volunteer salesman."[27] Female leaders offered a different perspective, reflecting the rising tide of feminist advocacy in the 1920s. Rather than accepting mere auxiliary roles as women's committee volunteers, some women advocated for their rightful place as paid professionals in the industry—albeit as sales staff for the residential consumer market where they would still be unthreateningly focused on domestic sales to other women. As Anna Steese Richardson, director of the Good Citizenship Bureau of the *Woman's Home Companion*, said in 1929, "Woman's Place is in the Homemaking Departments of the Industry."[28] Richardson's articulate speech is quite revealing of the time:

> WOMAN'S proper place in the social structure has been under discussion ever since Eve led that little procession of two from the Garden of Eden. Is it in the home? Try to keep her there! Has she found her niche in art or science? You will find her foot on the ladder. Is she a power in politics? Ask any candidate for office who has incurred the displeasure of a women's organization. Have business and the professions accepted her? Call the roll in your own office or store. . . . . You do not need women to climb poles, string wires, make surveys or experiment in laboratories, but you do need them to make contacts with homemakers—today a most important group to reach in building up your load. . . . In addition to their direct contact with consumers of power and light these women have the support of magazine editors who during the past few years have been preaching the gospel of less drudgery and more leisure for housewives, and of the women's organizations which are standing solidly behind the movement for better home equipment.
>
> Give [women] their place in the electrical industry and you will surely build up your load.[29]

Electrical sales promotion in Canada followed a similar path, although interestingly, BC Electric, which owned and operated the natural gas supply for the urban areas around Vancouver and Victoria, did not push electric cooking and heating as much as it did gas cooking and heating. The latter, of course, is a more efficient method for obtaining domestic heat, and since the company profited from gas consumption as well as electric consumption, it had little incentive to favor electric ranges. As company historian Cecil Maiden notes, the

1920s saw "many big improvements in cooking—especially gas cooking—and gas heating. The national trend itself was now towards the 'scientific approach' in housekeeping, and the 'B.C. Electric' was going before the public with the slogans 'Heat with Gas, the Scientific Way' and 'Cook with Gas, the Scientific Way.'"[30] The growth in BC Electric's power and light business focused more on residential lighting, commercial lighting, electric railways, mass transit, and industrial uses.[31] Promotions, affordable rates, and a predominantly urban population in British Columbia gave the province the highest electrification rate of any province in Canada: about 73 percent of B.C. households enjoyed domestic electric lighting by 1927, compared to a national average of 67 percent.[32]

Agricultural regions in arid climates that depended on pumped irrigation water were profitable markets for power companies, as were mass-production animal farms and food-manufacturing facilities. A U.S. Department of Agriculture survey of 1927, compiled at the request of *Electrical West*, reveals that pumped irrigation consumed 8 percent of all the power generated in the eleven western states. After California, the next three largest consumers of power for irrigation in the West were Idaho, Washington, and Oregon, respectively. In 1927 the Idaho Power Company sold 20.7 million kilowatt-hours of electricity in Idaho to pump water and another 12.5 million kwhs in Oregon. Other big players in the irrigation business were the federal government's Minidoka Irrigation Project in southern Idaho, which dedicated over 27 million kwhs to water pumping; the Washington Water Power Company, which dedicated 21.4 million kwhs to water pumping; and Pacific Power & Light, which dedicated 11 million kwhs to water pumping.[33]

But water pumping was only one key use of power in agriculture. Electrical engineers and manufacturing companies in the 1920s improved the performance and efficiency of many farm technologies, including milking machines, incubators for poultry farms, lifts and driers for grain elevators, and the like.[34] Utilities seeking to increase demand, therefore, focused a good deal of promotion efforts in agricultural communities in the Northwest. A great deal of large-scale commercial agriculture existed in the rural areas surrounding Puget Sound's major cities. Puget Sound Power & Light, which served these areas, in 1929 had a regular feature dedicated to "Farm Electrification" in its monthly *Puget Sound Electric Journal*. In fact, throughout the decade the journal included many articles on the topic.[35]

Significantly, the Northwest had higher rates of rural electrification than did most other rural areas of North America. The reasons are that farming in the dry but highly productive interior Northwest required irrigation, which often required large water pumps, and much of twentieth-century farming in the Northwest was industrial-scale and capital-intensive, more so than, for example, that of the humid prairie states and provinces. Besides improvements to existing electrical technologies for farmers, engineers also invented myriad new ways of applying electricity to agriculture, some of which bordered on the outlandish. Among the latter category was the "electric plow," which was in fact a method for killing pests in the field by electrocution. An article in the June 1928 issue of *Electrical West* described somewhat successful field tests for a tractor-drawn implement that charged plowed soil with "500,000 Cycle, 104-Kv. Energy" to kill insects and weeds. The writer noted that "The electrical discharges in the soil and air create considerable ozone. The smell behind the plow is very pungent."[36]

The biggest electricity loads came from mining, smelting, and manufacturing facilities. A great deal of technological innovation and sales promotion therefore focused on this sector of the market. As described in the previous chapter, on World War I, electrolytic metals processing consumed extraordinary amounts of power, dwarfing most other uses. Canada had set up an electrolytic zinc-processing facility in Trail, British Columbia, just north of the international border during the war, and it became the largest single load of West Kootenay Power and Light.[37] On the west side, BC Electric's largest new contract for power during the 1920s was the Britannia Mining and Smelting Company on Howe Sound. Just for this one contract in 1923, BC Electric built a new high-voltage transmission line from North Vancouver across thirty miles of rugged coastline to the company's property.[38] Across the international border in 1928, the Coeur d'Alene Mining District in northern Idaho installed in Kellogg a zinc-processing facility that used a new, experimental electrolytic process that required even higher voltages than previous processing methods. Washington Water Power supplied electricity to the district.[39]

One of the most important innovations of the 1920s was the development of functional, affordable electric air-cooling for industrial-scale cold storage. (Small-scale home air-conditioning units were still a few decades away.) In 1929 BC Electric provided power to new refrigerator plants at Pacific Coast Terminals, facilitating the storage and mass marketing of perishable food from this important port town. Innovations in electrical accounting and typing took off in this decade, too. The Burroughs office machine company manufactured a whole bank of new bookkeeping machines for BC Electric, which, according to Cecil Maiden, "itemized hundreds of thousands of bills, recorded a duplicate of each bill, and just by way of a final gesture of superiority, uncomplainingly added all the totals."[40]

Of course the electric power industry itself was constantly innovating and upgrading its own power-generation equipment. This is the main reason the cost of producing power declined so steadily in these years. Like many other industries, automation was a key element of electrical innovation, reducing the need for human labor while increasing efficiency. Automation came to hydroelectric power plants at this time, and the Pacific Northwest, with its prodigious water-power capacity, was at the forefront of innovation once again. The BC Electric Company, for example, installed Canada's largest automatic hydroelectric generating plant at its Alouette facility in 1928. It was said to be the second largest automated hydropower facility in the world.[41]

The economy in general and the electrical industry in particular were indeed booming on the eve of the Depression. A New Year's speech at the start of 1929 by the famous humorist Will Rogers alludes to the prevalence of "optimistic predictions by all prominent men who are doing well." If one is looking for optimistic predictions of the year ahead in early 1929, one could do no better than the February issue of *Electrical West*. Each February the journal published a lengthy summary of electrical development during the prior year and projections for the coming year. If any people in the industry suspected the good times were about to crash, their concerns were entirely absent from the pages of the journal. At the start of this grievously momentous year, optimism was the only emotion on display. Prefacing a series of statements from utility company executives about prospects for 1929, the editor of *Electrical West* observes, "Certainly the statements issued by these commercial executives cannot be considered as anything but forecasts of prosperity."[42] Few in the industry suspected that their own financial practices would contribute substantially to the coming economic depression, though the underlying value of electrical technology and the reliability of demand for electricity would help carry the industry through the rough times.

CONSOLIDATING OWNERSHIP TO CONCENTRATE PROFITS

For commercial utilities in the United States and Canada, a great deal of consolidation accompanied the 1920s boom as bigger and better capitalized companies bought and integrated competitors and neighboring systems. For the first time since the invention of electricity, the number of companies generating power in the United States actually declined in the 1920s, falling by over 40 percent from 1922 to 1927, even as electric generation nearly doubled in those same years.

There was so much consolidation in the United States that the number of electric light and power establishments in 1927 dropped below the number reported in the electrical census of 1902.[43] According to a study by U.S. senator Thomas Walsh's office, nationwide in 1925 and 1926 there were at least 1,500 utility company mergers.[44] That year the Federal Trade Commission (FTC) noted that a key characteristic of the industry in the mid-1920s was "the merger of these small local companies, operating in contiguous territory, into larger companies with larger areas of operation."[45] The most significant consolidation in British Columbia occurred in 1920, as mentioned in the previous chapter, when BC Electric purchased its main rival, the Western Canada Power Company, leaving just two large private companies dominating some 97 percent of the power supply of the province.[46]

Another hallmark of the decade was the growth of electric utility holding companies, which helped finance these mergers. Holding companies did not directly manage or "operate" utilities, but instead owned stock in the operating companies, lent them money for acquisitions and upgrades, and provided engineering, financial, and managerial advice—for a fee. The holding companies earned significant dividends, interest, and consulting fees from the operating companies under their financial control. In its 1927 study, the FTC identified sixty utility holding companies, twenty-two of which were subsidiaries of the others, so that thirty-eight holding companies exercised financial control over virtually the entire U.S.-based electric power industry. One holding company, Electric Bond and Share, controlled 10 percent of total U.S. generating capacity. The FTC bluntly explained that the holding companies proliferated at this time in order to "exploit the earnings" of the electric utility operating companies.[47] In essence, the success of the operating companies made them targets for predatory investors seeking to skim profits via a maze of pyramided financial relationships. The operating companies accepted these relationships as the cost of finance and security. Besides, they could pass on these expenses to customers in electric rates if the state utility commissions accepted them as reasonable costs of doing business.

The Puget Sound Power & Light Company (PSP&L) provides a graphic example of this consolidation phenomenon. In 1922 and 1923, PSP&L purchased three small electric companies: Granite Falls, Sumas, and Mutual. Then in 1924 it acquired six more railway, light, and power companies, plus two large conglomerates: Washington Coast Utilities and the North Coast Power Company. Washington Coast Utilities had merged together eighteen separate light, power, water, gas, and telephone utilities between 1910 and 1922, serving a comparable number of communities. The North Coast Power Company was a merger of thirteen separate traction, light, power, and water utilities acquired between 1905 and 1915. After absorbing these two, PSP&L in 1925 acquired four more independent utilities whose service territories stretched from eastern Washington to the town of Sequim on the Olympic Peninsula. In 1926 it acquired fifteen more companies serving rural counties straddling both the east and west sides of the Cascade Mountains, stretching from the Canadian border on the north to the Columbia River on the south. Five more acquisitions followed in 1927 and seven more from 1928 to 1930.[48] By 1930, PSP&L controlled a nearly uncontested service monopoly for the greater Puget Sound region, Cascade Mountains, and northern Olympic Peninsula. Only the municipally owned Seattle City Light and Tacoma Public Utilities provided any local competition for this company.

While the PSP&L annual reports and corporate history proudly touted these acquisitions as proof of the company's success, they were in fact orchestrated by Stone & Webster, a nationwide electrical engineering, management, and financial services firm that grew into a powerful utility holding company during the 1920s. Stone & Webster launched their Northwest operations in Seattle in 1899 and over several decades facilitated the mergers that created Puget Sound Power & Light and its acquisitions mentioned above. By 1927 the Stone & Webster group controlled 115 large utility companies, such as PSP&L, in fifteen states.[49]

This litany of consolidation and national holding company influence was not confined to the Puget Sound region. In fact, Stone & Webster's Northwest turf was rather small compared to the territory under the financial control of Electric Bond and Share, founded by J. P. Morgan in 1905 and still controlled largely by his son in the 1920s. Six of the U.S. Northwest's largest utility companies were part of this empire by the late 1920s: Pacific Power & Light, Northwestern Electric, Washington Water Power, Utah Power & Light, Idaho Power, and Montana Power. The Electric Bond and Share Co. (EBASCO) bought Washington Water Power in 1928. Most of the others were under the control of various EBASCO subsidiary holding companies, such as the American Power and Light Group and the Electric Power & Light Corporation. By 1929, EBASCO and its subsidiaries controlled 53 percent of the U.S. Northwest's electric load.[50]

Holding companies were constantly forming and merging in the 1910s and 1920s, mainly in order to concentrate control of operating companies and capture profits, although the more public rationale was to finance the interconnection of local systems into larger regional systems to take advantage of economies of scale and to improve service. In fact, service did expand and improve during this period while rates declined, so consumers were generally satisfied, with a few notable exceptions discussed below. Power system interconnections advanced quite a bit in the 1920s, too, in conjunction with holding company acquisitions. Construction of Rock Island Dam on the Columbia River, a Stone & Webster project started in 1928 and finished in 1931, spurred interconnection since utility executives expected the dam to generate large quantities of surplus power that could be marketed to other utilities. From 1928 to 1930, Utah Power & Light connected its system to Idaho Power, which connected to Pacific Power & Light, which connected to Northwestern, which connected to Washington Water Power. All of these were EBASCO affiliates. Finally, the whole group formed a power pool with Puget Sound Power & Light, the Stone & Webster affiliate building Rock Island Dam.[51]

As the control of electric utilities (operating companies) grew increasingly concentrated in the hands of a small number of holding companies, a chorus of anti-monopoly voices arose. One of the most vocal was Democratic senator Thomas Walsh of Montana, who introduced Senate Resolution 329, which was approved February 9, 1925, directing the Federal Trade Commission to investigate concentration in the electric power industry, especially in regard to the degree of control exercised by the General Electric Company. This quickly became one of the most publicized and sensationalized political issues of the late 1920s.[52] In fact, members of the Senate were so concerned about the growing financial power and political influence of the electrical industry that it refused to seat two newly elected senators after the 1926 midterm election (Frank L. Smith of Illinois and William S. Vare of Pennsylvania) because they had received inordinately large campaign contributions from utility companies.[53] Interestingly, Walsh's Senate resolution also directed the FTC to examine the degree to which General Electric and other commercial power companies sought to influence public opinion regarding public power. Advocates of public power were deeply critical of the extensive public relations efforts of the commercial companies to turn public opinion against municipally owned power systems. Enough sympathetic senators shared this concern to endorse this resolution asking the FTC to investigate the question. But before the FTC could act, opposing legislators added to the appropriations bills for 1925–1926 and 1926–1927 language barring the FTC from spending any public money investigating utility company propaganda.[54]

Despite this obstacle to researching power company propaganda, the FTC did go forward with its study of financial holding companies in the electric power industry. It released a 300-page report in 1927 with some rather troubling revelations. The findings of the report are worth reviewing in some detail here and below, as they shaped public debate and subsequent regulatory policies for the next decade. The letter of transmittal to the Senate from the chairman of the commission concludes with this expression of concern about concentrated control in the electric utility industry:

> One of the problems of public interest concerning some of the large electric-power groups is the extreme degree to which "pyramiding" has been carried in superimposing a series of holding companies over the underlying operating companies so that in one instance less than a $1 million investment in the majority of the voting stock of the apex holding company gave in 1925 full control of the entire organization of the group, having scores of underlying companies and several hundred million dollars of investment. Such pyramiding not only affects the financial stability of the electric-power industry, but also has a potential relation to the more general question of an undue concentration of control in the electric-power industry.[55]

The clause in the last sentence about risk to the "financial stability" of the industry was certainly prescient, given that the bubble would burst just two years later.

HIGH-FLYING FINANCE

> *A Holding Company is a thing where you hand an accomplice the goods while the policeman searches you.*
>
> —Will Rogers[56]

Because of the rapid growth of electrical consumption, corporate income for U.S. for-profit electric companies increased significantly between 1918 and 1924, as measured by their reported taxable income. This is especially notable since American businesses as a whole showed a decline in corporate income during this same period.[57] According to the U.S. electrical censuses, revenues for electric utilities nearly quadrupled between 1917 and 1927. (Detailed electrical censuses were taken only every five years at that time, including in 1917, 1922, and 1927.) Expenses grew at a slower pace than revenues in this same period, so profitability was rising.[58] Consequently, as noted above, an extraordinary infusion of capital flooded the industry in the 1920s. This resulted as much from the need to expand and improve service as from the desire of financiers to invest in a safe, growing business and capture a bigger slice of the electric revenue pie. In the United States, the capital value of central electric light and power stations tripled in the ten years between 1917 and 1927. Because state regulatory agencies set electric rates at a level designed to provide a fair return on capital investment, payback was virtually guaranteed, so money poured into the electric power industry—more money than necessary to meet electric demand, so for-profit utilities learned to pad their expenses in complicated ways that passed muster with the regulatory agencies but led to overcapitalization and financial abuses.

In the matter of excessive capitalization, a study published in 1924 by *Lake Shore News* (Lake Shore, Washington) compared the amount of debt per horsepower of three large commercial for-profit utilities and four large non-profit municipal utilities in the Northwest. The average debt per horsepower for the commercial utilities was $381; the average debt per horsepower for the four municipal utilities was $73.[59] This is a rather stark contrast, and the data source is not identified. Figures for the United States as a whole from the 1927 U.S. Electrical Census show a similar though less dramatic discrepancy between commercial utilities and municipal utilities. The capital value per horsepower of commercial utilities was $379, while the comparable figure for municipal utilities was $172.[60]

Municipal utilities experienced a virtually uniform threefold increase in electricity output, in the capital value of facilities, and in revenues from 1917 to 1927. In contrast, for commercial utilities in these same years, electrical output multiplied 2.9 times while capital investment in facilities and revenue from electric sales multiplied 3.6 times. In other words, debt and revenue were increasing much faster than actual electrical output. Commercial utilities were becoming less efficient in a cost-per-kwh basis while they were becoming more profitable. Publicly owned utilities, in contrast, produced more electricity per investment dollar and charged less per kilowatt-hour than investor-owned utilities.[61] Ironically, publicly owned utilities were not regulated by the states in the same way as for-profit utilities; they did not have to ask for rate increases from the PUCs. Nevertheless, they kept rates low and capitalization at reasonable levels because their primary purpose was to provide affordable energy services rather than profits for investors.

The story is quite different for Canada's electrical boom. At the start of the decade in 1920, the average cost per horsepower was $220 ($185 for British Columbia); by 1928 it was down to $200 ($173 for British Columbia).[62] Moreover, the difference in capital value per horsepower between commercial and municipal utilities in Canada was relatively small. Throughout the decade, Canada's output of electricity increased much faster than both capitalization and revenues. As the 1928 Electrical Census of Canada proudly noted, "per kilowatt hour of output, the industry has reduced the average investment between 1920 and 1928 by 23 percent [and] reduced average revenues by 24 percent."[63] The census explained this interesting record as primarily due to the growth of publicly owned nonprofit power systems in these years. Between 1920 and 1928, the publicly owned systems in Canada increased their share of electrical production from 24 percent to 30 percent of the total national output. As shown earlier in Table 7.1, the comparable figures for the United States starkly contrast with power output of publicly owned systems at 3 percent of the total in 1920 and 5 percent of the total in 1928. Clearly, on a cost-to-benefit basis, Canada was building a more efficient power system based on hydroelectricity with nonprofit government agencies steadily increasing their share of the market. British Columbia was a notable but temporary exception to the public power trend, as its municipal utilities controlled only 2.3 percent of the province's power output in 1928.[64]

In the United States, profit-seeking motivations were mostly responsible for the higher capital costs and higher electric rates of private utilities, and ironically the structure of the state regulatory system in the United States helped facilitate it. Because for-profit utilities could not unilaterally raise electric rates to increase their cash flow, they instead sought more indirect means of revenue enhancement. The most common method was to spend more money on facilities and show higher capital costs (loan fees and interest rates), which then justified higher electric rates to ensure a fair return on that investment. As the 1927 Electrical Census aptly notes, "Compared with most other industries, the value of plant and equipment in the electric light and power industry per dollar of annual revenue is large. . . . An unusually large fraction of the earnings is therefore consumed in meeting capital charges." What the census doesn't say is that the capital charges were usually profits for the same investment companies that owned the utilities, so they were essentially paying themselves.[65]

The average rate of return in the United States for the utilities under state regulation was about 8 percent per year—a healthy profit on a sustained basis. But for the holding companies organized one step above the reach of the state utility commissions, rates of return on investment were much, much higher. The 1927 FTC report documents annual returns on investment for the holding companies in 1924 and 1925 of 20 to 50 percent! Directors of these companies often served on more than one board and thus were raking in profits of this scale across multiple holding companies. The FTC report notes that interlocking directorates were "very numerous," especially among General Electric, Electric Bond and Share, and their 200 affiliated operating companies. Among the operating companies themselves, the FTC found that more than a thousand individuals served on two to four different electric utility boards of directors, while thirty-six men served on the boards of twenty or more separate companies.[66]

Thus, despite state efforts to suppress monopolistic price gouging, creative accounting and complex financing instruments allowed the industry to increase revenues and profits to levels beyond what most people considered reasonable. This effort to enhance profits made the commercial electric power industry in the United States an increasingly inefficient financial institution—though highly profitable for owners. The only protection against bloated capitalization and unnecessarily high electric rates was effective oversight of utility financial accounts. This is why the FPC regularly complained of its lack of adequate resources to accomplish these duties in the 1920s. The FPC's 1927 annual report, for example, specifically lamented its anemic budget for financial investigations and issued a threat that if more resources for evaluating utility company cost accounts were not forthcoming, it would divert resources to that task by placing a moratorium on processing all but the most essential hydropower license applications. Little changed, however, until after the election of Franklin Roosevelt in 1932.[67]

All these financial shenanigans occurred with the acquiescence, if not the blessing, of the state utility commissions, which likewise rarely had the resources or the organizational agility to uncover and challenge the complex financial arrangements and instruments invented by utility holding companies to enhance profits. Moreover, the state PUCs had the power to inspect the accounting books only of the operating companies doing business in their state. The financial records of the out-of-state holding companies were not subject to state inspection. Partly for this reason, Harvard economist William Ripley called holding companies "anonymous industrial autocracies."[68] The overburdened state public utility commissions tasked with protecting the public interest were outgunned and outmaneuvered by these highly profitable corporations, which were backed by some of the best legal and accounting consultants in the business. Consequently, ratepayers footed the bill for the pyramided financial structures surrounding and controlling the utility operating companies throughout this decade.

In the Northwest, some towns and businesses rebelled and forced the PUCs to re-examine utility financial records and rein in the unreasonable capital costs claimed by the utilities. Examples are discussed in the next section. Other communities turned to public power as a better alternative. But the road to public power was not easy. To buy or build a municipal power system required bonds, elections, taxes, and the hiring of technical expertise. Selling city voters on the idea of funding a city-owned power system could be tricky, and the commercial power industry went to great rhetorical lengths to make that task even more difficult, funding "citizens" organizations to attack public power in the press, editorializing against municipal power bond elections, and portraying public power as un-American. During the 1920s, public power proponents in Washington state repeatedly but unsuccessfully sought to expand public power authority through both legislation and citizen initiatives. Each attempt generated media battles in which the for-profit power industry vigorously defended capitalism and its record of service. To defeat a 1924 Washington state referendum sponsored by public power advocate Homer T. Bone, Puget Sound Power & Light helped establish and fund a group called the Voters' Information League, which generated what one Seattle newspaper called "canned propaganda," full of half-truths and distortions. The referendum was defeated, but later investigations by the FTC revealed the extent to which commercial utilities funded the misinformation campaign.[69] Commenting on a spate of public power bills and other proposed laws aimed at tightening regulations on utility companies pending in all three Pacific coast states in 1929, the editors of *Electrical West* complained that "the astounding part of it is that here on the Pacific Coast, where so much harassment is continually offered, the industry goes on rendering more and better service, advancing the interests of its customers and improving its own condition in the bargain."[70]

Despite these defensive justifications of their record, many communities, businesses, public interest groups, and politicians across the United States raised concerns about the concentration of utility company ownership, the financial and political power of the holding companies, and the excess profits wrung from consumers. (Things were considerably quieter in British Columbia about the dominance of BC Electric and West Kootenay Power and Light, most likely because Canadian electric companies were producing more power with less capital investment and collecting less revenue than the U.S. companies—in other words, they performed better.) Gifford Pinchot and the advocates of Giant Power railed against the dangers of concentrated wealth with their own ideological rhetoric while promoting publicly owned power systems:

> It is almost impossible to imagine the force and intimacy with which such a monopoly will touch and affect, for good or evil, the life of every citizen. . . . The unregulated domination of such a necessity of life would give to the holders of it a degree of personal, economic, and political power over the average citizen which no free people could suffer and survive.[71]

But not everyone who worried about utility consolidation, investment scams, and price-gouging favored public power. Businesses, of course, consumed electric light and power, too, and shared the broader public concerns about whether utilities charged fair rates and efficiently managed company finances. Some businesses actively supported public power, believing it would provide them lower rates. Many others, however, opposed public power on principle but staunchly supported effective public regulation. While favoring a strong role for private business in the American economy, large manufacturing and industrial customers of the electric utilities knew that more streamlined financial structures for the utilities would mean lower rates and, ultimately, big savings for them. Likewise, community leaders knew that cheap, reliable power served as an engine of economic development. Thus, concern about utility industry practices penetrated deep into American society among both partisans of public power and defenders of capitalism and for-profit utilities.

CONSUMER REBELLIONS AND "REASONABLE PROFITS"

The excesses of the profit-seeking companies in this decade left the investor-owned utility sector vulnerable to criticism from those who questioned the accuracy and necessity of the industry's high capital costs. Both public and private power companies constructed the same kinds of hydroelectric and thermal (steam-driven) generating plants, purchased equipment from the same manufacturing companies (e.g., G.E. and Westinghouse), used the same power plant control mechanisms and transmission materials, and had similar maintenance, service, and managerial requirements; so why were capital costs so different? Critics believed the answer lay in the unchecked profit ambitions of Wall Street financiers and their proliferating holding companies that issued watered stock, charged spurious supervisory fees, and made heavy demands for dividend and interest payments from their subsidiary electric operating companies—all of which siphoned off working capital from local electricity providers and transferred it to Wall Street.[72]

The famous Walla Walla, Washington, rate controversy of 1924–1925 provides an excellent example. Through a series of acquisitions after 1910, Pacific Power & Light (PP&L) had by the 1920s consolidated into a regional system serving communities in Washington's Yakima Valley and mid-Columbia Basin, the Walla Walla Valley, several eastern Oregon communities, and towns along Columbia River Gorge, along with its original home base in Astoria, Oregon.

It expanded and improved service in the region while reducing electric

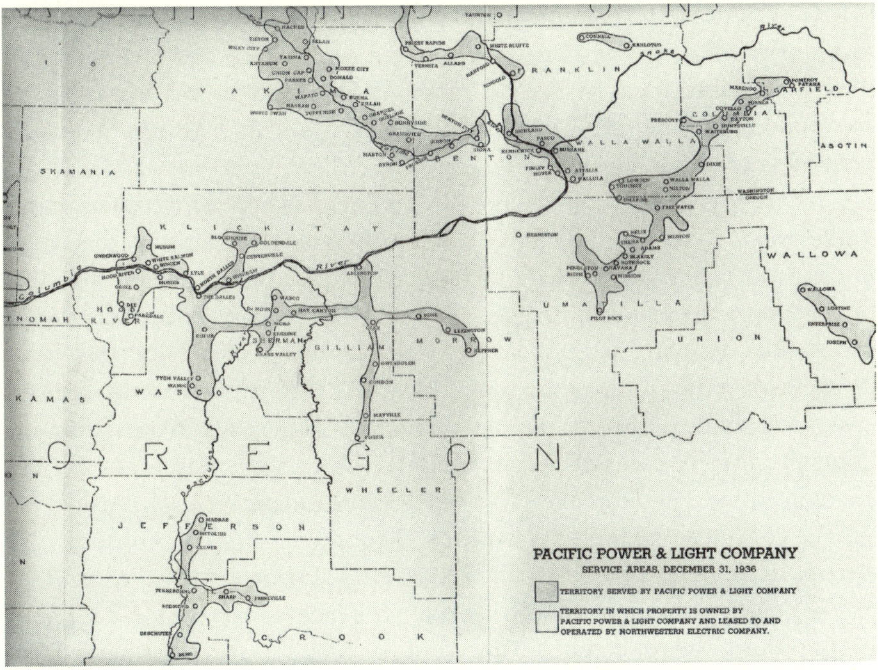

Figure 7.3. Pacific Power & Light Service Territory. Source: Pacific Power & Light Company, *Reclassification of Electric Plant: Statement A to I Inclusive* (Portland: Pacific Power & Light Co., n.d. [1940?]), 7.

rates,[73] yet became embroiled in a major ratepayer revolt mid-decade, the target of an organized rate protest by city leaders in Walla Walla in 1924. Even though Washington's public utility regulatory agency (at this time part of the Department of Public Works) had been scrutinizing PP&L's financial records and approving its rates for fifteen years, state regulators reopened hearings on the case in response to protests from Walla Walla. After a thorough investigation, PP&L's assets were devalued and its electric rates lowered. Investigators found that PP&L's financial directors had overvalued its property by 47 percent, sold $9 million worth of watered-down securities on a $6 million rate structure, inflated its business expenses with unallowed costs, and overcharged its small-town customers $600,000 between 1913 and 1923.[74] Obviously, the regulators had missed a lot of details over the years.

Pacific Power & Light did not act on its own. As a subsidiary of the giant holding company Electric Bond and Share, which itself was nested inside General Electric until 1925, PP&L was indirectly owned by the vast J. P. Morgan syndicate. Morgan's group began aggressively acquiring utilities in the Northwest after World War I. Holding companies earned substantial interest from loans to utilities for expansion and upgrades, but they also charged top dollar for "supervisory" services. Between 1910 and 1928, PP&L paid $1.5 million in these management fees to Electric Bond and Share.[75] This was all part of the cost of doing business for commercial utilities, a cost that, as critics regularly pointed out, did not burden publicly owned utilities.

Despite the Washington Department of Public Works' rate reductions for Walla Walla, PP&L continued to show healthy profits. A January 1929 article in *Electrical West* reported that PP&L's net profit for 1928 was triple than that of the company's previous best year—largely as a result of PP&L's sales force successfully doubling its gross sales of electrical merchandise in 1928. Due to its aggressive sales campaigns, annual electrical consumption for residential customers shot up from 322 kilowatt-hours per year in 1921 to 902 kilowatt-hours per year in 1928. This stunning increase in consumption per household guaranteed more revenue and profits even while electric rates declined.[76]

The continued high profits made excellent fodder for large consumers of electricity to lobby state PUCs to further reduce their electric rates and for public power partisans to tout the consumer benefits to be reaped from non-profit power systems. Investor-owned utilities fought these criticisms with extraordinary public relations initiatives, including public school textbook campaigns and scholarship programs that touted the virtue of private ownership, the effectiveness of state regulation, the financial soundness of the industry, and improvements in service and rates made by the investor-owned utilities. Utility executives hired publicists, paid to have editorials widely distributed in newspapers, and made large campaign contributions to key legislators. These aggressive campaigns to influence public opinion and policy makers angered public power advocates who lacked similar financial resources. Public power advocates consequently lobbied regulatory commissions to remove public relations costs from the list of allowed utility operating expenses. That utility companies could use revenue from ratepayers to shower their customers with anti–public power propaganda seemed perverse to public power proponents. In fact, after the Federal Trade Commission published its 1927 report on control of holding companies, Senator Walsh was able to get a supplemental resolution passed, again directing the FTC to investigate the public relations efforts of the commercial sector of the industry. This time other members of Congress did not block appropriations for the work, and the FTC went forward with a two-year investigation, putting the industry again on the defensive.[77]

In a December 1928 statement in *Electrical West* titled "We Have No Apologies to Make," the president of Pacific Power & Light Company in Portland, Guy Talbot, provided a typical, feisty response to the accusations of propaganda. After accusing the critics of commercial power of being malicious, self-aggrandizing, "amply subsidized propagandists" themselves, he played the slippery-slope card, associating public power advocates with those who "hope by the socialization of our industry to socialize all industry and thereby reach their objective of complete socialism." Talbot then admitted that a few companies may have been "over-zealous in their publicity and educational practices" but said they were responding to the "unscrupulous and envious attacks of government ownership propagandists." Talbot defended a monograph the commercial power industry sponsored for distribution in public schools, calling it "educational and not argumentative" and insisting, "the industry in the Northwest has no apology to make for any of its relations with educational institutions." Finally, he defended electric companies' political expenditures in 1926 to defeat a public power bill popularly named "the Housewives' Bill," saying,

> The money was legitimately and necessarily spent by these companies. They had a right to defend themselves against unfair competition. . . . They had the same right as any other industry confronted by similar conditions to protect the many persons who hold investments in their property through bond and stock ownership. . . . In fact, they not only had the right to do this, but it was their duty to do it.[78]

These public relations efforts helped the investor-owned utilities maintain their majority position in the region, but even more helpful was the fact that electricity became ever cheaper throughout the decade. Most customers cared less about ideological debates over capitalism versus socialism and more about the effects on their pocketbook. Thus, the private utilities best defended their position by lowering rates, improving service, and keeping publicly owned systems constrained and distant from their customers. Their relationships with the state utility commissions helped accomplish this, as the PUCs gave at least a nominal degree of legitimacy to the electricity rate structure. Samuel Insull, the head of one of the nation's largest utility holding companies, centered in Chicago, realized this would be one of the beneficial results of state regulation when he proposed it in 1898. Unfortunately, his foray into high-flying holding company finance in the 1920s would lead to his downfall and disgrace a couple of years later, when the wheels fell off the gravy train.

CHAPTER 8

The Bubble Bursts, 1929–1932

I am convinced we have now passed the worst and with continued unity of effort we shall rapidly recover.

—President Herbert Hoover, May 1, 1930[1]

President Hoover, quoted above, was sorely mistaken regarding a rapid recovery in 1930. The entire national economies of the United States and Canada spiraled downward for at least two more years and did not fully recover until a decade after Hoover's premature prediction. The utility industry survived this crisis well enough, better than most, in fact. Nevertheless, the Depression pushed many utilities to or over the brink of insolvency and ruined many fortunes. The Depression stalled development, led to workforce reductions and wage cuts, ended many dividend payments to investors, and ultimately led to major policy reforms in the United States after the election of 1932. The economy's tumble provided ample opportunities for reflection and reform, but government and industry accomplished very little toward solving the crisis during the first and worst three years of decline. In fact, they spent most of 1930 to 1932 denying or simply waiting out the crisis.

Booms and busts plagued the American and Canadian economies throughout the era of industrialization. Rapid expansions in economic activity were often built upon risky investments, speculative or volatile markets, corporate and political corruption, and cutthroat competition. Many people saw industrial capitalism as a game of chance where smart players took bets, risked millions, and amassed great fortunes. "Casino Capitalism," in the memorable words of economist Susan Strange,[2] thrives in laissez-faire or poorly regulated economies and emerging markets. While a few well-positioned individuals can get fabulously wealthy, most people do not appreciate or benefit much from erratic economic cycles. Nor do nations as a whole. Modern governments, therefore, have developed a variety of mechanisms to ameliorate these wild swings and promote steady, reasonable growth, either through incentives and restraints on the private sector or through direct government intervention. That's what Progressive Era economic policies were mostly about. But these emerging twentieth-century efforts to rein in overexuberant capitalism had limited success. National corporations wielded as much power and influence as did governing institutions—often more—and public sentiment still retained a considerable distrust of centralized political authority, at least in the United States. Corporations therefore expanded faster and were more nimble than the limited and underresourced regulatory bodies created to ensure that corporations acted fairly and responsibly in the public interest. And, of course, government spoke with many voices. Corporations were well represented among the constituencies that influenced policy making. If government agencies failed to adequately ensure fairness and economic stability in the markets or justice and equal opportunity in social relations, it was often because many policy makers and administrators danced to different tunes. No consensus prevailed regarding what constituted the public interest or what distinguished greed from a reasonable return on investment. Laws, therefore, embodied contradictions often, and compromises always.

Times of crisis focus attention on public policy and the public interest, offering a unique opportunity to evaluate problems and restructure institutions to solve them. Initial responses to crises, however, tend to be chaotic, contradictory, and contentious. During the first three years of economic decline that followed the North American stock market crashes in late 1929, the primary targets for blame were Toronto and Wall Street speculators, bankers, the big holding companies, and their allies and representatives. Leaders of the commercial utility industry met the crash first with denial, then with defensiveness, then with introspection. It was a hard fall from wealth and grace to bankruptcy and villainy. Despite all, the commercial industry remained firmly committed to private control of utilities and firmly opposed to the expansion of public power and stricter state regulation. Reforms had to be imposed against the will of industry leaders, and only after years of economic hardship and embarrassing revelations sufficiently damaged the commercial industry's public support and political standing. In the meantime, electric utility operating companies at the local level muddled through the early years of the Depression with almost no government assistance, fighting against predation by the free-falling holding companies, but buoyed by more or less reliable demand for power from customers who had become dependent on electricity and were willing to pay for it despite the hard financial times.

THE WALL STREET DEBACLE AND AFTERMATH

If ever there was a financial bubble that burst, it was Wall Street in 1929. Aggregate values on the New York Stock Exchange had nearly tripled from 1924 to 1929. Electrical industry stocks rose even faster. During just the eighteen months before the crash of October 1929, General Electric stock prices rose from $128 per share to $396 per share. Electric Bond and Share stock prices rose from $90 to $187 per share in that year and a half. An incredible amount of speculative wealth was created, and investors scrambled to capture a piece of the expanding pie. U.S. Federal Reserve chairman Alan Greenspan seventy years later referred to such a market condition as "irrational exuberance."[3] Contemporaneous critics called it a speculative orgy. One popular means for exploiting this bubble beyond one's immediate financial capacity was to borrow money at modest interest rates to buy stocks whose values were rising rapidly. This was called buying on margin, and it generally required only 10 percent down from a borrower who was assured of great profits as long as stock values continued to rise. Savvy investors knew the roller coaster would drop after reaching some as yet unknown pinnacle, but so much money could be made so quickly that no one wanted to get off the ride until stock values reached their peak—which they did in October 1929. As soon as the overblown market showed clear signs of deflating, everyone who had bought on margin and everyone playing the market for speculation tried to cash out. This accelerated the decline, which caused even more investors to pull out. A vicious cycle continued unabated for about three weeks, erasing more than a third of the market wealth created in the previous five years. Falling even harder than aggregate stock values, General Electric stock was down 42 percent by mid-November.[4]

Most everyone assumed (or hoped) this was just a temporary glitch. Stock prices stabilized and even climbed a little through the first few months of 1930, about the time President Hoover was predicting imminent and rapid recovery, but then began falling again and kept falling for two more years with only periodic brief rallies until the stock market bottomed out in July 1932. Ninety percent of the aggregate value of stocks on the New York Exchange evaporated between October 1929 and July 1932. Again, falling even harder than the rest, a share of G.E. stock was worth only $9 when the market hit bottom—just 2 percent of its pre-crash value. The stock value of the American Power and Light holding company, a subsidiary of the Electric Bond and Share empire that controlled Pacific Power & Light, fell from $175 a share before the crash to $3 a share in July 1932—an erasure of more than 98 percent of its stock value.[5]

Harvard economist John Kenneth Galbraith's best-selling 1954 classic, *The Great Crash, 1929*, identifies the pyramided utility holding companies as one of five main causes of the economic collapse. Decrying the human greed at the foundation of the crisis, Galbraith observes,

In November 1929, a few weeks after the crash, the Harvard Economic Society gave as a principal reason why a depression need not be feared its reasoned judgment that "business in most lines has been conducted with prudence and conservatism." The fact was that American enterprise in the twenties had opened its hospitable arms to an exceptional number of promoters, grafters, swindlers, impostors, and frauds. This, in the long history of such activities, was a kind of flood tide of corporate larceny.[6]

Of course this was not just a crisis in the stock market. The Wall Street crash exacerbated weaknesses and vulnerabilities in the North American economy, triggering the worst, most prolonged economic depression in American history, a financial and human tragedy that quickly spread worldwide. In the United States, industrial production fell by 25 percent the year following the crash. By 1932, unemployment had reached 25–30 percent, farm income had fallen by half, and workers' wages plummeted from $53 billion to $31.5 billion. In Oregon, 866 manufacturing facilities shut down in the years 1929 to 1932, while jobs in the timber industry declined by half. Thousands of banks, overly leveraged in the speculative bubble, ran out of cash and closed their doors, extinguishing the life savings of many citizens at a time in American history when there was no Social Security safety net. As Portland General Electric (PGE) historian Craig Wollner aptly puts it, "People lost their homes, their savings, their insurance, their optimism, and their faith."[7]

Canada, which had boomed at least as spectacularly as the United States in the 1920s, experienced a similar crisis, starting, however, with a crash in farm commodity prices (particularly wheat) in 1928. The prairie provinces were especially hard hit. The Wall Street debacle spilled over into Canadian markets, which degenerated into widespread national economic decline. Canadian gross national production fell precipitously from $6.1 billion in 1929 to $3.5 billion in 1933, unemployment reached 30 percent, and national income declined by half. Globally, the fall in Canada was second only to that of the United States in severity.[8]

Oddly, the trade journals and public relations materials from the electrical industry showed little sign of concern in the aftermath of the crash. No mention was made of the stock market crash in the November 1929 issue of *Electrical West*. Only one article in the December issue mentioned the crash, and did so only to deny its significance. In "Stock Market Will Not Affect Western Conditions," the journal's editorial staff called the historical event merely a "near panic" and claimed, "In so far as the electrical industry is concerned, business will be good in 1930. If the huge deflation of market values in securities has shaken the confidence of anyone, it is not apparent. Rather, a distinct feeling of optimism prevails."[9] This denial of the crisis continued well into 1931. Not until August of that year, nearly two years after the crash, did any articles in *Electrical West* acknowledge a depression. When the editors did acknowledge the crisis it was only to cheerlead industry leaders back to optimism: "Of all businesses, that of supplying electric light and power has been least hard hit by the depression. . . . The electric light and power companies, through their executives, can play a prominent role in leading business out of the doldrums." Only two more articles that year mentioned the Depression, both in the October issue and neither evincing any concern or sympathy. Instead, one complained about how government spending didn't seem to be affected by the Depression, while the other mentioned the Depression only in order to challenge its existence. Titled "Who Said Depression?" the latter article crowed about record sales of ranges and water heaters by Pacific Gas & Electric in California, claiming that "economists to the contrary, there is no buying slump in the West; at least when a wide-awake utility and aggressive dealers join forces to wreck bogeys." Indeed, though the holding companies and bankers were in crisis, the operating companies had real assets, stable sales, and honest revenues—all of which helped maintain hope in the industry despite the widespread hopelessness elsewhere.[10]

As noted by the optimistic articles in *Electrical West*, a healthy demand for electric power persisted despite the crisis in financial markets, rising unemployment, and falling wages. Unlike the lumber industry, for example, sales of electricity and electrical appliances remained relatively stable in the United States as a whole and in the Northwest during the first two years of the Depression. Those who had not lost their jobs apparently had not lost their interest in electrical living either. Among the eleven western states, as reported in the February year-in-review issue of *Electrical West*, the number of customers and kilowatt-hours sold increased slightly in 1930 over the levels reported in 1929. In contrast, at the national level, U.S. energy output in 1930 declined marginally.[11] Utility executives cancelled or postponed construction of many of the planned new hydroelectric and steam-driven power plants due to the slowdown in demand, but many new projects nonetheless moved forward, as executives did not expect the economic crisis to last.

Finally showing the effects of the Depression by 1931, energy generated by utilities in the eleven western states declined 3.8 percent from 1930 levels. Planned capital investments for 1932 were down 18.4 percent from actual expenditures in 1931. The *Electrical West* retrospective on 1932 was gloomier, as that year marked the depth of the Depression. Energy output fell about 13 percent from 1931 to 1932 in the eleven western states, 10 percent in the United States as a whole. Even with the persisting crisis, though, construction of new facilities did not halt. Eleven hydropower facilities were under construction in the West in 1932, seven of them in the Northwest states. While utilities launched fewer new power plants in these years, many of those under construction were very ambitious projects. In fact, the Stone & Webster Company and its Northwest subsidiary Puget Sound Power & Light in 1931 completed construction of the very first dam on the main stem Columbia River.[12] Considered too big a river to dam prior to this, Stone & Webster's Rock Island Dam opened the floodgates, so to speak, to Columbia River development in the coming decade. But the federal government subsequently led the charge (see chapter 9). In the Southwest, the Bureau of Reclamation started Hoover Dam on the Colorado River in 1931. The era of megadams had arrived, right in the middle of the Depression.[13]

In Canada, signs of weakness appeared in the economy prior to the stock market crash. While electric power output for the nation as a whole increased by 20 percent in both 1926 and 1927, it increased only 12 percent in 1928 and 10 percent in 1929. In the first year after the stock market crash, power generation was almost flat, going up less than 1 percent. The next year, 1931, power generation fell 8 percent across Canada. Showing a little more resilience, British Columbia's electrical output increased 7 percent from 1929 to 1930 and 2 percent the next year. A few hydropower projects under construction continued even as demand slipped, with 25,000 more horsepower added to British Columbia's generation capacity from 1930 to 1931. While British Columbia was especially hard hit by the Depression due to its heavy reliance on primary exports, particularly timber, the electrical industry in British Columbia fared better during the first years of the crisis than provinces back east. But, as in the rest of Canada, economic conditions turned distinctly sour in 1931 and thereafter, until World War II. Ridership on Vancouver's streetcars declined, so BC Electric lowered rates to try to entice customers back, with little success. The rail system operated at a loss by 1931, and twenty employees of the streetcar system lost their jobs. The hard times kept coming.[14]

Like President Hoover in the United States, British Columbia premier Simon Fraser Tolmie was a pro-business conservative who believed the proper response of government in hard economic times was retrenchment. As historian Robin Fisher reveals, Tolmie convened a group of prominent businessmen to advise him on fiscal policy during the Depression and, predictably, their advice was to cut government back drastically and slash social programs (health, education, and welfare). Tolmie followed their recommendations for the most part, which of course only made a bad situation worse for those suffering from the stagnating economy. It is perhaps no surprise that leaders of for-profit corporations would think it best to cut costs during times of declining revenue. That is usually a prudent strategy for organizations that must earn profits to survive. But it is a counterproductive strategy for governments during economic crises. When markets fail and capitalists retrench to preserve their accumulated wealth, governments are the only remaining institutions available to fill the breach. Unfortunately, Conservative Party leaders in the B.C. legislature held to the same principles as Premier Tolmie, so it accomplished very little to ameliorate the Depression. The Conservatives in British Columbia had swept overwhelmingly to power in the 1928 elections, winning three quarters of the legislative seats after more than twenty years of Liberal Party rule. But the Depression quickly reversed the fortunes of British Columbia's Conservative Party. The province's Liberals won the 1933 election by a margin similar to that of the Conservatives' victory in 1928.[15]

The low point in Canada's Depression came in 1933, the last year of more or less laissez-faire conservative rule and retrenchment policy. From 1929 to 1933, gross production as well as per capita income in the province declined by half. More than a quarter of B.C. workers were unemployed, leading to widespread social unrest.[16] According to BC Electric's chronicler Cecil Maiden, "private fortunes which had not already crashed were falling thick and fast." A growing percentage of the B.C. population was on welfare relief, exacerbated in the Pacific province by refugees from the prairie states (the counterpart to the U.S. Dust Bowl refugees flocking to California, Oregon, and Washington). So many people were out of work, according to Maiden, that BC Electric's "collection of accounts, in 1933, was something of an ordeal." As lumber mills and other large customers in British Columbia closed shop, BC Electric cancelled new capital expenditures and building programs, focusing just on maintaining essential services. The only exception was a two-year-old dominion government–sponsored project to build better-quality coke-manufacturing plants (the fuel, not the soft drink). With subsidies from the central government in Ottawa, this gas plant was the only BC Electric project under construction in 1933. On the east side of the Cascades, West Kootenay Power and Light continued construction projects after 1929, including the Corra Linn Dam on the West Arm of the Kootenay River, completed in 1932, but after that it postponed or cancelled other planned projects on the Kootenay and Pend d'Oreille rivers.[17]

Soon after the stock market crash, the financial institutions in the United

States set up to exploit the profits of utility operating companies began to unravel. Indebtedness undergirded much of the utility holding company structure. As revenues declined, debts could not be paid and foreclosures ensued. An illustrative example is the collapse of W. B. Foshay's modest utility empire. From 1924 to 1929, Foshay and a business partner launched two holding companies and acquired control over dozens of utilities in twelve states, two Canadian provinces, Nicaragua, and Honduras. In the Northwest, the operating companies he controlled served thirty communities in Oregon, nineteen in Washington, and fifteen in Idaho—mostly under the name of the West Coast Power Company. (He did not own any utilities in British Columbia.) Foshay built corporate offices in Portland and Seattle and an extravagant office tower in Minneapolis—something like the Washington Monument on steroids. During the giddy years of growth after 1924, securities for one of Foshay's holding companies, People's Light & Power, increased in value from $2 million to $30 million. But after the crash, his house of cards collapsed quickly. His second holding company, Public Utilities Consolidated, was in receivership already by November 1, 1929, sued by several corporations to whom it owed money. The Federal Trade Commission found examples of "shocking financial mismanagement" of his enterprises, and within a year Foshay was indicted on mail fraud. Foshay was eventually sentenced to fifteen years in prison, but President Franklin Roosevelt pardoned him in 1935.[18]

In this panicky environment, the financial markets became predatory. Shareholders sued directors, business partners sued each other, and creditors everywhere called in debts. Hemorrhaging holding companies sought transfusions from the utility operating companies to stanch their losses. This exacerbated the problem of excessive capital costs that the state utility commissions were investigating and increasingly denying. Operating companies, on the other hand, tried to sever their relationships with the holding companies like passengers fleeing a sinking ship. Some investors saw the crisis as an opportunity and helped push over the cliff competitors on the brink of disaster so their remaining assets could be acquired for a song. The J. P. Morgan syndicate did this to Samuel Insull's utility empire (more on this below).

Since all the major commercial electric utilities in the Northwest were subsidiaries of the holding companies, the financial crisis and predatory practices harmed Northwest power companies and their local investors, directors, and customers. For example, the Portland Electric Power Company (PEPCO), serving most of the Portland metro area, had been managed ably since 1913 by Franklin Griffith. By the 1920s, PEPCO and its partner corporations (Portland Railway, Light and Power and Portland General Electric) were part of a larger regional holding company named Pacific Northwest Public Service Corporation, which Samuel Insull's top-of-the-pyramid Central Public Service Corporation in Chicago swallowed up in 1929. Desperately in need of revenue after the crash, Central Public Service saddled PEPCO with unnecessary debts and questionable management fees, even though the utility was widely known to be well run and not in need of advice from Chicago. Griffith tried to work things out with the investors back east, but the situation deteriorated as the Depression deepened. For the first time in company history, power output, and therefore power revenues, declined in 1932 over the previous year's output. Suffering from reduced revenue and parasitized by Central Public Service, Griffith travelled to Chicago in 1932 to wrest control of PEPCO from the Insull conglomerate. If they did not cooperate and cut PEPCO loose, Griffith threatened, the directors of Central Public Service would likely be playing bridge with him in the state penitentiary in Salem, Oregon. They relented, but too late. PEPCO's parent company, Pacific Northwest Public Service, retained debt that it could not service and went into foreclosure in March 1934. Even after restructuring its debt, it still could not repay the boom-era loans as the Depression dragged on through the 1930s. Finally, in 1938 Pacific Northwest Public Service declared bankruptcy.[19]

In the meantime, the whole Insull empire collapsed. Considered one of the architects of the utility holding company, with an enviable reputation as one of the ablest financial managers in the business, Insull nevertheless met his own Waterloo in these years. At the moment of the crash, Insull controlled the third largest utility conglomerate in the United States. He served as president of eleven utilities, chairman of the board of sixty-five utilities, and board member of eighty-five utilities. He created Central Public Service to consolidate more utility holdings in a bid to fend off predatory intentions from the J. P. Morgan syndicate, which had formed a super–holding company, the aptly named United Corporation, aimed at consolidating the entire industry. Morgan's United Corporation formed in 1927 with a capitalization twenty times that of the gigantic U.S. Steel Corporation, which Morgan also controlled. Fairly quickly, Morgan was able to absorb under United's wing most of the significant U.S. holding company systems, except Central Public Service. Insull had decided to fight back. But he had not counted on the stock market crash. Overly leveraged from his competition with Morgan, Insull barely hung on through 1931. Worth $3 billion before the crash, Insull's empire was worth only a quarter of that by the end of 1931. In June 1932 he reluctantly resigned his directorship in sixty corporations. Six months later, Central Public Service Corporation went into receivership and was bankrupt by March 1933.[20]

Like W. B. Foshay before him, Insull was charged with mail fraud: using the U.S. mail to promote securities with overinflated values. Since there were no laws at the time against the kind of speculation that led to the collapse, such charges were about the best the government could do to bring responsible parties to justice. Unlike Foshay, however, Insull was acquitted of the charges in 1935. Nevertheless, his reputation was ruined, and whether deservedly or not, he became a poster child for what went wrong with American capitalism in the 1920s. As soon as Insull's Central Public Service Corporation fell, J. P. Morgan and partners swooped in to collect the booty for pennies on the dollar. As with the briefer economic depression of 1893–1896, financiers with deep pockets could pick up bargains and further consolidate the industry, which is what Morgan and his associates did. By 1932 just sixteen holding companies controlled 75 percent of the U.S. power supply. Crisis for some was indeed opportunity for others.[21]

THE TIDE TURNS

The year 1932 marked a turning point in the United States. The opposition party and a new president won control of the legislative and executive branches in the November elections, resulting in a spate of new policies designed to reform systemic problems and abuses. Even the industry itself called for reform by 1932. Notably, Congress passed the banking reform Glass-Steagall Act in 1933, designed to limit speculation and set up a depositor's insurance program: the FDIC. Two years later, Congress approved the Public Utility Holding Company Act to restrain abuses in the utility-financing sector. But reforms also emerged at the state and local levels even earlier, starting with the elections of 1930. In Oregon, the public furor over the economic crisis revitalized the progressive wing of the Republican Party. A dark horse candidate not favored by the Republican establishment, Julius Meier, won the governor's office in a bitterly contested election. He was a strong public-power proponent, and he appointed a crusading consumer advocate as the Oregon Public Service commissioner responsible for regulating utilities. In the same election, Oregon voters finally approved a ballot initiative that permitted the formation of "people's utility districts," or PUDs, for developing nonprofit power systems at the county level. The enabling law, however, required two votes to establish a public-power system: one to authorize it and another subsequent vote to approve the bonds to fund it. That gave the commercial power industry two chances to defeat every PUD initiative, and they were quite successful in doing so in subsequent years. In Washington, voters also approved a public utility district initiative after many previous failures at the ballot and in the legislature. This more liberal law required only one vote to authorize and fund a PUD. Consequently, public-power advocates in Washington successfully set up many more PUDs than advocates in Oregon during the next decade.[22]

The commercial power industry spent a lot of money trying to defeat these voter initiatives and editorialized regularly against them in regional newspapers and the pages of *Electrical West*. One editorial in *Electrical West* warned readers about "Another Crop of Anti-Utility Legislation." Dismissing the initiatives and their backers, the author charged that the measures were "backed by extremists or misguided political groups and fomented by misunderstanding or imagined evils in the economic system." The president of Washington Water Power, Frank Post, warned that the Washington initiative was "filled with dynamite." U.S. senator Clarence C. Dill, a proponent of public power in Washington, countered "If the Power Trust beats this bill it will have the country people by the throat."[23] After the Washington public-power bill prevailed, however, an editorial in *Electrical West* denied any significance to the election and the public opinion behind it, chalking up the results to "utility baiting . . . prohibition, hard-times, [and] socialistic schemes" that swept "many a political opportunist into office."[24] But the tide was turning, and utility executives sensed it.

In the mind of the editorial writer above, one of those so-called opportunists was surely the new Oregon Public Service sommissioner, Charles M. Thomas. After the 1930 election, the new Oregon legislature in 1931 abolished the Public Service Commission, composed of three commissioners, and reorganized it as a public utilities commission, with a single commissioner with broader powers. After his appointment by Governor Julius Meier as the sole commissioner in 1931, Mr. Thomas went after the commercial utilities with missionary zeal, acting on many previously filed complaints and initiating many other investigations and actions on his own. The 1931 and 1932 annual reports of the commissioner reveal an aggressive effort to trim bloated capital and valuation claims and reduce rates whenever possible within the confines of the law that still guaranteed reasonable returns on investment. In his first year, he reduced residential and commercial electric rates charged by Eastern Oregon Light & Power Co. and Pacific Power & Light; he reduced farm service rates charged by the Yamhill Electric Co.; and he reduced street railway fares charged by PEPCO. The next year, he continued his crusade, reducing electric rates charged by Northwestern Electric and disallowing many "service charges" paid by Northwestern to "affiliated companies under common corporate control." This action was designed to exclude from operating expenses parasitic charges by holding companies when those fees were excessive or not based on actual services performed. The order by Commissioner Thomas in this case, number U-F-526, was directly aimed at charges by Electric Bond and Share of about $55,000 per year for "general services." Commissioner Thomas found them unjustified and excluded them from the company's acceptable operating expenses that counted toward determining rates.[25]

Reflecting on his mission and philosophy, Thomas gave a lengthy speech at the Portland Civic Auditorium in June 1933, which has been quoted liberally by several historians due to its bluntness and passion. He called holding company financiers "pillagers on the high seas of frenzied finance," and detailed many of the abuses committed by them and their minions in the Northwest, including some of Portland's most prominent citizens. He also explained in embarrassing detail how Oregon ratepayers subsidized excessive pay and benefits for executives and directors. All the executives' memberships in high-society clubs and commercial associations like the Portland Chamber of Commerce were paid by the utilities. One executive had nine different club memberships paid for, including three country club memberships. Such revelations during the depth of the Depression surely did not sit well with the audience.[26]

In contrast to the persistent optimism expressed in the pages of *Electrical West* through 1931, the January issue of 1932 finally exhibited some soul-searching and humility. An editorial that month titled "A New Ticket for 1932" uncharacteristically acknowledged that "we have been shaken up, shaken down, a lot of conceit has been knocked out of us, most of our pet theories proved to be nothing more than theories, our ego has been deflated and even our pride has been humbled. All of which means that our feet are back on the ground, our coats are off, and we are ready to write a new ticket for 1932."[27] Following this preamble, the editorial proposed five New Year's resolutions for the industry: (1) balance individualism with an ethic of cooperation, (2) help solve the unemployment problem, (3) reform utility financing, (4) break up the concentration of ownership and drive out corruption, and (5) improve production efficiency and find alternative jobs for workers displaced by mechanization.[28] This was a remarkable transformation for the editorial pages of *Electrical West* and a decidedly reformist agenda being embraced by core partisans of for-profit power. And it came at an opportune time. As mentioned, 1932 would be a critical turning point in the United States.

On the campaign trail in September 1932, presidential candidate Franklin Roosevelt, then governor of New York, took a swing through the Pacific

Northwest. Well informed on the subject of public utilities and sympathetic to Cook's and Pinchot's Giant Power concept, Roosevelt delivered a major campaign speech in Portland on the subject of public utilities regulation and hydroelectric power. It foreshadowed most of the significant developments of the next decade in the U.S. Northwest's hydropower system, so it is worth reviewing and contextualizing the key elements of the speech.[29]

First, the president-to-be asserted his philosophy of democratic governance that would define the coming New Deal. "The object of Government," he said, "is the welfare of the people. The liberty of the people to carry on their business should not be abridged unless the larger interests of the many are concerned. When the interests of the many are concerned, the interests of the few must yield." Roosevelt then asserted that electrical development had become a "national" question by the 1920s. Many electric power operating companies had become interstate enterprises, buying and selling power across state lines. Washington Water Power, Idaho Power, Utah Power & Light, and Pacific Power & Light all exemplified this. More to the point, financial control of operating companies was extensively vested in interstate, national, and international corporations like Electric Bond and Share. Because of that, Roosevelt said, it was time for the federal government to involve itself in promoting, shaping, and regulating this critical technological system at the heart of manufacturing, commerce, and daily life.

Roosevelt articulated another principle regarding public ownership of rivers and their water-power potential. Just as navigable rivers were used but not owned by private interests, water-power sites on America's rivers, Roosevelt argued, should be leased but not owned by private interests. Because the great majority of excellent water-power sites in the West resided on publicly owned federal lands, Roosevelt believed that hydropower facilities built on those public lands must serve the public interest. In classic candidate rhetoric, Roosevelt announced: "I promise you this: Never shall the Federal Government part with its sovereignty or with its control over its power resources, while I am president of the United States."[30]

To underscore his desire for the federal government to take a more active role, Roosevelt argued that industry financiers had escaped the control of the state utility commissions. He cited the Federal Trade Commission reports on watered stock, overcapitalization, and other holding company abuses that took money out of the pockets of consumers to enrich Wall Street. "It is an undoubted and undeniable fact that in our modern American practice the public service commissions of many States have often failed to live up to the very high purpose for which they were created." Roosevelt criticized President Hoover for opposing federal regulation of electric utilities and turning a blind eye to the "non-enforcement" of much of the Federal Water Power Act of 1920.

To remedy the problems, Roosevelt called for a new regulatory philosophy on the part of the states, which he promoted in New York while governor. The state PUCs, he said, should not merely be an arbiter between consumers and utilities, but should serve as "agents of the public" in guaranteeing adequate service and reasonable rates. "This means, when that duty is properly exercised, positive and active protection of the people against private greed!" He also called for expanded regulatory authority to be vested in the Federal Power Commission to deal with interstate power transactions. Roosevelt then laid out an eight-part proposal for specific regulatory reforms at both the state and federal levels, three of which involved greater transparency in public utilities financing (of this, he said, "Let us turn on the light!"). The others included federal regulation of holding companies, which were currently out of the reach of the state commissions; greater control over the issuance of stocks, bonds, and other utility securities; greater cooperation between state PUCs and the Federal Power Commission; and a new law "making it a crime to publish or circulate false or deceptive matter relating to public utilities."[31]

The last point was aimed at the five-year-old brouhaha regarding anti-public-power propaganda generated by the commercial utility sector. Not mincing words, candidate Roosevelt noted that the Federal Trade Commission had revealed "a systematic, subtle, deliberate and unprincipled campaign of misinformation . . . if I may use the words, of lies and falsehoods. . . . It has permeated the schools, the editorial columns of newspapers, the activities of political parties, the universities and the printed literature of our bookstores." Clearly, he sympathized with the public-power advocates, although his position was nevertheless quite moderate: "I do not hold with those who advocate Government ownership or Government operation of all utilities. I state to you categorically that as a broad general rule the development of utilities should remain, with certain exceptions, a function for private initiative and capital." Then he laid out the exception to the rule, which has come to be known as the "yardstick" principle, or as he called it more colorfully in his speech, "a 'birch rod' in the cupboard to be taken out and used only when the 'child' gets beyond the point where a mere scolding does no good." Roosevelt asserted that citizens have "an undeniable basic right" to democratically establish a government-owned and -operated electric service if they are not satisfied with the service or rates provided by their privately owned utility. These public-power systems would serve as a "yardstick," he concluded, "to prevent extortion against the public and to encourage the wider use of that servant of the people—electric power." Promising federal investment in an electric power yardstick for the Northwest, Roosevelt announced that "the next great hydroelectric development to be undertaken by the Federal Government must be that on the Columbia River."[32] And so it was.

Interestingly, as indicated in the quote above, Roosevelt's final goal behind both the regulation of commercial utilities and development of public-power systems was to keep rates as low as possible to encourage greater electrical consumption. Studies repeatedly showed that lower rates led to greater consumption. Many utilities realized this and lowered their rates intentionally to stimulate consumption with the expectation that overall profits would increase as per capita consumption increased. For example, Puget Sound Power & Light's *Electric Journal* in 1931 had an editorial titled "Our Job" that explained how recent rate reductions were "earned through the increased use of electricity." "Our job," the editorial stated, "is to acquaint our customers with our low rates . . . so they will use the maximum amount of electricity and therefore get the maximum return for their electric dollar, and our earnings will continue to increase."[33] Likewise, Oregon Public Utilities commissioner Thomas specifically stated in nearly every case when he lowered a utility's electric rates that "the reduction will bring additional patronage and increase the volume of business to the utility."[34] It usually worked. But, as Roosevelt noted, "many selfish interests in control of light and power industries have not been sufficiently far-sighted to establish rates low enough to encourage widespread public use." Roosevelt and many policy makers of his day believed that widespread use of electricity was central to prosperity, that access to affordable, reliable electric power was even a basic human right. If the object of government is the welfare of the people, as Roosevelt argued at the start of his speech, then ensuring widespread use of electricity should be an important function of governing institutions.

HEADING TOWARD REFORM

The financial crisis following the stock market crash pressed the reset button on the whole global economy as well as the international electric power industry. The North American electric power system entered a transitional era of greater openness to change in the 1930s. The shock of the Depression altered established presumptions and patterns of development. Most citizens came to understand that unchecked greed was destructive and unregulated capitalism untrustworthy; they saw the need for institutional reform, yet deep political divisions persisted over how best to reform the crippled economy. Conservatives wanted government retrenchment and economic incentives to big business. Liberals and progressives wanted government to provide at least temporary jobs and economic stimulus programs, stricter regulation of business, and a social safety net for the vulnerable while preserving a modified, responsible form of capitalist enterprise. Socialists wanted to abolish capitalism entirely as corrupt, unjust, and oppressive. As usual, the U.S. and Canadian ships of state charted a zigzagging middle course between these contending positions. Despite the disagreements and the unprecedented nature of the economic challenges, a broad commitment to reform led to many new and more rigorous regulatory policies in the Northwest as well as a minor boom in the development of public-power systems, especially in Washington state.

This trend of the 1930s was more a shift in trajectory than a fundamental restructuring of the economic system. It cannot be accurately called a revolution. A mix of public ownership and private ownership continued, but with more growth in the former and less growth in the latter for the next several decades. The financial establishment's extraordinary profit-taking from the public utilities during the 1920s was significantly restrained during the 1930s, but not ended. Holding companies were more carefully regulated but not entirely abolished. The U.S. government accelerated its involvement in power systems during the Depression, as a generator and distributor of hydropower from old and new federal dams, as a regulator of both private and public hydropower facilities built on the nation's rivers, and as a promoter of public power. But none of those roles were novel. The U.S. government had been building hydropower dams and distributing their power to local customers with a marketing preference for public-power systems since the first decade of the century. It created a federal counterpart to the state utility commissions in 1920: the Federal Power Commission (FPC). It promoted comprehensive regional planning and coordinated river basin development throughout the 1920s. New Deal initiatives of the 1930s were simply an extension and acceleration of these trends. As Jay Brigham argues compellingly in his book *Empowering the West*, "the election of 1932 did not represent a complete political break with the past. Instead, a strong political sentiment for government action to prevent economic abuses and modernize the country had existed several decades before the New Deal."[35] The same can be said of the election of 1933 in British Columbia, as the reforms attempted there after 1933 followed many precedents from decades past, as the next chapter reveals.

CHAPTER 9

A New Deal for the Northwest, 1933–1938

To whatever strange ports we are wafted on the warm breeze of the New Deal, we shall not see the lotus land of laissez-faire again. That is probably the major historical fact of our generation.

—William Orton, *Atlantic Monthly*, 1935[1]

The Great Depression focused public attention not just on problems but also on their solutions. By the twentieth century the majority of North Americans no longer supported laissez-faire economic policy during an economic crisis, nor by 1932 did they wholly trust business leaders to lead an economic recovery on their own. Most Americans and Canadians in the 1930s wanted their democratically elected governing bodies to engage in problem-solving. Progressives at the turn of the century had pioneered a new kind of activist government that sought goals broader than mere economic boosterism. Increasingly, North Americans expected elected officials and governing institutions to actively pursue social well-being along with economic growth and to invest public money in that duty when necessary. British economist John Maynard Keynes advocated this philosophy, arguing that governments had a responsibility to use fiscal policy to ameliorate the highs and lows of economic cycles. Though experimental and by no means the predominant economic theory at that time, Keynesian philosophy advanced in both Canada and the United States from 1933 onward, though less so in Canada and more so in the United States during the 1930s.

In the 1930s, Canadian politicians at both the federal and many provincial levels followed a more cautious course regarding government intervention. The dominion government in particular did not embrace public works programs and deficit spending on the same scale as did the United States government, though many provincial governments strongly urged it to do so. The Conservative Party in Canada won a majority in the national elections of 1930, defeating the Liberal Party, which had held power since 1896. But the new

Conservative prime minister Richard Bennett, who served from 1930 to 1935, held to traditional laissez-faire fiscal policies similar to those of U.S. president Herbert Hoover. Bennett hoped to pull Canada out of the Depression mainly by promoting exports and taxing imports, with little effect. His government sponsored relatively few programs to help the jobless, homeless, and hungry. After five years the electorate booted his party out and returned the Liberals to power, with MacKenzie King back as prime minister. King is considered one of the architects of the Canadian welfare state, although his embrace of the welfare state started too cautiously to be of much help in ending the Great Depression. He served as prime minister from 1921 to 1930 and again from 1935 to 1948, but despite his social welfare commitments, he kept a tight rein on federal and provincial spending during the Depression.

In British Columbia, the Conservative Party came to power in the election of 1928 but promptly confronted the stock market crash and fell from favor after providing no effective remedies for the crisis. Although the Liberal Party swept the B.C. elections of 1933, its new premier, Thomas Dufferin "Duff" Pattullo, had difficulty securing the funds needed to put people back to work and jump-start economic development. His platform was remarkably similar to the New Deal platform in the United States. Pattullo advocated a modified capitalism with the state playing a vigorous role to restrain abuses by the nation's privileged class and provide a social safety net. Unlike President Roosevelt, however, Premier Pattullo had few financial resources to implement his vision. He lobbied the Canadian government vigorously for public works and welfare funds of the type western states in the United States were receiving from the U.S. government's New Deal programs, but the Canadian government provided comparatively little. Mostly British Columbians had to wait out the Depression, while watching with a mix of consternation and desire the progress of U.S. investments south of the border. With the provincial government in debt, the dominion government tightening its belt, capitalists retrenched, and risk avoidance endemic in the private sector, British Columbia's economy remained adrift in a Sargasso Sea of stasis through most of the 1930s. As in the United States, World War II kicked Canada's economy into gear, including British Columbia's.[2]

In the United States, newly elected president Franklin Roosevelt was determined to wield the federal government as a tool for social relief, economic recovery, and institutional reform—the "three Rs" of the New Deal as it is referred to in most history textbooks. Because this kind of philosophy had been applied only cautiously and piecemeal in the past and because the current economic depression exceeded in scope and persistence every other economic crisis in national historical memory, Roosevelt and his advisors were emboldened yet at the same time unsure how to proceed. They knew only that inaction was not an option. In May 1932, just six months before his election at a commencement speech at Oglethorpe University, candidate Roosevelt made clear his intention to do something about the persistent economic crisis if elected:

> The country needs and, unless I mistake its temper, the country demands bold, persistent experimentation. It is common sense to take a method and try it: If it fails, admit it frankly and try another. But above all, try something. The millions who are in want will not stand by silently forever while the things to satisfy their needs are within easy reach. We need enthusiasm, imagination and the ability to face facts, even unpleasant ones, bravely. We need to correct, by drastic means if necessary, the faults in our economic system from which we now suffer.[3]

This commitment to action by the Roosevelt administration and a willing U.S. Congress wrought profound changes throughout the nation, including far-reaching transformations in the Pacific Northwest and its electric power regime. During the first five years of Roosevelt's tenure (1933–1937), the federal government started construction on two colossal dams on the Columbia River (finishing one in 1937 and the other in 1942), proposed a regional Columbia Valley Authority modeled after the TVA but after vigorous opposition settled for a more limited Bonneville Power Authority, enacted the Public Utility Holding Company Act, created the Rural Electrification Administration, expanded the regulatory authority of the Federal Power Commission, and stimulated the creation of dozens of rural electric co-ops and other public power projects in the Northwest. And this list includes only the highlights. It was a remarkable period of development, innovation, and change led by federal investment and leadership. The U.S. economy incrementally recovered as a result but did not really boom again until World War II, when war production, massive deficit spending, and trade with allies in Europe finally catapulted the U.S. economy out of its decade-old slump.

The architects of all of this were driven by a broad social vision of coordinated regional development, with widely dispersed and more equitable prosperity, yielding self-sustaining productive and democratic communities living peacefully on the bounty of local natural resources that they would use and conserve wisely for the long term. Easier said than done, of course. In the face of chronic contention over New Deal programs and how to allocate their costs and benefits, the inevitable political compromises resulted in much more limited and pragmatic reforms. While some New Dealers dreamed of social transformation, they eventually settled for incremental but notable improvements in wages, profits, public services, and public amenities while stimulating general economic growth.

BOLD GESTURES AND GRAND VISIONS

Uncle Sam took up the challenge in the year of 'thirty-three,
For the farmer and the factory and all of you and me,
He said, "Roll along, Columbia, you can ramble to the sea,
But river, while you're rambling, you can do some work for me."

—"Ballad of the Great Grand Coulee" by Woody Guthrie, 1941

At his pre-election speech in Portland, Oregon, Franklin Roosevelt had promised that if he was elected, "the next great hydro-electric development to be undertaken by the Federal Government must be that on the Columbia River."[4] He said "next" great development because the federal government was already building the world's largest dam on the Colorado River: Boulder Dam (later renamed Hoover Dam). Authorized in 1928 and finished in 1935, the eyes of the nation, indeed the world, were focused on Hoover Dam. Not only was it the largest concrete structure on the planet at the time it was built, it was also the largest producer of hydropower in the world. But not for long. Within months after being sworn into office, Roosevelt fulfilled his campaign promise by launching not one but two precedent-setting Columbia River projects: Bonneville Dam and Grand Coulee Dam.

Because the Columbia River is the second largest river in the United States in terms of volume—200 million acre-feet flow through its channel per year, more than fifteen times the flow of the Colorado River—just the thought of damming it boggled the minds of engineers. But Boulder/Hoover Dam emboldened the government's dam-building ambitions. Though FDR and his advisors initially planned in 1933 to build a low dam at the Grand Coulee site, by 1935 the government committed to a high dam that when finished literally dwarfed Boulder Dam. It was 550 feet tall (equivalent to a fifty-five-story building), and stretched a mile in length across the Columbia floodplain, almost 4.5 times the length of Hoover Dam. Grand Coulee required a mind-boggling 12 million cubic yards of concrete. The Bureau of Reclamation estimated that if that amount of concrete was laid out as a sixteen-foot-wide road, it would stretch from New York to Seattle, then south to Los Angeles, and back to New York.[5]

The extraordinary size of the dam and the extraordinary volume of water in the Columbia River meant unprecedented potential for power generation—a kilowatt capacity from one source almost beyond comprehension at the time. Understanding what this meant in terms of the Northwest power supply is crucial to understanding some of the objections raised against the dam in the 1930s. The dean of the Engineering School at the State College of Washington, H. V. Carpenter, put Grand Coulee power in perspective for a meeting of scientists in 1935 affiliated with the National Resources Committee. The big commercial electric utilities in Washington, Idaho, and northern Oregon, he said, had an installed capacity of 1,112,000 kilowatts in 1935. In contrast, the high dam at Grand Coulee was projected to generate 1,890,000 kilowatts. With an additional 432,000 kilowatts planned for Bonneville Dam, these two federal projects alone would triple the power supply in the region at a time when consumption was flat or declining due to the Depression.[6] No wonder Washington Water Power adamantly opposed building Grand Coulee Dam, fearing it would depress electric prices the way overproduction of wheat depressed crop prices in the Great Plains. But the optimistic Carpenter assured his audience, "The possibility of absorbing this rather startling amount of power is not as hopeless as it may appear." He then argued that consumption before the Depression had climbed annually at an impressive rate and would do so again once the economy improved. Moreover, power markets in the region were far from saturated, he said, and the prospect of reduced power rates would very likely stimulate increased consumption. The low power rates would also likely attract energy-intensive industries to the region. All these factors, Carpenter believed, would drive demand up to the level of supply in a reasonable amount of time. He was presciently correct on all counts.[7]

Getting the two dams authorized was not a cakewalk. With Congress overwhelmed with legislation during the famous "First 100 Days" of the new administration, Roosevelt had to launch these projects on his own authority. Moreover, despite the phenomenal electoral sweep of the Democrats in 1932, a great deal of contention remained over federal hydropower development. As a consequence, neither Bonneville nor Grand Coulee secured congressional approval in 1933. So Roosevelt used existing funds and authorities available in the Public Works Administration, the Army Corps of Engineers budget, and the Bureau of Reclamation budget to get the two dams started. Two years later, Congress finally authorized both projects in the Rivers and Harbors Act of 1935.[8]

Neither of these projects emerged out of thin air in the summer of 1933, however. Portland-area boosters had been advocating for many years for a dam on the lower Columbia River, to improve navigation through what was known as the Cascade Rapids. They also hoped the hydropower from such a dam would increase the power supply and reduce electric rates in Portland. Similarly, boosters in central Washington's rural Columbia Basin had been advocating for more dams to provide irrigation water to the arid lands there with the additional hope that the hydropower from such dams would lower the cost of irrigation and bring down power rates for the agricultural communities on the east side of the Cascades. These dreamers were assisted by two acts of Congress in 1925 and 1927 that directed the Army Corps of Engineers (CoE) to study and report on comprehensive multipurpose development options for the nation's major river systems. The 1927 River and Harbor Act specifically directed the CoE to study the Columbia River. After more than three years of preparation, in 1931, the CoE's Major John Butler completed a report, nearly 2,000 pages in length, recommending eight dams on the Columbia and its major tributaries, to control the river from the Canada border to Pacific tidewater. Grand Coulee and Bonneville were among the eight. (The CoE soon added two more dams to the plan, bringing the total to ten.) Hoover was still president at the time, though, and he opposed additional federal dam development during the Depression. He felt it was important to balance the budget during the economic crisis, and he supported the private power industry, which did not want competition from federal dams selling power at low rates. Hoover also worried about the current glut of crops and low farm commodity prices, so he was particularly opposed to launching a Columbia Basin reclamation project supported by Grand Coulee Dam. Consequently, his administration sat on Major Butler's report for about a year before releasing it to Congress. It was not printed for public distribution until 1933, during a much more favorable political climate. Nevertheless, its contents leaked out after 1931, and supporters of the two big dams made good use of Butler's recommendations. Butler had become convinced that hydropower was the primary benefit to be derived from Columbia River development and that electricity revenues from the dams would easily repay the costs of construction while subsidizing navigation improvements and irrigation.[9]

These two dams formed the cornerstone of an eventual Federal Columbia River Power System, transforming Northwest livelihoods and landscapes in general and the regional electric power system in particular. But in 1933 this future was far from clear or certain. River and hydropower development during the 1930s was contested and compromised, a result of what was politically feasible to accomplish at each moment in time. In this time of continued economic crisis and political transition, the Columbia River became the symbol of regional opportunity, and electric power the expected engine of regional recovery. As the government embarked on a challenge to build the "World's

Greatest Engineering Wonder," Grand Coulee Dam raised ambition and hope to a new level at a time when neither found much fertile ground in the developed world.[10]

Bonneville and Grand Coulee dams embodied a variety of hopes to a diverse cross section of society. At the most immediate and visceral level, these projects offered the promise of employment at a time when one in four able and willing workers could not find jobs. During Roosevelt's inaugural address in March 1933, he said, "Our greatest primary task is to put people to work."[11] Unwilling to wait for the slow gears of Congress to act on these two projects, FDR decided for the sake of immediate employment to get them under way and let Congress catch up later. Applicants flooded the employment offices, trying to get jobs at the two dam sites. At its peak the Grand Coulee project alone employed 7,000 workers.[12] In addition to the jobless poor, many of the middle class and wealthy in the Northwest saw hope in the two projects. Workers' wages would be spent on food, clothing, rent, and entertainment, stimulating commercial activity for small businesses and farms. The purchase of construction materials for the dams would stimulate larger businesses and manufacturing companies in the region. Beyond the immediate economic boost from construction activities, however, was a general hope for future development and prosperity from the hydropower, irrigation, flood control, and navigation improvements facilitated by the two dams—a hope undergirded by a prevalent technological optimism and the realization that northwesterners were building "The Biggest Thing That Man Has Ever Done," in Woody Guthrie's iconic words.

New Deal dreamers sought more than mere economic recovery, though. A significant number of intellectuals, politicians, writers, artists, and opinion leaders reacted against the grinding poverty and economic dysfunction of the Depression years by analyzing the flaws in the economic order and imagining a better future. They proposed many alternative forms of social organization, and Columbia River hydropower often played a crucial role in the material foundation of those hopeful visions in the Northwest. A number of professional associations in North America in the 1930s promoted a quasi-utopianism in which cities and industry would be reorganized into regional configurations unified by climate, resources, and culture. The new regimes would harmonize city and countryside; revitalize rural life; conserve natural resources; provide cleaner, safer living and working environments; and enable more decentralized democratic control of business, government, and daily life.[13] The Regional Planning Association of America and the Technical Alliance were two main articulators and disseminators of these ideas. Leading intellectual lights such as Lewis Mumford, Stuart Chase, and Benton MacKaye belonged to both organizations. In his book *Technics and Civilization*, written in 1930–1931, Mumford places special emphasis on the important role of electricity in supporting this social transformation. His colleague Stuart Chase best summarized the view of electricity's predicted beneficial effect on society in his often-cited 1933 *Fortune* magazine article, "A Vision in Kilowatts":

> Electricity can give us universally high standards of living, new and amusing kinds of jobs, leisure, freedom and an end to drudgery, congestion, noise, smoke, and filth. . . . It will shift population, change the map . . . profoundly affect both the numbers and the skills of workers, revise—upward—the quantity and variety of commodities, break down the division between country man and city man . . . it promises a world replete with more freedom and happiness than mankind has ever known.[14]

The Tennessee Valley Authority (TVA) was the U.S. government's first attempt to put this regionalist vision into practice. David Lilienthal served on the board of the TVA in its first year of operation. He had previously served on the Public Service Commission of Wisconsin, regulating public utilities. Lilienthal enthusiastically embraced Mumford's and Chase's vision of regional social reformation founded on cheap, abundant public power. In dedicating the TVA in 1933, Lilienthal expressed the following hopeful vision, plagiarizing in part Chase's "Vision in Kilowatts":

> This valley will be the first to enjoy the full fruits of this new age, the Age of Electricity. Those who have its blessings in abundance will come into a new kind of civilization. New standards of living, new and interesting kinds of jobs, totally new industrial processes, and end to drudgery, congestion, waste. . . . For in this valley in another decade, electricity will hardly be recorded in cost, so cheaply can your communities then supply it.[15]

Geographer Patricia Kellner finds this "rhetoric of the electrical sublime" fairly common in the Pacific Northwest from the 1920s through the 1940s.[16] Likewise, Northwest historian Wesley Arden Dick documents the prevalence and significance of this regional culture of electrical optimism in his 1973 dissertation, "Visions of Abundance." In an article published in 1989,"When Dams Weren't Damned: The Public Power Crusade and Visions of the Good Life in the Pacific Northwest in the 1930s," Dick quotes Democratic congressman Knute Hill of Washington praising President Roosevelt's plans for Columbia River development in 1933:

He has a dream contemplating the development of the stupendous waterpower resources of our country for the benefit of all its citizens. We are living in the electrical age and its blessings should be distributed equitably among all the people, as they are common inheritors of this great resource provided us by a bountiful nature.

As this quote shows, big dams during the Depression offered not just economic but also psychological hope. When Roosevelt's decision to build Bonneville and Grand Coulee dams hit the newspapers in the late summer of 1933, it generated a wave of emotion (not all of it positive) throughout the region. People gathered on street corners in Spokane to discuss the news, businessmen in Portland marched through downtown, politicians who had advocated for the dams praised Roosevelt's vision and heralded the coming of a new, prosperous era for the Northwest. Washington's senator Clarence Dill effused, "With cheap electricity on the farms, in the homes, in the stores, in the factories, and on the highways and the railroads, we shall build the greatest electrical empire here in the Northwest the world has ever known." When Roosevelt came to the Northwest in the summer of 1934 to celebrate the start of construction on the two megadams, Senator Dill introduced him to a crowd of 20,000 cheering people at Grand Coulee, saying Roosevelt was ushering in "a new America in which all will have a more abundant life." The *Seattle Star* editorialized that Roosevelt's speeches on the significance of Bonneville and Grand Coulee embodied "The Great American Dream—a dream of a happier, cleaner, more fruitful country."[17]

Not everyone cheered on the projects or shared the vision of economic and social rejuvenation through public development of hydropower. Most of the private utility executives were skeptical or downright hostile to federal power development. However, if they thought they would be able to purchase large blocks of cheap power, they sometimes welcomed the call for federal development, as in the case of the Portland General Electric Company and Bonneville Dam. But federal policy had a "preference" clause in which public power systems enjoyed priority access to electricity from federal dams, so many commercial utility executives were not confident they would get reliable access to federal power. More troubling to them was the fear that the hydropower from Bonneville and Grand Coulee would flood the market and depress electric rates so much that the commercial utilities would struggle to earn profits. Consequently, many of the big privately owned utilities, such as WWP in Spokane, expended extraordinary energy criticizing the projects and trying to get them cancelled. Many business-class allies of the utilities in the banking sector, in chambers of commerce, and in the media joined the fray, challenging the need for and benefits of federal Columbia River development.

A 1935 article in *Electrical West*, "WANTED: Buyers for One Billion Kw/hr," provides a good example of the private industry's skepticism and worry about Bonneville and Grand Coulee dams. The author starts provocatively: "To find a market for 190,000 kw. of new generating capacity that is being dumped by the government in the next few years on top of a present surplus of 440,000 kw. in the three major markets affected is one of the major economic problems immediately facing the people of the Pacific Northwest." After eight pages of analysis, he concludes that the Northwest market "will be insufficient to absorb more than a fraction of the power made available by the new projects under construction."[18]

Another example of the skeptic's view of Grand Coulee was a July 1935 article by Robert Ormond Case in the popular national magazine *Saturday Evening Post*, "The Eighth World Wonder." While admitting in deft and evocative prose that the proposed dam was indeed a marvel of engineering, exceeding the monumentalism of Cheops's Great Pyramid, Case nevertheless suggests that it was an expensive boondoggle to produce power and irrigated farms where neither was needed. He predicted that ultimately the dam, a "memorial to the vision—or egotism, or folly" of national leaders, would end up a useless monument like Cheops's pyramid.[19] Journalist Jim Marshall calls Bonneville a "Dam of Doubt" in a June 1937 article in *Collier's* magazine, criticizing among other things that the dam threatened the multimillion-dollar annual salmon fishery of the Columbia River system. Other critics argued that the costs were too high, that opening new farms in the Columbia Basin would exacerbate the glut of crops and the problem of low prices, that the flood of cheap power into the electricity market would harm the profitability and investments of the commercial utilities, and that publicly funded, government-owned hydropower systems were un-American. As Bonneville Power Administration (BPA) historian Gus Norwood notes, there was plenty of bad press about Grand Coulee and Bonneville dams in those years.[20]

Roosevelt and his allies had answers to many of these charges. New Dealers hoped to get farmers off the marginal lands in the nation's Dust Bowl and resettle some of them in the Pacific Northwest and California, where federal reclamation projects would provide a more stable environment for agriculture. Thus, FDR expected a net reduction in farm acreage as a result, and more healthy and sustainable farm communities. Fish ladders would be built into Bonneville Dam to allow upstream passage of spawning salmon. (No mitigation for the young smolts headed back seaward was planned, though.) On the power question, the dam advocates remained convinced that markets for the prodigious amounts of electricity would develop once the power became available. This was a game of prediction without certainty, though. Skeptics said the power from Grand Coulee would go unused for decades. The Bureau of Reclamation (BoR) predicted that it would take only about fifteen years to absorb the power capacity at Grand Coulee. In fact, it took half that amount of time. Surprising everyone, power demand to meet the needs of the war industries that sprang up in the Northwest during World War II, the aluminum plants and plane- and ship-building factories, absorbed every kilowatt of power these two federal dams could produce (see next chapter). By 1943, the *Saturday Evening Post* was singing a different tune about Grand Coulee Dam. An article by Frank Taylor, "The White Elephant Comes into Its Own," concludes that "the scoffers" of several years earlier were dead wrong. Reviewing the two dams' contributions to the war effort, Taylor claims they were "indirectly winning the war for us. . . . Now, with their turbines roaring full tilt, the aluminum pots are hungry for more power." With cheap and abundant federal power, the "sparsely settled Northwest" might become "the light-metals capital of the world."[21]

One reason opponents of these two federal dams lost the battle was that many business interests expected to benefit from the cheap power and the economic stimulus derived from construction of Bonneville and Grand Coulee during those glum years of the Depression. For example, Franklin Griffith, president of Portland General Electric (PGE), expected to get access to cheap power from Bonneville Dam since it was located so close to Portland. His company had considered building its own hydropower dam at the site a few years before but abandoned the idea because it would have cost too much and taken too long to build. Tens of millions of dollars borrowed at commercial interest rates would have been tied up in construction for years before any power could be sold to recover the investment. In an article he wrote for *Electrical West* in 1938, Griffith explains why construction of Bonneville Dam was "impractical for private capital to undertake," and why "the government's low-cost Columbia River energy should benefit the Northwest."[22]

Another reason opponents lost the battle was that one of the key objections to federal development was an ideological argument that had lost much of its luster during the Depression: the argument that business enterprise should be undertaken by the private sector, not the government, and that unfettered supply-and-demand markets should determine the timing and location of economic development. Such endorsements of the virtues of capitalism and the evils of government rang hollow to most people during the Depression. The weary and desperate majority looking for hope and change found a ray of optimism in these grandiose make-work projects launched by the New Dealers in 1933 and 1934. But support for these projects could not be taken for granted. The defenders of capitalism and critics of government planning, paternalism, and public power fought tenacious public relations battles and retained loyal and astute defenders in state legislatures and the U.S. House and Senate. Roosevelt and the New Dealers had to fight for every new program and policy and compromise continuously during the approval and implementation of each project or regulatory reform.[23]

While the Liberal Party in Canada entertained similar grand visions of economic recovery and social transformation, especially in British Columbia, government gestures toward implementing those visions were markedly modest in contrast to those in the United States, to the disappointment of Canadians suffering from the Depression. The Canadian government was especially hesitant, compared to the U.S. government. Most of Canada's provinces were clamoring for more federal aid during the Depression, often led by British Columbia, whose newly elected premier in 1933, T. Dufferin Pattullo, had fully embraced a platform almost identical to Roosevelt's New Deal. Pattullo did not meet Roosevelt until 1937 and always claimed that his political philosophy was homegrown. Indeed, his orientation toward governance and his views on the responsibilities of the state to actively pursue and protect the public welfare had been fairly consistent throughout his political career, which stretched back to the first decade of the twentieth century. Nevertheless, the two had a lot in common. British Columbia political scientist Martin Robin describes Pattullo in terms that would be an equally apt description of Roosevelt: "Pattullo's major virtues were not theoretical exposition and philosophical eloquence. Instead, he thrived on incurable optimism, boundless self-confidence, and bold pragmatism. Pattullo was an energetic problem-solver, willing to apply a myriad of solutions to the host of immediate problems which attended his premiership."[24]

In leading the B.C. Liberal Party in the 1930s, Pattullo put his province at the Canadian vanguard, embracing Keynesianism, promoting aggressive public investment to ameliorate the Depression, and advocating systemic reforms intended to establish a modern capitalist welfare state—what he called "socialized capitalism." Pattullo felt that in hard economic times government should expand credit to finance public works and thereby boost employment. He believed that, even in good economic times, the government should guarantee a minimum standard of subsistence and health care, just as it provided a minimum standard of education. B.C. historian Robin Fisher credits Pattullo's government for offering "what was probably the most vigorous response to the social and economic problems of the depression of any government in Canada."[25]

The campaign platform the B.C. Liberals ran on in 1933 stated, "No person in British Columbia should be allowed to want for food, clothing or shelter through inability to obtain employment." The platform specifically called for a federal system of unemployment insurance; an expanded and improved old-age pension system; a provincial economic council to promote and coordinate economic development; an agricultural marketing board to assist farmers; a highway commission to plan and build a better transportation system; a public utilities commission to regulate electric, gas, water, and other utilities; and a national health insurance program. The platform also, ironically, promised to cut provincial taxes and expenditures.[26] Pattullo's party hoped that the dominion government would fund welfare and stimulus programs since the economic crisis was a national problem. It no doubt also realized that tax cuts would be popular and that voters always expected fiscal prudence from government.

To Pattullo's eternal frustration, federal leaders in Ottawa did not share his vision. The Conservatives remained in power in the national government until the federal elections of 1935, and their prime minister, Richard Bennett, adhered to a rigid monetary policy during the crisis, cutting expenditures and relying on foreign trade and the private sector to solve the economic crisis. Bennett's government did provide relief payments for the large camps of unemployed workers scattered across the country, and after years of economic decline, it finally initiated some modest financial incentive programs for businesses, much like Hoover's in the United States, but its first goal was to keep debt and deficits down as tax revenues plummeted. Pattullo wrote to Prime Minister Bennett in December of 1933, describing the rising social desperation and the need for government action: "Unemployment is destroying self-reliance and self-respect . . . and is creating more and more dependence upon the state and ominous disrespect for the present order of our economic and political structure. Private endeavour alone cannot reasonably meet the situation. There must be governmental intervention."[27]

Despite these pleas, very little additional federal aid came to British Columbia or any other province from 1933 to 1935. Consequently, the B.C. Liberal Party worked to implement its policies with its own meager resources. Handicapping the province was a provincial debt of over $165 million (for a population of only 700,000), which the Liberals inherited in 1933 from the Conservative Party. Worse, the government's banker had closed the province's credit line because of British Columbia's unhealthy financial condition, so British Columbia could no longer borrow to cover obligations or fund public works. The Canadian banking system was private, and provincial finances were regulated by the dominion government, so British Columbia could not raise its own capital to accomplish its goals. Pattullo felt hemmed in by stingy, self-interested bankers and an unsympathetic national government. Nevertheless, his party made some modest progress. In the 1934 legislative session, British Columbia created an economic council and a crop-marketing program for farmers, raised the minimum wage, and lowered the maximum workweek to forty-eight hours. In the next few years the B.C. Liberals also established an industrial relations board, aided the mining and fishing industries, increased funding for welfare and schools, created a public utilities commission and a highway commission, designed Canada's first government health insurance plan (authorized in 1936 but scrapped in 1938), and started a few modest public works projects (a new City Hall and a hotel in Vancouver, a toll bridge across the Fraser River, and some highway construction). This fulfilled just a small portion of the B.C. Liberal Party's ambitions, though. Pattullo and his allies therefore hoped that a Liberal win in the national elections in 1935 would finally kick welfare and stimulus programs into high gear with more generous infusions of capital from the dominion government in Ottawa. They were once again disappointed.[28]

After five years in power in Ottawa, the Conservative Party lost the 1935 election, and MacKenzie King of the Liberals returned to his former seat as prime minister. A longtime family acquaintance of Pattullo's and an icon of early twentieth-century Canadian Liberalism, King nevertheless shared few of Pattullo's views on the proper role of government and the proper relationship between the provinces and the dominion. Contrary to Pattullo's hopes, King adopted almost the same retrenchment policy as Bennett after discovering to his shock in 1935 that he had inherited from the Conservative Party a $3.25 billion debt. Like most politicians of his era, King felt the responsible thing to do under the circumstances was to cut expenditures and raise revenue. He immediately cracked down on deficit spending, including borrowing and debt incurred by the provinces, and insisted on recovering all revenue from income taxes and tariffs collected in the provinces that were owed to the dominion. (British Columbia hoped to keep more of these revenues for itself to fund relief and recovery programs.) So, at the same time that Pattullo and other provincial leaders were demanding loans, grants, and reimbursements from Ottawa, the prime minister was tightening the federal belt and exhorting the provinces to do the same. British Columbia had to continue pursuing its "Little New Deal" on its own and in anemic proportions.[29]

SYSTEMIC REFORM

Like Roosevelt, Pattullo was not just interested in providing relief to the vulnerable during the economic crisis; he had a plan for structural economic and social change that he believed would avoid such crises in the future and create a more just and prosperous society. Many of his party's policies were designed to achieve systemic reform, but most of them were either compromised or largely unfulfilled in the 1930s. He got nowhere with his proposal to restructure relations between the provinces and the dominion. The B.C. legislature developed a provincial health care system in 1936, approved by public vote in 1937, but it remained so controversial that Pattullo shelved it after the referendum. His party did finally create a Public Utilities Commission in 1938, but it was weak, politically controlled by the B.C. cabinet, and did not undertake any serious reforms. Private corporations remained firmly in control of the power supply in British Columbia for several more decades. Pattullo's reformist plan was emphatically not radical, just a bit ahead of its time. He saw the Liberals as the moderate party situated between the Conservatives and the Socialists represented by the Cooperative Commonwealth Federation (CCF). The CCF gained a great deal of electoral ground in the 1930s and even outperformed the Conservative Party in British Columbia's 1933 election. Pattullo did not like either the Conservatives or the CCF, however, and refused to join governing coalitions with them, which contributed to his loss of the premiership in 1941.[30]

Though Pattullo proposed transformative structural reform, the Depression, along with limits to provincial authority and the lack of financial assistance from the federal government, ultimately hampered his plans. Consequently, Pattullo's reform vision increasingly gave way to pragmatic adjustments. After the provincial election of 1937, when the Liberals lost four seats, Pattullo and his party substantially retreated from their commitment to social reform. While dramatic changes were taking place in the U.S. Northwest, folks in British Columbia had to content themselves with slow and incremental change. With for-profit companies totally dominating the power industry in British Columbia and consumer demand remaining depressed, British Columbia's electric utilities stayed in a retrenched mode, and the power system there remained in relative stasis until 1939, after the war in Europe had stimulated trade and manufacturing in Canada, boosting economic activity and power demand.[31] Notably, the federal government of Canada, even under Liberal rule, could not justify extraordinary expenditures to relieve human suffering during the worst economic depression in the nation's history but found the will and justification to spend extraordinary sums of money on war.

South of the border, Roosevelt and the New Dealers went far beyond immediate assistance for the unemployed and idle industry. Like Duff Pattullo in his original platform, the New Dealers also sought to fix the problems that led to the Depression and create a sustainable and just economic system whose benefits distributed widely across society. Unlike in the B.C. Liberal Party in the 1930s, though, reform of the electric power industry loomed large in Roosevelt's agenda. After launching a remarkable spate of relief and recovery programs in the first congressional session of his first term in office (1933–1934), Roosevelt, his key advisors, and his allies in Congress turned more assiduously toward systemic reforms in what historians call the Second New Deal.

The year 1935 was particularly significant. Many Americans know the U.S. Congress passed the Social Security Act in 1935, creating the nation's first universal old-age pension program. Also that year, Congress passed a Banking Reform Act that restructured the Federal Reserve; passed the National Labor Relations Act, which created the National Labor Relations Board; enacted a wealth tax that sought to redistribute concentrated wealth; created the Resettlement Administration to help farmers ruined by the Dust Bowl (it became the more broadly focused Farm Security Administration in 1937); and created the Works Progress Administration, which put two million people to work. There was more. In regard to the electric power industry, 1935 was probably the most important year of the New Deal. Congress passed the Public Utility Holding Company Act (PUHCA) to end holding company abuses, passed a closely related Federal Power Act (FPA) that expanded the responsibilities of the Federal Power Commission, and passed the Rivers and Harbors Act that authorized Bonneville and Grand Coulee dams and many other dams across the nation. In that same year, Roosevelt established by executive order the Rural Electrification Administration to bring power to areas of the nation not yet served by electric utilities.

In PUHCA, the new Federal Power Act, and the REA the New Dealers sought to transform the electric power system to achieve several fundamental goals: eliminate or greatly restrain holding company influence over public utilities, provide universal access to electricity at rates affordable to everyone; regionally coordinate electric transmission grids on the Giant Power/Superpower model, reorient "public service corporations" toward public service rather than corporate profit, and enable state and federal agencies to more effectively guide the industry toward these goals. Policy makers need agents to implement policy, and the New Dealers chose the new Securities and Exchange Commission (SEC, established in 1933) and the Federal Power Commission (FPC, established in 1920) to implement their national power policies. PUHCA empowered the SEC to restructure or abolish public utility holding companies that did not meet certain criteria. To simplify the arcanely complex and obfuscating pyramid structure of the holding companies, PUHCA basically allowed only one layer of ownership over the utility operating companies and required that the relationship between them be demonstrably direct and beneficial. If a holding company controlled multiple utility operating companies, those operating companies had to be regionally contiguous to facilitate coordination and improve the likelihood that one or two adjacent state public utility commissions would be able to effectively regulate them. To eliminate the predatory management and consulting fees that holding companies charged operating companies, and to reduce the inflated capital costs and profits generated by sweetheart deals between them, PUHCA required that any sales of goods or services between affiliated utility companies be on an at-cost basis only—no profit allowed. Importantly, PUHCA also authorized the SEC to regulate the issuance of interstate utility securities to eliminate many of the worst forms of watered stock and irresponsible speculation. If a public utility corporation did not meet these and other requirements, the SEC could break it up, restructure it, or dissolve it.[32]

Title II of PUHCA and the related Federal Power Act gave the Federal Power Commission a different but complementary set of responsibilities. Created fifteen years earlier by the Federal Water Power Act (see chapter 6), the FPC's authority was largely limited to licensing hydropower dams on public lands and navigable rivers. Congress also authorized the FPC to perform hydropower facility valuations and regulate rates for interstate transmissions of wholesale hydropower, but it rarely performed those functions due to a lack of funding and a lack of political support from the Republican presidential administrations of Harding, Coolidge, and Hoover from 1921 to 1933. New Dealers wanted to energize those neglected duties and expand the agency's responsibilities. Title II of PUHCA, covering new FPC duties, along with a set of revisions to the 1920 Federal Water Power Act codified in the Federal Power Act of 1935, gave the FPC responsibility over all interstate exchanges of wholesale electricity whether from hydropower or not. New Dealers wanted to give the FPC stronger authority over interstate power sales because the U.S. Supreme Court had ruled in the 1927 case *Public Utilities Commission v. Attleboro Steam Company* (273 U.S. 83) that interstate sales of electricity were beyond the authority of the states to regulate. Congress directed the FPC to ensure that interstate power sales were kept at reasonable rates and were nondiscriminatory. The latter policy prohibited interstate power companies from burdening customers in one state with excessive rates in order to advantage customers in another state with artificially low rates. The FPC still had no authority to regulate local rates except in the absence of a state-level regulatory commission.[33]

As noted above, the New Dealers had to fight and compromise for every reform. Objections to PUHCA led to a lot of deal-making and adjustments before the law passed. The National Association of Railway and Utility Commissioners (NARUC—now called the National Association of Regulatory Utility Commissioners—same acronym) mounted vigorous opposition out of fear that the federal government would usurp their responsibilities and jurisdiction. Their concerns were varied and inconsistent. Virtually all chafed at the implication that state-level regulations were inadequate and ineffective. While state commissioners often lamented their inadequate resources in house, they did not like the national, public accusation that they were not up to the job. Many feared that local control and local interests would be lost or diluted in a federal regulatory agency, a reasonable concern. Some leaders of the utility industry with multistate operations actually supported federal regulation because they would have to deal with only one regulatory agency, but their enthusiasm for federal regulation depended on the degree to which they felt federal regulation would actually benefit them. Under a presidential administration like Calvin Coolidge's, who famously said, "the business of America is business," federal regulation could easily work to the advantage of the captains of the power industry. But under Roosevelt, who believed "the object of government is the welfare of the people," federal regulation might not be as advantageous to the wealthiest class, especially during a time of economic crisis in which capitalism seemed to have failed the people.

NARUC had additional concerns. Many members objected to the idea of the FPC prescribing a uniform system of accounting for determining utility valuations and rates because it might require them to change their own long-established procedures. Some feared that a federal agency would be more susceptible to influence from national corporations that had no stake, other than profit, in a state-level operating utility. Others feared the opposite: that crusading federal regulators would curtail the profitability of locally based utility corporations, harming the pecuniary and competitive interests of the state. Consequently, Washington, D.C., policy makers crafted language in the bills to address these and other objections from NARUC. Cooperation with state commissions became a major feature of the legislation and of the FPC's responsibilities. Mostly the law empowered the FPC to fill the gaps outside the reach of the state commissions rather than usurping their authority. While the FPC did establish a uniform system of accounting, it applied only to interstate utilities reporting to the FPC. State utility commissions did not have to adopt the standards. Because a state commission could not compel an interstate utility to open its records regarding the costs of power produced out of state, the new federal law authorized state commissions to petition the FPC to investigate and report on those costs. The law also authorized states to file complaints with the FPC to address adequacy and reliability problems with interstate power transmissions. The law gave the FPC authority, after a complaint and an investigation, to compel "proper and adequate interstate service." The Federal Power Act thus created multiple opportunities for federal and state regulatory commissions to voluntarily collaborate to achieve more effective regulation of rates and service. Nevertheless, the leadership of NARUC felt the law impinged too much on state jurisdiction and joined the critics of the bill during debate in 1935.[34]

One facet of the new Federal Power Act advanced Roosevelt's desire to promote regional coordination of electric power systems—one of the key principles of Giant Power/Superpower and the foundation behind the Tennessee Valley Authority (TVA). The Federal Power Commission summarized its new responsibility concisely in its 1935 annual report:

> A primary objective of this act is a planned coordination of the power facilities and resources of the nation. The Commission is directed to divide the country into the most economical regional districts and to accomplish, either by voluntary action of the utilities themselves, or by compulsory Commission action . . . the interconnection and coordination of power facilities within such districts.[35]

This, too, would be easier said than done. Nevertheless, it was a clear statement of the administration's commitment to regional power pools and coordinated power planning (more on this in the next chapter).

Though controversial at the time, this policy of regional coordination of power supplies was simply part of a general national trend developing steadily since World War I. The increased role of the federal government in promoting and directing this trend, however, attracted the special ire of industry executives. Leaders of the commercial power industry obviously preferred to control the process themselves. Greater regional interconnection and coordination advanced significantly in the Northwest and elsewhere in the United States during the 1930s as a result of New Deal policy, despite the fact that Congress never authorized a Columbia Valley Authority (CVA). World War II greatly accelerated that trend just as the previous world war had done.

PUHCA and the Federal Power Act went a long way toward reforming flaws and inefficiencies in the national electric power system, but they did not address a crucial equity issue of deep concern to Roosevelt: the fact that over 85 percent of Americans who lived in rural areas still had no electricity in 1930. America remained very much a rural nation at that time, with 44 percent of the population living in communities with fewer than 2,500 residents. That meant that over a third of all Americans lived without electricity in 1930. Roosevelt would need more than the SEC and FPC to address this problem.[36]

The low rate of rural electrification in the United States contrasted sharply with that of many other developed nations, to the slight embarrassment of U.S. policy makers. For example, New Zealand, primarily an agricultural nation with a population density of about a third that of the United States, had two thirds of its farms electrified by 1935, compared to 14 percent in the United States. New Zealand had a primarily public-power system. Even more impressive were many Western European nations. Denmark had 85 percent of its farms electrified in 1935, with much of the impetus for electrification coming from self-organized farmers' cooperatives. Germany reported 90 percent rural electrification that year. Germany had 5,000 co-ops serving rural areas, with 53 percent of the nation's electricity generated by publicly owned power plants. The government of France launched a national program of electrification after World War I and raised the percentage of the population with electric service from 20 percent at the end of the war to 93 percent by 1935. Switzerland reported nearly 100 percent electrification in 1935, with 65 percent of its power publicly owned and derived from hydropower. In electrification, the United States had a comparatively poor record, and Roosevelt was determined to do something about it. Clearly, Europe's success in the 1920s and 1930s founded on a preponderance of public-power systems served as a model for the New Dealers.[37]

GETTING POWER TO THOSE WITHOUT IT—RURAL ELECTRIFICATION IN THE NORTHWEST

The prominence of the Rural Electrification Administration in history textbooks and public memory can leave the erroneous impression that rural electrification was a rare occurrence before the New Deal, that farmers all over the United States lacked electricity before the REA. But the degree of rural electric service in fact varied widely by region and even within regions, with many rural areas getting electrical service well before the 1930s. In fact, rural electrification was patchy everywhere in North America, with some farm regions getting hooked up early in the century and others waiting until the 1960s to get power.[38] In the United States at the start of the Depression, farm electrification was lowest in the Southeast states, ranging from 2 to 5 percent, and highest on the Pacific coast, with California at 71 percent, Washington at 50 percent, and Oregon at 35 percent. The national average, however, was 14.3 percent. Thus, an overwhelming majority of farms lacked service in the 1930s, and most of those farmers wanted service.[39]

The situation in Canada was starker, where rural electrification proceeded at a slower pace than in the United States. Provinces with public power systems like Ontario made a more concerted effort to extend electric lines outside cities and towns than did provinces with predominantly commercial utility companies, but electrical penetration of rural areas nevertheless remained minimal in all provinces.[40] Canadians who had electricity consumed a lot of it, but except for those living in cities, relatively few enjoyed electrical service even as late as 1930. The Canadian Census of Electrical Industries did not distinguish rural from urban customers in 1930 (or prior to that), so reliable rural electrification statistics are hard to come by. But some rough estimates can be derived from the censuses. The Canadian Electrical Census of 1930 indicates that about 75 percent of provincial households in British Columbia had electric service. Revealingly, the same percentage of British Columbia's population lived in the greater urban and suburban areas centered in Vancouver and Victoria, which were well served by BC Electric. If most households in these developed areas had electrical service, along with medium-size towns like Kelowna, Kamloops, and Chilliwack, then probably few small towns or rural areas had any central-station electrical service.[41]

Many people saw the poor penetration of electricity into rural areas as a lamentable social inequity worthy of attention from policy makers. Canadians made few improvements to this situation in the 1930s, though. In contrast, after 1935 the REA in the United States moved rural electrification forward significantly while articulating a new principle that all Americans, regardless of income, class, or residence, should have electrical service. But the changes REA wrought were not as revolutionary as some have claimed, and they did not happen quickly.[42]

Another misconception is that commercial for-profit utilities did not care about farmers and would not extend rural lines to them. This is only partially true. Where a commercial utility believed rural demand was sufficient to generate profitable revenue, it usually extended rural lines without much prodding. Where profits were questionable, utilities needed incentives. Where the investment obviously would not pay off, the commercial utilities refused to provide service. Only a government agency or a nonprofit utility driven by social welfare policies could afford to build distribution lines to rural regions containing only scattered customers using small amounts of power.

Before the Depression, commercial utilities generally provided rural electrification only to customers who happened to live near existing service areas or distribution lines or in irrigation districts where the large electricity demands of irrigation pumping made building new transmission lines profitable. Historian David Nye notes that hydropower-based electric systems often were among the first to serve rural areas because the power source was typically in more remote locations (river valleys in mountainous regions). Fuel-based power plants could be built near load centers (cities and industry), but hydropower plant locations were determined by physical geography, and those places best suited for dams were often far from cities but not far from farms. This mixed record regarding the willingness of commercial utilities to bring power to rural areas explains why some farmers had few grievances against the commercial utilities while others were so aggrieved they called for a government takeover of utilities. It also explains why commercial utilities and their political allies could approvingly tout their fine record in rural electrification while detractors and their allies could cite damning statistics about the failure of commercial utilities to serve rural areas.[43]

Evincing an early interest in rural markets, the National Electric Light Association (NELA) formed a Committee on Electricity in Rural Districts in 1911. NELA served as the research and public relations arm of the investor-owned sector of the industry. This committee mainly investigated opportunities and reported them to its members. But as the market for power applications in agriculture grew over the next decade, the investor-owned utilities paid increasing attention to it and developed major promotional campaigns aimed at farmers (see chapter 7). By the 1920s, rural districts—at least significant portions of them—were a viable market, and many utilities sought to add them to their service territory in rosy anticipation of future revenue. Prodded by Wisconsin Power and Light, NELA established a more active Committee on the Relation of Electricity to Agriculture (CREA) in 1923. It recruited universities, government agencies, farm organizations, and equipment manufacturers to join it in advancing farm electrification. CREA had a good run for about a decade, establishing a number of rural electrification projects around the United States. But its effects were limited and experimental, more promise than progress. Extensive (not to mention universal) rural electrification was simply not feasible for commercial utilities in many regions because farmers could not pay the rates necessary to repay costs and provide a reasonable profit to the utilities. The Depression hampered CREA's activities and industry enthusiasm for rural service. Not able to expand beyond its experimental projects, CREA sank into obscurity and finally disbanded in 1939. As a result, public power systems assisted by federal incentives dominated rural electrical development in the 1930s.[44]

Ironically, the privately owned Stone & Webster affiliate Puget Sound Power & Light (PSP&L) built one of the first rural distribution lines in the Northwest. As early as 1913, this company expanded service into the farming areas around Puget Sound and generally succeeded because farms in that area were small and clustered near urban markets that PSP&L served, reducing the overall cost of extending electrical distribution lines. The company established a farm-electrification department in 1924, and in subsequent years its monthly *Puget Sound Electric Journal* often included articles on rural and agricultural electrification. It regularly sent representatives and exhibits to county and state fairs to reach that particular market. In 1928, for example, PSP&L set up 140 linear feet of demonstration booths at the Western Washington Fair in Puyallup, attended by 200,000 people. The booths featured an electric "stump-burner" (a useful implement in heavily forested western Washington!), electric egg incubators, milking machines, sterilizers, butter churns, dairy refrigerators, water pumps for garden hoses and farm irrigation, electric water heaters, an electric oat sprouter for chicken feed, "hotbeds" for germinating plants early in the season, paint sprayers, and a variety of electric shop tools and motors. That was just the section of the exhibit dedicated to farming. At one end of the exhibit, PSP&L had assembled a full-size cottage with all modern electric appliances for lighting, cooking, laundry, bathing, sanitation, and even an electric fireplace for heat. As the profiler of this exhibit notes, "The whole display was enough to make any city man or woman want to tear loose and get a farm home, raise chickens by electricity, milk cows by electricity, raise feed and gardens with irrigation pumped by electricity, and have a few acres of stumps around to burn out with electricity."[45] By June 1929, PSP&L claimed to serve 30,000 farms in eighteen Washington counties supported by 4,000 miles of transmission line.[46]

As shown above, farmers near urban centers like Seattle and Tacoma were among the most likely to get electric service before the REA. Another lucky group was residents of farm communities near hydropower dams, especially those that needed to pump large amounts of water to irrigate crops in the dry inland Northwest summers. As discussed in earlier chapters, federal agencies and utilities often electrified these regions after the dams were completed. The Yakima Valley, Snake River Valley, and Washington's Okanagan Valley are prime examples. Idaho Power's main customers included rural communities since the founding of the company in 1915. It aggressively sought to bring power to irrigated farms in the Snake River Plain. Idaho Power's chronicler Susan Stacy claimed that throughout the 1920s, Idaho ranked among the top ten states for rural electrification. The percentage of farms electrified in Idaho when Congress formally established the REA in 1936 had already reached nearly 30 percent, double the national average though still significantly less than Washington (54 percent) and Oregon (44 percent).[47]

Not all relationships between utility companies and farmers in the Northwest were as cooperative as the ones noted above. Many residents of rural areas felt frustrated and neglected by the industry, such as the farmers of Franklin County in eastern Washington, who signed right-of-way easements over to Pacific Power & Light (PP&L) in 1916 for a high-voltage transmission line running across the county—and across their land. PP&L was connecting its transmission system with Washington Water Power's to buy surplus hydropower from the latter's Long Lake Dam. One local landowner, Lloyd Howton, recalled that the PP&L agents sent to acquire rights-of-way promised that electricity would be supplied to area farmers as well as to the rural towns of Kahlotus and Connell. But, Howton claims, "after most of the easements were signed, the company stated that there wasn't enough revenue to support a distribution system. There were enough easements hanging fire to force an agreement to serve Kahlotus, but the farm service was dropped." This experience apparently politicized Howton, who later became a public power advocate and a commissioner of the Public Utility District of Franklin County.[48] Rural residents complained constantly in the 1920s and 1930s about high rates or the lack of service in rural areas. Those who had power were generally happy (if rates were reasonable); those who did not were unhappy.

To secure electric service or get redress for unreasonable rates, farmers could file a complaint with the state Public Utilities Commission (PUC) asking it to order service extensions and/or reasonable charges. Utilities often required rural residents to pay extra fees up front to cover the cost of extending power lines to their residences. These could be several thousand dollars a mile, an exorbitant sum for a family in 1930. Alternatively, if enough customers were available, utilities charged a high minimum monthly bill for several years, sufficient to recoup power-line construction costs. On top of all this, utilities often charged higher electric rates for rural service.[49] Of course, this all made business sense, but it kept farm regions largely out of service. Franklin Roosevelt, Morris Cooke, and David Lilienthal hoped to ameliorate this through New Deal programs. Throughout the 1920s and early 1930s, farm organizations like the Grange regularly advocated for service extensions and helped farmers appeal for assistance from the state PUCs. For example, the Washington Division of Public Utilities handled over fifty complaints in 1929 regarding electric service extensions. As a result, it fully revised its rules that year regarding special fees and minimum bills for service extensions.[50] If a group of farmers could not get satisfaction from either the utility or the PUC, they could cooperate to build a power line themselves. But until the REA came along with low-interest, long-term loans, this was usually impractical. High costs, difficult loan terms, and technical requirements often beyond the farmers' means inhibited informal cooperatives. Getting power to energize the electric lines posed another set of challenges. Thus, electricity remained out of reach for the majority of farmers.

Because Canada's rural population was even less densely settled than the United States, rural service there was more expensive and more difficult to acquire. No national program of rural electrification comparable to that in the United States existed in Canada during the Depression. Rural electrification efforts in Canada occurred more in the 1950s than in the 1930s.[51] But the lack of government assistance did not stop farmers and rural residents from trying to get power to their properties during the Depression. One way was to appeal to the public utilities commission. As noted earlier, British Columbia created a public utilities commission in 1938 and authorized it to evaluate and act upon complaints from citizens who wanted utilities to extend service to their residences, just as PUCs did in the United States. The B.C. Public Utilities Commission report of 1940 included a section listing all the formally filed requests for service extensions and its decisions regarding each. Even in the Liberal-controlled province of British Columbia with its "Little New Deal," farmers and other rural residents got no immediate assistance from the new commission. Revealing strong support for the bottom line of profitability, the commission sided with the utilities in eighteen of twenty complaints by citizens that year, who wanted an extension of electrical service at more reasonable rates. The commission report states, "most of the complaints received arise out of a misunderstanding on the part of the complainant as to why he should be required to make any contribution to the cost of the extension." The commission's summary findings for each complaint settled in the utilities' favor simply stated, "insufficient revenue to justify extension" or "revenue requirements of company not unreasonable."[52] BC Electric continued to focus on urban and industrial service through the 1940s and ignored most outlying communities. People living in northern British Columbia, far from the more densely populated southwest corner of the province, had little hope of receiving electric service.

Despite the scattered progress in rural electrification in the United States during the decade or so before the Depression, the great majority of rural districts and dwellers still had no electricity in 1930—more than 85 percent were still in the kerosene age. As noted earlier, however, the three Northwest states were far ahead of the national average in farm electrification: Oregon was at 35 percent, Washington just over 50 percent, and Idaho 33 percent.[53] But even this highly progressive record left most Northwest farms without electric service when FDR came into office. New Dealers considered this inequitable and unconscionable, so they marshaled some modest resources and created a cooperative national program to help farmers get affordable electric service. It became one of the most popular New Deal programs. New Deal historian William Leuchtenburg claims, "Perhaps no single act of the Roosevelt years changed more directly the way people lived than the president's creation of the Rural Electrification Administration in May, 1935." America, he said, was divided "into two nations: the city dwellers and the country folk," with the former enjoying the fruits of modernity while the latter "toiled in a nineteenth century world."[54] This divided social order personally concerned the president, and he worked assiduously to promote programs aimed at revitalizing rural America, to stem what he considered an unhealthy and unnecessary flight from the country to the city. The regional planners profiled earlier in this chapter shared those concerns and sought ways to better integrate city and country and equalize standards of living between the two. Electrification seemed one of the best opportunities to raise rural quality of life as well as productivity of the farm. Considering the record of the previous decade, it was clear that government involvement would be necessary to accomplish this goal in a reasonable time frame.[55]

Historians mainly credit Morris Cooke for getting Franklin Roosevelt interested in a federal program for rural electrification. Cooke coauthored the Giant Power proposal of 1925, which envisioned coordinated regional power development, including an ambitious plan for rural electrification and a preference for publicly owned power systems, making it somewhat of a blueprint for New Deal power policy. In 1933 Roosevelt appointed Cooke to a leadership position in the Public Works Administration—one of the New Deal relief and recovery agencies. Most interested in advancing rural electrification, Cooke repeatedly lobbied Roosevelt and his cabinet to create a national program to promote it. A moderate and a collaborator by inclination, Cooke initially proposed government incentives to the commercial utilities to accomplish the work, especially because nearly 90 percent of the electric power business in the United States was controlled by for-profit utilities in 1932. But ideological differences and mutual distrust between the New Dealers and the commercial power industry stymied cooperation. Industry leaders hated Roosevelt's power policies, doubted the need for and feasibility of rural electrification, and adamantly opposed any programs that would strengthen the publicly owned utility sector. Cooke met with private industry leaders numerous times in 1933 to promote a cooperative program but ended up with little to show for it. Harold Ickes, the imperious and curmudgeonly secretary of the interior, stymied Cooke's efforts on the other side. Ickes oversaw a vast number of New Deal programs. He hated the commercial power industry for its anti–public power propaganda and financial abuses of recent years and unequivocally told Cooke he would not work with "the scoundrels." Thereafter Cooke, with Ickes's blessing, developed a proposal for a purely federal government program administered within the Department of the Interior, which he delivered to Ickes in early 1934. It was based largely on a plan to provide low-cost financing and technical advice to nonprofit organizations (cooperatives) formed for the explicit purpose of building power lines and small generating stations to serve rural needs. This became the model for the REA.[56]

Roosevelt wrapped Cooke's proposal into a much larger unemployment relief bill he sent to Congress in January 1935. It faced some stiff opposition but finally passed in April 1935, with $100 million earmarked for rural electrification. Congress created no special agency to administer this program, so on May 11, 1935, FDR created by executive order a temporary Rural Electrification Administration to coordinate projects. Shortly thereafter, FDR authorized the new REA to serve as a lending agency. He also appointed Morris Cooke to run the agency. Both public and private utilities as well as qualified farmers' cooperatives could apply for funding, so Cooke once again sought cooperation from the private sector to get as many projects as possible going as quickly as possible. But industry leaders were not enthused about the program, did not like the rules for participation drawn up by the administration, and were not interested in helping make the REA a success. After many meetings with mixed success, Cooke decided to focus his attention on public utilities and farmers' cooperatives, but the opportunities in those sectors were limited. Consequently, the REA got off to a slow start. Roosevelt was disappointed that the program after a year had not made a dent in the unemployment work-relief rolls, so he threatened to disband the REA if Cooke could not ratchet up its accomplishments.[57]

To solve the problem, Cooke sought congressional assistance in turning the REA into much more than a relief agency. He wanted to enlarge the REA's scope, authority, and funding in order to make real progress on rural electrification. Senator George Norris and Representative Sam Rayburn came to Cooke's rescue in 1936, crafting a bill to give a broader purpose and mandate to the REA. As expected, a fight broke out over its provisions in the House and Senate. Norris and representatives from the American Farm Bureau and the Grange wanted a not-for-profit government program of ambitious scope that would electrify six million farms in the next decade. The commercial power industry wanted a much smaller program, which they would implement, with federal financial incentives. Anti–New Deal fiscal conservatives wanted to kill the bill entirely. After several months of wrangling, a compromise won enough votes to pass, and President Roosevelt signed it into law on May 21, 1936. The law authorized the new REA to loan up to $40–$50 million a year for ten years to finance power lines and power plants for rural areas that lacked electrical service. It also offered small loans for home and business wiring and electrical appliances. To protect existing utilities, recipients of REA loans could not organize electrical service in areas that already had it. Importantly, the law was not designed as a subsidy. REA borrowers had to pay back the loan over twenty-five years at the same interest rate the government paid for its securities, at the time about 3 percent annually.[58] Congressional opponents of the program insisted on these provisions. Congressional advocates of commercial utilities also insisted that for-profit companies, not just public agencies and farmer co-ops, be allowed to apply for REA loans. Senator Norris desperately wanted to cut the commercial sector out of the program, but Cooke and Rayburn felt the for-profit sector could be helpful, so the final bill allowed the commercial utilities to apply for loans. This last compromise secured the bill's passage. The REA law also stipulated that funds be allocated nationally by need; in other words, states with the lowest rates of rural electrification would get the lion's share of REA funds each year. By the summer of 1936, the new REA was busy helping rural co-ops get established and qualified for loans.[59]

Probably the first rural electric co-op established in the Northwest in response to the REA was the Northern Lights Co-op in the Idaho panhandle. It was Idaho's first REA co-op and the sixth in the nation to receive a loan from REA. It launched in 1935 under the awkward name Northern Idaho Rural Electrical Rehabilitation Association, but changed to Northern Lights in 1936. With two loans from REA it built power-distribution lines in 1936 to dozens of small farms scattered across rugged north Idaho and adjacent areas of Washington and Montana. Six other REA co-ops formed in Idaho in the years before the United States entered World War II.[60]

Eight rural electric co-ops formed in Oregon between 1936 and 1942. The first to be approved and funded by REA was the Blachly–Lane County Cooperative Electric Association launched in 1936 and formally incorporated in April 1937. In the pastoral Willamette Valley between Eugene and Corvallis, the farmers there for years had been unable to convince Mountain States Power Company, the local utility, to provide them service. A few years after Oregon passed its people's utility district law in 1930, the farmers in Lane County organized to create a PUD, one of Oregon's first. But they had trouble securing a power source and never succeeded in getting funding authorization for the PUD. When Congress established the REA in 1936, the leaders of the PUD switched tactics and formed a co-op. REA quickly approved their loan application, including funds for a small hydropower plant on nearby Lake Creek because Mountain States Power would not agree to provide the electricity at a reasonable wholesale rate. This was one of REA's first loans for power generation. By November 1938 the co-op had built its power plant and forty-six miles of distribution lines serving 125 members. Three years and several more loans later, the co-op more than doubled its customer base to 300.[61]

Mountain States Power responded by building sixty miles of what were commonly called "spite lines."[62] The REA law allowed co-ops to serve only areas that had no pre-existing service, so to stymie co-op expansion, commercial utilities often hastily constructed electric lines in areas where the co-ops planned to expand their service. The commercial utilities would choose the areas with the greatest density of potential customers both to help ensure eventual profitable service for themselves and to make the co-ops ineligible for REA loans. The REA had a rule that co-ops needed at least three customers per mile of distribution line. When a commercial utility captured the most financially viable areas, the co-op was left with the most remote and expensive areas, raising the cost of the project for all the members of the co-op and risking falling below the three-customers-per-mile threshold. Despite these spite lines, the Blachly-Lane co-op found ways to continue bringing power to customers that Mountain States did not or would not serve. A contract for super-cheap power from Bonneville Dam in 1940 greatly enhanced its ability to do so.[63]

One of the nation's largest and most successful rural electric co-ops, Inland Power & Light, got its start in northeast Washington in 1937 as the Inland Empire Rural Electric Association. It adopted its abbreviated moniker in 1954. By the 1970s it had become the largest co-op west of the Continental Divide, serving an area larger than the states of Connecticut, Delaware, and Rhode Island combined—10,000 square miles of farm country in eastern Washington and northern Idaho, stretching from the Canadian border to Oregon. The story of Inland Power & Light actually began before the REA. Farmers in the wheat country surrounding Spokane wanted electric service and sent a delegation to the privately owned Washington Water Power Company in Spokane, meeting with WWP president Frank Post. One of the delegates later recalled that Post summarily turned them down. "It was absolutely fruitless. We accomplished nothing." West of Spokane in the Palouse Hills, several other small communities seeking electric service also contacted WWP representatives in 1934 and 1935 and met with the same result. In frustration a group of farmers near Spokane tried to establish a public utility district under the 1930 Washington state law that authorized PUDs. Because PUDs had taxing and condemnation authority, they required a formal vote of approval from the people living in the proposed utility district. Not wanting any competition from a publicly owned utility in its territory, WWP sent agents door to door to convince people to vote no. They told voters that their taxes would go up, that their service might be curtailed, and a suite of other true, half-true, and untrue claims. The PUD vote was narrowly defeated. Afterward, WWP lowered its rates.[64]

About the same time, the REA got full congressional authorization, and Grange leaders and county agricultural extension agents started advising the eastern Washington farmers to pursue their goals through a co-op rather than a PUD. Because co-ops had no taxing or condemnation authority and required only a voluntary membership, they were much easier to establish. After a few months of negotiation, several groups of farmers and small communities joined together and incorporated the Inland Empire Rural Electric Association in March 1937. Because their proposed service area was rather extensive, they applied for a large loan of $455,000 to build 463 miles of power lines to serve 1,530 members. There was one problem: they did not have a power supply. They had hoped to purchase bulk power at wholesale rates from WWP, but not surprisingly, the company refused to cooperate. The power rates WWP offered were simply too high to make the project financially feasible. In Washington, D.C., REA administrators held up approval of the loan pending resolution of the power-supply problem. The chief engineer for REA, Benjamin Creim, came out to Washington state to negotiate personally with WWP. In a closed-door meeting with President Frank Post, Creim got him to agree to sell the co-op power for five years at the wholesale rate of 1.4 cents/kwh. The agreement was signed, and Inland got its loan from the REA and its power from WWP. In the spring of 1938, after some fast and furious construction, the first building served by the new co-op received electricity: a schoolhouse in the bucolically named White Rose Prairie. A witness at the gathering to see the schoolhouse's initial illumination recalled, "What a celebration! It's impossible for people today to realize what getting electric service for the first time meant to people."[65]

Once the first group of co-op members got electricity, their neighbors who had not joined the co-op suddenly wanted in on the opportunity. A cascade of new members quickly grew Inland's customer base and required constant expansions. This troubled WWP deeply, as it wanted to control all the profitable markets in its region and did not want potential new customers deciding to join the co-op instead of buying service from WWP. The enthusiasm for electricity among the farmers caught WWP by surprise, and the company decided to capture as much of that potential market as possible. So WWP sent agents out into the field to identify rural areas where the density of farms might be sufficient to support profitable service. Once the company identified the most profitable places to run distribution lines, it quickly started power-line construction in order to keep the co-op from serving those areas. Inland formally complained to the REA about one of these WWP spite lines, and the REA ruled against WWP, finding that the co-op had contracted to serve the area first and was thus permitted to continue construction. By the end of 1939, Inland had built hundreds of miles of power line. In 1940 the recalcitrant Frank Post died, and WWP subsequently offered to buy the co-op lock, stock, and barrel. Inland's directors turned down the offer because WWP would not commit to extending service to all the farmers in the region. The co-op directors felt an obligation to complete the job and get 100 percent rural electrification. By 1941 they had built 1,500 miles of distribution lines and were contracting for Grand Coulee power at a wholesale rate one fifth the price they paid WWP for power.[66]

The stories of the birth and growth of Northern Lights, Blachly-Lane, and Inland Power & Light are similar to those of the several dozen other co-ops established at this time in the Northwest. (See Figure 9.1 for a map of rural co-op service areas in Washington and Oregon circa the 1940s.) Members of these co-ops were generally proud of their accomplishments. Moreover, they usually had chips on their shoulders about the commercial utilities, not just because the companies originally refused to extend them service, but because the success of their co-ops showed unequivocally that electric service was feasible. To them, stubbornness or greed seemed the only real obstacle to serving their farms and communities. As Norwegian immigrant farmer Olaf Rise of the Okanagan Valley of northern Washington told journalist Marquis Childs in 1951, "Except for the pressure of REA's program the farmers today would still be sitting in the dark listening to promises. Like anything else, the power business needs competition and our co-ops have supplied that competition."[67]

Figure 9.1. Rural Electric Co-ops in Washington and Oregon. Source: Gene Tollefson, *BPA and the Struggle for Power at Cost* (Portland: Bonneville Power Administration, 1987), 173.

Though the Northwest states already had relatively high rates of rural electrification in 1937, the REA made a big impact on the region. Over the next ten years, REA helped electrify most of the remaining areas of the Northwest. After 1941, war-related materials rationing (especially of copper) temporarily shut down rural electrification projects. But impressive strides had already been made. Rural electrification in Oregon jumped from 44 percent in 1937 up to 74 percent in 1942. In Washington in those few years, it jumped from 54 percent to 76 percent.[68] After World War II ended, the program picked up steam again, Congress reauthorized the agency for another ten years, and it facilitated another doubling of the co-op customer base across the nation.[69] Several other New Deal programs, including Bonneville and Grand Coulee dams and the new Bonneville Power Administration (BPA), created in 1937, accelerated electrification in the Northwest in the 1930s and 1940s. The cheap and abundant federal power from the two dams together with BPA's construction of electric transmission lines, REA's loans, and a policy of "preference" for public power by both REA and BPA brought about a crescendo of public power development in the Northwest, which will be a major topic in the next chapter.

CONCLUSION

In his book *Electrifying America*, David Nye argues that several key principles prominent during the New Deal undergirded the REA: that technology had the power to transform lives, that government can and should be a force for positive social change, and that the rural cooperatives promoted democracy and self-sufficiency rather than hierarchy and dependence.[70] The same principles stood behind most other New Deal programs, including the decision to build the Bonneville and Grand Coulee dams. Though Nye admits that things did not always turn out as rosy in rural America as New Deal planners hoped, the historical experience of those whose lives were affected by the REA and by construction of the federal dams on the Columbia did validate the New Dealers' faith in these principles. Every study of rural electrification acknowledges that the program transformed lives, greatly advanced the nation toward universal access to electricity, and vitalized small communities during the Depression. Everyone also agrees that without the REA, rural electrification would have proceeded much more slowly. Likewise, every study of the history and legacy of Bonneville and Grand Coulee dams acknowledges that they were transformative projects that provided extraordinary benefits to the region and the nation during World War II.

But at what price came this progress? Leaving aside the question of the environmental impact of the dams for now, critics of the REA and the big federal dams, then and now, have argued that these benign-seeming aid projects were unnecessary, expensive, and nefariously imposed "outside" federal government authority over local and state control, ultimately curtailing individual freedom and independence. Historian Brian Cannon takes a close look at the implementation of REA projects in the West to test these claims and finds little to support them. Farmers without electricity certainly thought the projects were necessary. The program had modest funding and required repayment by the beneficiaries, so it was not much of an economic burden. As for the concern about Big Government, Cannon concludes, "The history of rural electrification in the northern and central Rocky Mountain region during the 1930s demonstrates that westerners at the grass-roots level played key roles in the implementation of New Deal programs and disposition of funds. In many cases, including electrification, agencies empowered westerners rather than ran roughshod over them."[71] REA federal administrators offered financial and technical assistance and regulated the flow and conditions of that assistance, but local recipients of REA loans self-organized their co-ops, proposed projects that suited their needs, implemented them, and locally administered them over the long term. While REA staff exercised a good deal of oversight during the loan-approval and project-construction phases, they did so to ensure program success, not to wield power for the sake of it. After construction was complete and the power lines energized, REA barely maintained a presence unless asked for advice or needed to resolve a problem. As Cannon summarizes, "Westerners have not been the passive objects of state power."[72]

There are two basic flaws in the common criticisms of the New Deal. First, the critics imply that federal programs and federal authority itself are somehow external, not "local"; that the federal government is distinct from and antithetical to "the people." This is little more than political rhetoric. As shown in this chapter and revealed by Cannon and many other scholars, New Deal programs were conceived and launched by people with state and local affiliations and commitments. Local advocates put Grand Coulee and Bonneville dams on the federal agenda. Senators and representatives in the U.S. Congress advocated for New Deal programs to benefit their states and congressional districts as much as or more than for national interests. As a recent Speaker of the House, Tip O'Neill, famously said, "All politics is local." Roosevelt's appointees to head agencies all came from somewhere—Cooke from Pennsylvania, Lilienthal from Wisconsin—and were shaped by their experiences in those states. Most New Deal agencies had decentralized administrative structures, which is the literal definition of a "federal" system of government. There were regional, state, and even local offices of most federal agencies, staffed primarily by local people.

Not only is the U.S. government equally local and national in its loyalties and sympathies, but it must be understood fundamentally as an instrument of the people, especially if one understands the directors of powerful corporations as people. While some persons and interests obviously get more and better representation than others—political influence is certainly not equitable—any argument that the democratically elected U.S. government is separate from and hostile to the people is simply polemical. New Dealers sought to modestly reallocate costs, benefits, and wealth in American society. Some interests did not like the policies and programs, while others applauded them. But those policies and programs themselves derived from people seeking change at all levels—local, state, regional, and national. In that sense, the New Deal programs, as exemplified by the REA, were as much a bottom-up as a top-down effort.

The second key flaw in the charges of New Deal critics is their representation of the federal government as monolithic, as a kind of unitary force from Washington, D.C., intent upon aggrandizing power to pursue a specific agenda. This, too, cannot be supported by the facts. As shown in this chapter and legions of New Deal political histories, the policies coming out of Washington, D.C., in the 1930s, indeed at all times in the nation's history, were bitterly contested, compromised, inconsistent, and constantly changing—reflecting the diversity of interests and opinions represented by political leaders, who themselves came and went with political alliances reshuffling every election cycle. The U.S. government is far too fragmented and incoherent to wield anything like hegemonic authority or intent. While critics of the New Deal on the right claimed the government was tyrannically oppressing liberty, critics on the left claimed it wasn't doing nearly enough to liberate the people from capitalist oppression. In fact, the messy suite of government policies in the 1930s concertedly sought benefits for wealthy capitalists as well as for workers, small businesses, farmers, and the poor. Whether New Deal programs and policies accommodated the demands of the people or oppressed the people depends entirely on which people and contexts one considers.

The same debate enlivened British Columbia's politics during the 1930s. There, Pattullo's Liberal government after 1933 did what it could to provide a better social safety net during the economic crisis, to stimulate industry and job creation, and to provide infrastructure and business regulations to improve quality of life and economic performance. In response, the Conservative Party railed against what they called the Liberals' socialistic tendencies, while the socialist CCF Party railed against the Liberals' support for capitalism and its failure to ameliorate poverty and unemployment. As B.C. citizens muddled through the Depression, their government adopted an inconsistent collection of compromised laws and programs reflecting the fragmented and conflicted nature of the people and interests of the province.

Because government is heterogeneous, its policies inconsistent, and its resources limited, the outcomes of government programs likewise go in multiple, unexpected, and sometimes contradictory directions. While New Dealers succeeded in getting much of the countryside electrified, thereby improving quality of life and economic production for formerly off-the-grid farmers, it did not alter certain trajectories in American agriculture that troubled the New Deal planners and regionalists. The U.S. population continued to trend rapidly urban, family farms continued to decline while large corporate farms increased, rural communities languished, and, as David Nye notes, overproduction, low crop prices, and soil exhaustion remained serious challenges. Technology and government initiatives indeed had the power to transform lives, but not always as planned and not always for the better.

CHAPTER 10

Integrating the U.S. Northwest Power System, 1937–1941

What cotton is to the South, what wheat is to the Great Plains, what automobiles are to Detroit, what steel is to Pennsylvania, power is to the Columbia basin.

—Richard Neuberger, 1938[1]

In its ability to solve the social and economic problems of the area rests the enduring value of the vast hydroelectric servant so long dormant in the waters of the Columbia.

—Bonneville Power Administration, 1938[2]

While New Deal programs brought significant change to the U.S. Northwest during Roosevelt's first term, the subsequent five years—1937 through 1941—literally transformed the region. Publicly owned power systems experienced a crescendo of grassroots and government support, especially in Washington state, aided by a windfall of cheap federal power from the Columbia River dams distributed by the newly created federal Bonneville Power Administration. Economic activity in the United States and Canada accelerated late in the decade as conflicts in Europe and Asia presaged the start of a second world war. The private power sector and finance capital recovered from the post-crash doldrums and regained confidence by the late 1930s. With the upswing in production and profits from war preparations and with the public focused increasingly on patriotic rhetoric, the commercial power industry vigorously and often successfully asserted itself against the rising tide of public power enthusiasm in the U.S. Northwest. A fevered pitch of hostile rhetoric and political struggle marked the years just before World War II. Foreshadowing important changes to come, however, both the public and private utilities became slowly integrated into a regional transmission system sharing an increasing dependence on cheap federal electricity supplied by the Bonneville Power Administration (BPA) from the federal dams on the Columbia River. This new federal agency quickly grew to become the dominant force in the electric power system of the U.S. Northwest. In essence, the federal government's involvement in power production and distribution in the region stabilized both nonprofit and commercial utilities and launched a decades-long trend toward institutional integration and mutual acceptance and accommodation. Prior to World War II, the two competing utility types—for-profit versus nonprofit—captured their own service territories and built separate and competing systems. They rarely cooperated, saw the other's advance as their own loss, and did not interconnect or share power. They only cooperated with their own kind while working against the other's interests. Once BPA began selling large blocks of wholesale power to both public and private utilities on its own expanding regional transmission grid, the two sides would eventually find themselves integrated into a single interstate power system, which remains to the present day.

Importantly, during this decade a new institutional and technological infrastructure developed in the Northwest that substantially universalized and democratized electrical service. This occurred due to persistent political effort and popular pressure in support of such a transformation, and its accomplishment required significant government involvement and federal investment. This institutional transformation met committed resistance from those who favored weak government and unfettered capitalism, but the massive failure of the private sector during the Great Depression had weakened popular support for this position and strengthened support for government interventions to stimulate and restructure economic institutions.

Fewer institutional and power system changes occurred in British Columbia in these years, although a major transformation was just around the corner, in the postwar era. British Columbia's economy boomed as the war heated up, but public power did not advance to the same degree as south of the border because neither the Canadian government nor the B.C. government invested significantly in developing rivers or hydropower facilities at this time. That task was left primarily to the privately owned companies and the market.[3] A semblance of an integrated system had already developed in British Columbia, however, because the great majority of the power supply to the municipal utilities came from the commercial utility sector. Thus, the two were much more interdependent and much less conflicted in British Columbia than in the United States. This situation was similar to the relationship that developed in the United States between rural electric co-ops and the commercial utilities. Outside the Pacific Northwest and the Tennessee Valley, the rural co-ops mostly got their power from commercial utilities. The former depended on the latter, and the latter saw the co-ops as a small but valuable wholesale customer.

As long as they did not compete for the same customers, the commercial utilities in most regions of the United States agreed to sell wholesale power to the co-ops.[4] Likewise, the commercial utilities that dominated the power system in British Columbia did not feel threatened by the little municipal utilities in British Columbia's small towns and were willing to sell them bulk-rate power when profitable.

Changes in the Northwest electric power system in the 1930s wrought profound changes in the physical, economic, and cultural landscape as well. In the nineteenth and early twentieth centuries, every region of the United States and Canada was known by its primary economic drivers. For generations, the Pacific Northwest was represented in promotional literature as a land of giant trees, leaping salmon, bucolic farms, and mineral wealth—a cultural landscape dominated by images of lumber mills, fish canneries, and apple orchards framed by emerald mountains and clear-running rivers. In the 1930s, hydroelectric power captured Northwest image-making, becoming the most prominent new feature of regional identity formation, especially among those envisioning the Northwest's future. As Oregon journalist Richard Neuberger wrote in 1938, the new gospel of hydroelectric-driven prosperity echoed in every corner of the region:

> It is evangelism of a sort, with facile orators sketching lavish word-portrayals of a land of prosperity, pay rolls, commerce and industry. . . . At weekly luncheons, Rotary club members in Seattle and Portland and a hundred lesser communities listen in a roseate glow of enthusiasm, as chamber of commerce spokesmen describe the glorious era when every acre of ground within sound of the Columbia's booming hymn of power will be covered by a factory.[5]

In 1939, the Bonneville Power Administration produced a public relations film titled *HYDRO! The Story of Columbia River Power*, touting in purple prose the regional benefits to come from federal Columbia River power. Two years later, as Grand Coulee neared completion, the agency redoubled its propaganda efforts by hiring Woody Guthrie to compose his iconic verses about Uncle Sam's Pacific Northwest dams.[6]

After 1937, hydropower development increasingly transformed the Northwest's largest rivers, colliding destructively with another key economic driver and cultural icon of the region: the salmon fisheries. While incrementally vanquishing the wild salmon runs in the U.S. Northwest, each new hydropower dam boosted other economic drivers, opening new agricultural regions in the Columbia Basin, enhancing urban and industrial development, creating a whole new electrolytic metals- and chemicals-processing industry, and rearranging geographical alliances. Regret over what was lost remained muted as the new hydroelectric-industrial order took shape in the U.S. Northwest, banishing the Depression once and for all.

BONNEVILLE POWER ADMINISTRATION—MANDATE AND EARLY ACCOMPLISHMENTS

With Bonneville and Grand Coulee dams authorized and under construction in 1933, President Roosevelt and members of Congress scrambled to come up with a plan for marketing the extraordinary amount of power they would produce. Bonneville power would be available by 1937 and Grand Coulee power by 1942. Roosevelt and progressives in Congress like Republican senator George Norris of Nebraska wanted a Columbia Valley Authority modeled after the Tennessee Valley Authority. But conservative opposition to any more TVAs proved insurmountable. The legislative wrangling over a compromise solution for marketing Bonneville power began in earnest in 1935 and didn't end until August 1937, when Bonneville Dam was virtually finished and ready to generate power. The Bonneville Project Act itself was a last-ditch compromise that satisfied no one. The crux of the debate centered on four simple issues that had profound policy and outcome implications: who would market the power from Bonneville (and later from Grand Coulee), how would it get to customers, how would its price be determined, and who would be able to buy it?[7]

Regarding the first question, there were three options for who would generate and market the power. The Army Corps of Engineers (CoE) was building Bonneville Dam. As an agency dedicated to improving navigation and flood control, the CoE by the 1930s had also become a major producer and marketer of hydropower from the dams it built and managed. It naturally wanted to retain its control over the power from Bonneville. Allied with the CoE was the Portland Chamber of Commerce, investor-owned Portland General Electric (PGE), the governor of Oregon, and other movers and shakers in the greater Portland region. Less than forty miles from the Bonneville site, Portland stood to gain a huge advantage over other urban rivals in the Northwest if it could capture the bulk of the power from Bonneville. Over the years, the Portland elite had developed a good working relationship with the CoE and felt their interests would be served best by the CoE's retaining control of Bonneville power. This same question arose with Grand Coulee Dam, which was being built by the Bureau of Reclamation (BoR). Would the BoR or some other agency market Grand Coulee power after the dam was finished? The Federal

Power Commission (FPC) represented the second option. Congress had just expanded the FPC's authority in 1935 (see last chapter), and the agency was deeply involved in breaking up utility holding companies, promoting regional power coordination, and regulating interstate power sales. However, the FPC did not own or manage hydropower facilities or directly market power, so this option would require additional authorities and expertise. The third option, favored by public-power enthusiasts and New Dealers, was to create a new federal agency like the TVA to distribute Bonneville power regionally rather than locally, on terms favorable to public-power systems and with a mandate to promote rural development and the widest possible use of electricity at the lowest possible price. In other words, an agency with a social mission focused on spreading prosperity where it was most needed rather than concentrating wealth where it already existed.[8]

The second issue of contention closely related to the above was who would actually transmit the power to willing customers and markets. In most cases the CoE and BoR put their power up for sale at the dam (what is technically referred to as the "bus bar"), although recently the BoR had been building more of its own transmission lines to reach markets. Mostly, the CoE and the BoR sold wholesale power to nearby utilities, which then transmitted the power to their customers. This gave the federal agency little or no control over distribution or consumer rates. It also permitted just one or a few utilities to monopolize the power from the federal dam. Such a policy for Bonneville would again advantage PGE since it could easily and quickly build a transmission line to the dam. Indeed, PGE hoped to capture most if not all the power Bonneville would generate. Conversely, this policy disadvantaged rural co-ops, small PUDs, and distant municipal utilities that lacked the capital to build long-distance transmission lines to Bonneville. They would be cut out from direct purchase of the low-cost power. It would also disadvantage other investor-owned utilities whose markets were significantly more distant from Bonneville than Portland. Bonneville was Portland's plum that city boosters hoped to keep for themselves. A large array of interests in Puget Sound and in eastern Washington and Oregon, along with public-power advocates and regionalists seeking distributed development, all advocated for the creation of a power marketing agency that would build high-voltage transmission lines to bring Bonneville power to more distant parts of the Northwest. This would cost more up front and complicate the task of marketing Bonneville power, but it would also spread the benefits to a wider population and better facilitate regional development goals.[9]

The price of Bonneville power was the third major question under debate, and its resolution would profoundly shape the character of regional development. Part of the reason Portland interests and the CoE got along so well in the 1930s is that the CoE felt that if Bonneville power was going to be transmitted to far-flung markets by a federal agency, then the rate schedule should reflect transmission costs, with lower rates for markets near the dam and higher rates for markets farther from the dam. Their reasons were economically sound. Not only did long-distance transmission lines cost a lot to build, but the farther electricity is transmitted from its source, the more energy is lost through the transmission lines. From a strictly economic calculus, more distant markets should pay higher rates for power. This obviously would benefit Portland, because it was much closer to the dam than any other major Northwest city. It also would concentrate manufacturing and population in areas reasonably close to the dam, where the power would be cheapest. Others in the region and their allies in Congress wanted a single, uniform rate for Bonneville power regardless of distance. While not an economically efficient option, that position promoted regional equity. A great deal of government water and power policy divides over this tension between efficiency and equity.[10] Just like the issue of transmission, resolution of the rate controversy would determine whether the benefits of Bonneville power were concentrated near the dam or spread throughout the region.[11]

Finally, Congress wrangled over who would be able to buy Bonneville power and under what conditions. Strong partisans of public power wanted all of Bonneville's electricity reserved for publicly owned utilities. The commercial utilities just as strongly fought for equal access to Bonneville power. The compromise position, not surprisingly, was access for both public and commercial utilities but with an expressed preference for publicly owned systems—the same "preference" policy that had been in place for power sold from federal dams since the first decade of the twentieth century.[12]

The contending interests and their congressional allies introduced several bills in 1935 and 1936; all failed to pass. More attempts at compromise bills in 1937 similarly led to impasse until Roosevelt personally intervened and lobbied for a settlement that reflected his own preferences and the recommendations he had received from his National Resources Committee and the Pacific Northwest Regional Planning Commission the year before.[13] The basic institutional framework these two affiliated organizations provided Roosevelt became the template for the Bonneville Project Act. As passed and signed by the president on August 20, 1937, the Bonneville Project Act embodied New Deal principles and favored public-power interests far more than the interests of the commercial power industry and the Portland boosters. But the president and his loyal New Dealers did not get all they wanted. The act temporarily created a new federal authority called the Bonneville project administrator, who would be responsible for marketing the power from the dam. This settled the matter as to who would control and sell Bonneville's power. The CoE retained control of the dam and operated the generators but did not control or dispose of the power.[14]

To bridge the impasse between those who opposed a new agency and those who wanted a powerful new CVA, the act made the Bonneville administrator a limited and temporary authority to deal with just the electricity question. New Dealers hoped to get a CVA approved in the next session of Congress, while their opponents hoped to eventually kill the temporary authority altogether. Neither happened. In 1940, the Bonneville administrator became a standalone federal agency called the Bonneville Power Administration (BPA).[15] It did not have the broad authority that the TVA did, but it became a powerful and permanent fixture in the region. The 1937 act authorized the administrator to construct high-voltage transmission lines "to interconnect the Bonneville project with other Federal projects . . . now or hereafter constructed." This was a vague but undeniable reference to Grand Coulee, which Roosevelt and the Pacific Northwest Regional Planning Commission wanted in the new authority's portfolio. In another compromise, the law left open the question of who would control and market Grand Coulee power. Within a couple of years this, too, was settled in favor of the BPA.[16]

The 1937 Bonneville Act directed the administrator to build and manage a backbone transmission grid connecting the power from federal dams to willing markets "in order to encourage the widest possible use of all electric energy that can be generated . . . and to prevent the monopolization thereof by limited groups." The BPA had authority to design a master transmission plan for the region, which it did in its first year, and purchase rights-of-way to get the lines built. The authority to determine when and where to build its transmission grid gave BPA unrivaled influence over the direction and character of future electrification in the region.[17]

On the question of whether rates should be uniform or vary by transmission distance, the 1937 act equivocated. It did not exclusively prescribe a uniform rate but made it clear that the administrator was to promote outcomes that a uniform rate was best suited to accomplish. Here is the relevant passage: "The said rate schedules may provide for uniform rates or rates uniform throughout prescribed transmission areas in order to extend the benefits of an integrated transmission system and encourage the equitable distribution of the electric energy developed at the Bonneville project."[18] So the rates could vary, but the goal remained equitable access and an integrated system. The first Bonneville administrator established two rates: one for sales within fifteen miles of the dam, for which the customer built the transmission line, and the other for everyone else anywhere else in the region. This was nearly the uniform rate New Dealers and public power interests wanted. To further facilitate equity and discourage profiteering, the act prohibited individuals and for-profit companies, except for regulated public utilities, from buying and reselling Bonneville power. For public utilities, the administrator was required to include stipulations in the power-sale contract to ensure that resale rates to customers reflected the low cost of Bonneville power and that retail rates were "reasonable and nondiscriminatory."[19]

The 1937 act strongly reinforced the "preference" policy. Section 4 of the act opens with the following unequivocal statement:

> In order to insure [sic] that the facilities for the generation of electric energy at the Bonneville project shall be operated for the benefit of the general public, and particularly of domestic and rural consumers, the administrator shall at all times, in disposing of electric energy generated at said project, give preference and priority to public bodies and cooperatives.[20]

The act also directed the administrator to ensure that at least 50 percent of Bonneville's power was reserved for publicly owned utilities and rural electric co-ops until at least 1942. It further directed that if more power was needed to meet demand from public-power systems, the administrator could reduce or rescind the power-sale contracts with for-profit companies with five years' notice. Power-sale contracts were for twenty years and renewable, with the stipulation that public bodies and co-ops had priority if demand exceeded supply. Because many public-power organizations (PUDs, rural co-ops, and municipals) were trying to get launched at this time to take advantage of Bonne-ville power, sections 4(c) and 4(d) of the law stated that applications for power contracts from public-owned utilities should not be denied just because voter approval and financing for the public utilities was not yet secured. The law directed the administrator to allow "a reasonable time" for these public bodies to get their ducks in a row.[21]

A final feature of the bill is worth noting, as it crucially determined the price utilities paid for Bonneville power. While everyone expected the electricity to be cheap, no one knew how cheap. The thought of unprecedentedly low wholesale power flooding the market with the government giving preference to public-power systems understandably alarmed commercial utility executives. That is why they fought so hard for access to Bonneville power. To mollify critics of federal spending on social programs, the law required that the costs of the Bonneville project be repaid in a reasonable period of time at a reasonable interest rate—but not the whole cost. The dam was originally proposed and substantially designed as a navigation improvement. One half of the dam was dedicated to transportation, with a huge navigation lock for ships and barges, the largest "single-lift" lock ever built at that time. The dam buried several rapids of the Columbia River and made 187 miles of the lower river navigable for "full-sized ocean freighters," according to the First Annual Report of the Bonneville administrator. Advocates of cheap power thought it was unfair for electric ratepayers to subsidize the navigation components of Bonneville Dam, so the 1937 act directed the FPC to determine how much of the cost of the dam was attributable to navigation and how much was attributable to power generation—a somewhat arbitrary calculation, as the two functions were significantly integrated in the dam's design. Nevertheless, the percentage of the cost of the dam the FPC allocated to hydropower was the investment the Bonneville administrator had to pay back to the U.S. Treasury through power-sale revenues. This substantially determined the cost of Bonneville power. The FPC made its financial decision in February 1938, allocating just over 40 percent of the initial cost of the dam to hydropower. (The FPC also determined that after the staged installation over several years of all the electrical generators, hydropower generation would be responsible for 57 percent of the ultimate cost of the dam.) The administrator elected a forty-year repayment period at 3.5 percent interest annually, which was 1 percent higher than the U.S. government was paying on its own debt at that time. This allowed the administrator to adopt a wholesale rate for Bonneville power of just $17.50 per kilowatt-year (a kilowatt of energy, 24 hours a day, 365 days a year)—by far the lowest wholesale electric rate in the United States. That translates to 1/5 cent per kilowatt-hour. At the time, most Northwest residential consumers were paying from 2 to 5 cents per kilowatt-hour—ten to twenty times the Bonneville wholesale rate.[22]

While the Bonneville Project Act of 1937 contained numerous compromises and disappointed most of the legislators who worked on it, there were nevertheless some clear winners and losers. It was a loss for the Portland interests who hoped to monopolize Bonneville's cheap power for themselves but only a minor setback for the investor-owned utilities, which got Bonneville power at the same rate as the publicly owned utilities while having to settle for second place on the preference scale. The CoE (and shortly the BoR) lost a significant asset when Congress gave control over the electricity from these dams to another agency. No doubt they hoped the temporary authority for the Bonneville administrator would expire and marketing authority revert to them, but this was not to be the case. The winners included electric power consumers of all types: "public bodies and co-ops" in particular; boosters of small towns across the region, who hoped cheap abundant federal power would stimulate business and improve quality of life; regionalists and advocates of decentralized economic development; and the architects of New Deal policy. In fact, though significantly compromised, the 1937 act substantially fulfilled many of Roosevelt's policies articulated in his "Portland Speech" of 1932 when he was on the campaign trail (see chapter 8). The act clearly served the "interests of the many" rather than the interests of the few; it asserted a large federal role in shaping regional electrical development; it kept the ownership of the Columbia River's power in public hands rather than privatizing it; and it advantaged publicly owned utilities while not directly harming or excluding the commercial utilities. As Roosevelt had proposed in 1932, Bonneville would serve as a "yardstick . . . to prevent extortion against the public and to encourage the wider use of that servant of the people—electric power."[23]

Passing a law is one thing; implementing it is quite another. Having authority to appoint the Bonneville administrator, the Roosevelt administration was in a position to set the future agency's course right from the start. Tellingly, Secretary of the Interior Harold Ickes appointed James Delmage Ross (popularly known as "JD") as the first Bonneville administrator. Ross had been the superintendent of Seattle City Light since 1911 and a hero of national stature for the public-power movement. An engineer by training, Ross gained a reputation as a visionary leader of one of the largest and most successful municipal utilities in the United States. No stranger to politics, JD Ross's status ballooned in 1931 after Seattle mayor Frank Edwards abortively attempted to fire him. Edwards was allied with Stone & Webster, the holding company that controlled Puget Sound Power & Light, and in March 1931 he decided to get rid of this thorn in the side of the commercial power industry. But Ross and City Light were so well liked that Seattle residents immediately launched a recall of the mayor. In just two months they held a special election and booted out Mayor Edwards. The replacement mayor, Robert Harlin, immediately reinstated Ross. Franklin Roosevelt heard about the hullabaloo all the way across the country while he was governor of New York, thanks in part to the fact that his son-in-law, John Boettiger, published the Seattle *Post-Intelligencer* newspaper. Roosevelt and Ross became friends, and the president eventually appointed him to the Securities Exchange Commission. But Ross was frustrated with that assignment and longed to focus his attention on the Northwest. His appointment as Bonneville administrator in the fall of 1937 surprised few people, though it alarmed many because it clearly signaled that the Roosevelt administration wanted a public-power advocate on the job.[24]

Ross was one of the optimists who felt Bonneville's power could and would quickly be absorbed by an expanding market. First, even though the economy remained sluggish in its recovery, consumption of electricity in the Northwest states had been steeply rising since 1933, especially in Washington. Ross documented that rise in his first annual report as administrator (see Figure 10.1). Second, the U.S. government was beginning to ramp up its defense preparations in response to hostilities growing in Europe and Asia. As with World War I, the Pacific Northwest would experience a great deal of war-related production. Ross saw this coming, but neither Ross nor anyone else in 1937 foresaw just how deeply involved the United States would become or how revolutionary the war's impact would be on the Pacific coast states. Third, as a public-power system manager, Ross knew that lower rates led to greater consumption (see chapter 8). In his first annual report, Ross lays out his argument that all the key sectors of the Northwest economy would be consuming much more power in the near future, as low-cost federal power came on the market. At one point he cited studies by the Oregon Public Utilities Commission, showing that from 1933 to 1936, as the price per kilowatt-hour steadily declined in Oregon, the average annual consumption per household increased at an even higher rate. Oregon utilities in those years earned less per kilowatt-hour yet raked in more total revenue. Their customer base grew by about 15 percent while household consumption rates increased by about 25 percent. This and similar examples convinced Ross that when cheap Bonneville power hit the market, electric rates would fall even further, and consumption would correspondingly grow rapidly. Though he wouldn't live to see it, his prediction was correct.[25]

Ross's twenty-page argument in his first annual report about likely increases in electric consumption in the Northwest supported two key objectives that he pursued in his first year as administrator, regarding Bonneville's wholesale electric rate and its construction of transmission lines. The Bonneville Project Act left it up to the administrator to set wholesale electric rates, and Ross wanted a very low universal "postage-stamp" rate. As noted above, the law required electric revenues to repay about 40 percent of the initial cost of the dam in a reasonable time frame. A fiscally conservative approach to the rate question would predict less aggressive consumption growth than Ross did and set higher electric rates to ensure BPA would meet target revenue goals. Ross instead argued that a lower rate would actually generate more revenue. His argument about rates and markets apparently carried the day, as the FPC

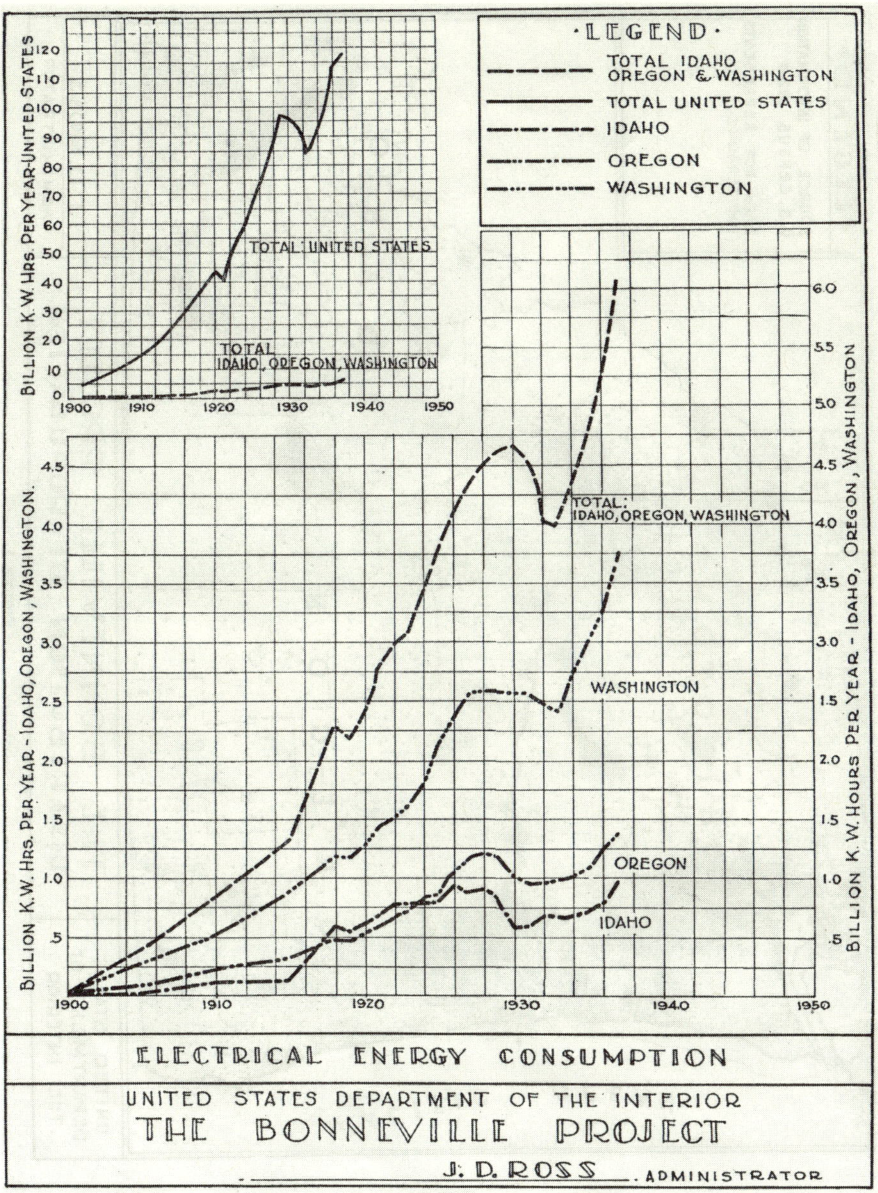

Figure 10.1. Electricity Consumption in Washington, Oregon, and Idaho, 1900–1937. Source: Bonneville Power Administration, *First Annual Report of the Bonneville Administrator* (Portland: BPA, 1939), 8.

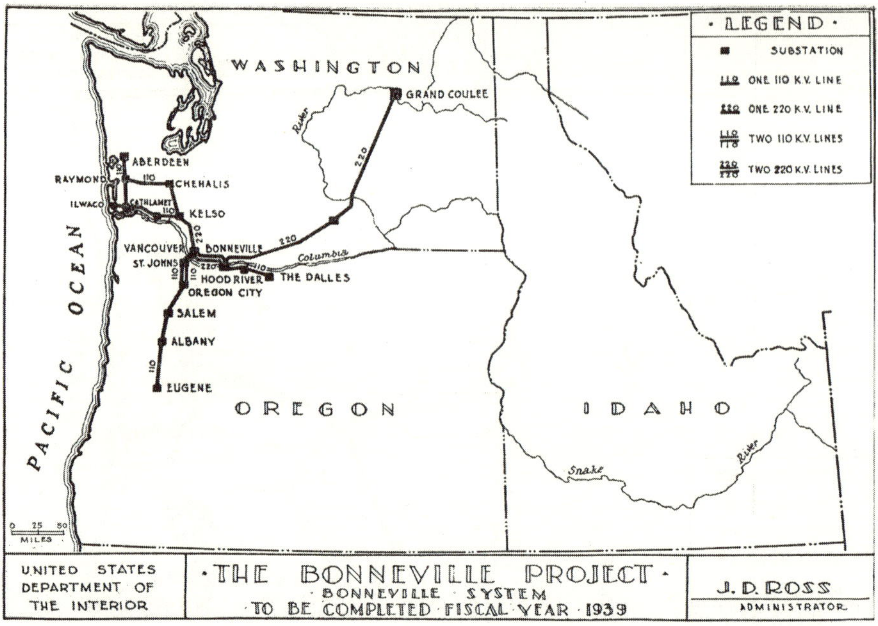

Figure 10.2. BPA Transmission Master Plan: Proposed Transmission Grid for 1939. Source: Bonneville Power Administration, *First Annual Report of the Bonneville Administrator* (Portland: BPA, 1938), 45.

approved his unprecedentedly low, "postage-stamp" wholesale rate of $17.50/ kilowatt-year.[26]

The transmission issue again pitted Ross against fiscal conservatives. Many of those opposed to federal construction of Bonneville and Grand Coulee dams had argued that there would be little need for the electricity from the two dams and that much of their power potential would be wasted for years if not decades (see chapter 9). Consequently, a conservative approach to planning transmission lines would start small and build incrementally as markets appeared. Ross did not want to go slow and wait for demand. He wanted to quickly build the infrastructure necessary to create new demand. Where others saw a satiated market, he saw a huge potential waiting to be served. An aggressive high-voltage transmission construction program to bring Bonneville power to the far corners of the region would unlock this potential, he believed, but without federal transmission that incipient market would lie dormant. Congress had provided no funds for building transmission lines in the agency's first-year budget and had been rather stingy in the next fiscal-year budget. So Ross and his staff put together a transmission grid master plan, based on the grid proposed by the Pacific Northwest Regional Planning Commission in

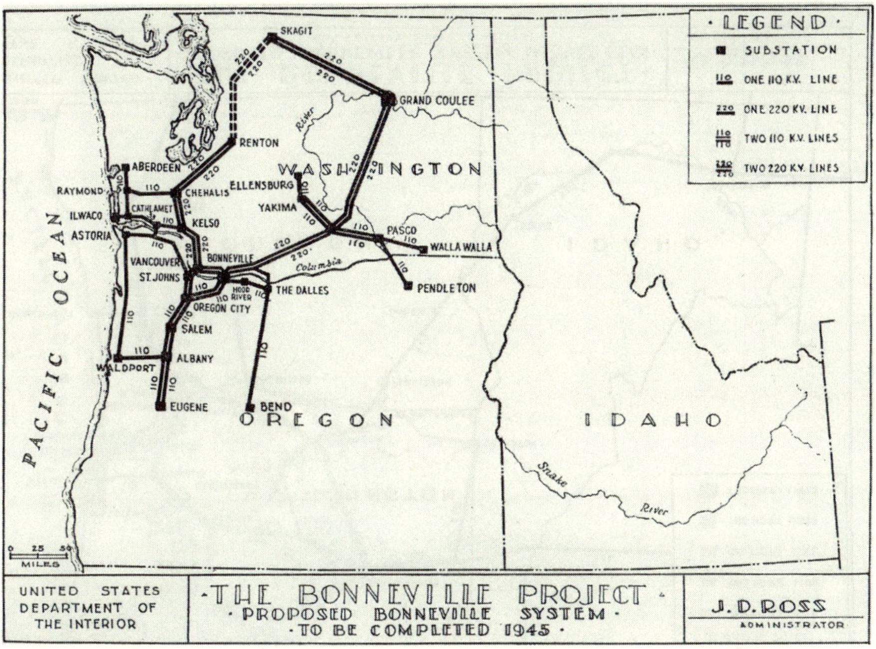

Figure 10.3. BPA Transmission Master Plan: Proposed Transmission Grid for 1945. Source: Bonneville Power Administration, *First Annual Report of the Bonneville Administrator* (Portland: BPA, 1938), 47.

1935. He published the master plan (see Figures 10.2–10.4) in his first annual report and made his strongest pitch for federal funds for an ambitious construction program.

Support from Roosevelt and several New Deal agencies got his construction program going. The Public Works Administration under Ickes's control provided additional funds, and the Works Progress Administration provided workers to clear the rights-of-way for the transmission lines. (Unemployment in the Pacific Northwest remained at about 20 percent in 1938, so construction jobs were desperately needed and appreciated.) Preparations began in the summer of 1938 for the intertie with Grand Coulee Dam, and in the spring of 1939, BPA started construction of the large transmission line connecting Bonneville to Vancouver, Washington, and Portland, Oregon. The day after BPA formally launched the Bonneville-Vancouver line, J. D. Ross died at the Mayo Clinic in Minnesota while undergoing abdominal surgery. This ended his short career as Bonneville administrator, though his policies lived on.[27]

After a short and troubled stint with interim administrator Frank A. Banks, Ickes appointed Dr. Paul Raver as Bonneville administrator in August 1939.[28]

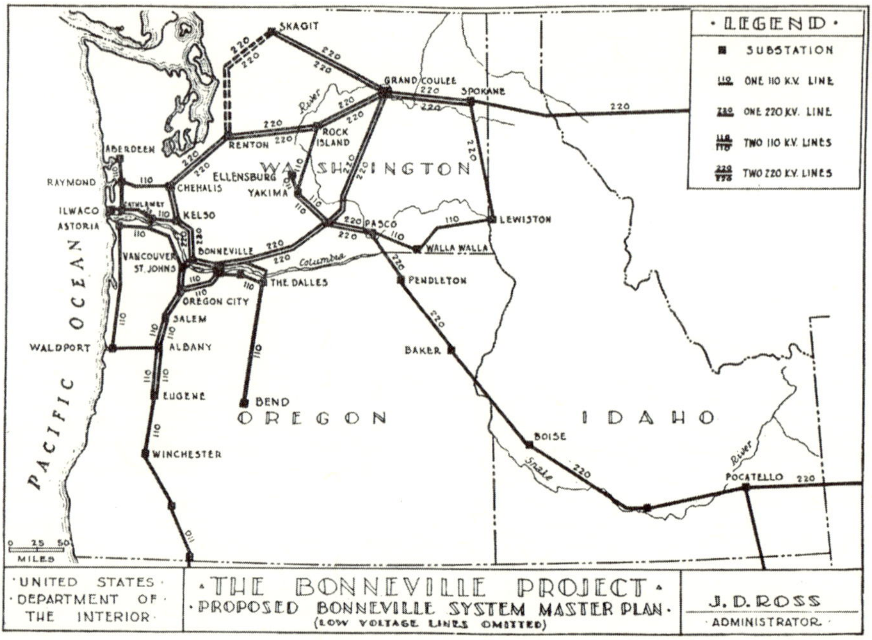

Figure 10.4. BPA Transmission Master Plan: Proposed Transmission Grid Full-Build-Out. Source: Bonneville Power Administration, *First Annual Report of the Bonneville Administrator* (Portland: BPA, 1938), 48.

Raver served until 1954. His background as a professor of public utilities at Northwestern University, a longtime member of the Illinois Commerce Commission, and a proponent of rural electric co-ops in Illinois prepared him well for the job. Shortly after Raver assumed his new position, he gave a speech at Reed College (where well-known public administration scholar Charles McKinley taught from 1918 to 1960) on BPA and public power. The topic was particularly timely, as Portland voters would decide that month whether to create a publicly owned utility to serve the city. Raver's predecessor, Ross, had raised the ire of the commercial electric industry and its political allies by aggressively promoting public power in the region. He helped PUDs and rural co-ops secure financing, supported public-power voter initiatives, and personally joined the propaganda battles, penning pro–public power newspaper editorials and giving numerous speeches. In advance of the 1940 election, the commercial electric utility sector put up a vigorous defense of its record serving Portland. Both sides in the public-/private-power debate wanted to know whether Raver would follow in Ross's footsteps or take a more moderate stance. When he gave his speech at Reed College, Raver knew his words would be intensely scrutinized. He started out implicitly distancing himself from Ross, saying,

> This month the people of Portland will vote on whether they wish to set up a people's utility district to distribute electricity under public ownership instead of through the two privately owned systems operating at present. Today I am not going to advise how Portland should vote. Whether it should have public ownership . . . is for the people to decide. I do not believe that I am free to enter this campaign.

Immediately after withdrawing himself from the political contest, however, Raver acknowledged that the Bonneville Project Act imposed on him three main duties: to encourage the widest possible use of electricity, to prevent limited groups from monopolizing Bonneville power, and to give preference to publicly owned utilities. The implications were not lost on the audience. Raver would help to build a very different power system for the region than that created by and preferred by the commercial utilities. Over the next few years, Raver followed nearly all the same policies as did Ross, just in a less overtly partisan manner.[29]

Things moved astonishingly quickly from 1939 to 1941. The Corps of Engineers installed the first two generators at Bonneville by 1938, with two more scheduled to be in service by the end of 1940. The four represented about half of Bonneville's capacity of 518,400 kilowatts. The much bigger Grand Coulee Dam was slated for many more generators with much higher capacity: eighteen generators of 108,000 kilowatts each were planned at the time. The BoR installed the first two in 1941 and a third at the start of 1942. To get this power to market, BPA completed the 37-mile, 230,000-volt transmission line to Vancouver by the end of 1939, with a huge electrical substation at Ampere, Washington, to serve the first set of large aluminum-manufacturing facilities setting up shop in and near Vancouver. BPA completed the 234-mile, 230,000-volt Bonneville–Grand Coulee line in July 1940, more than two years ahead of schedule. Anticipating this intertie in 1940, at Raver's request, Roosevelt had issued Executive Order 8526, giving BPA the authority to market Grand Coulee power, leaving the BoR to manage only the dam and its generators—just as the CoE did at Bonneville Dam. The Bonneville–Grand Coulee transmission line at the time was the longest in the region, and it used six of the world's largest transformers. It connected the Spokane and Portland markets and allowed BPA to serve dozens of smaller communities and utilities in between. By 1941 BPA had also built a 230,000-volt intertie to the Tacoma-Seattle area, connecting with the region's two largest and most successful municipal utilities, as well as several smaller municipals, PUDs, and co-ops in western Washington.

Also in 1941, transmission lines were completed or under construction from Bonneville east to The Dalles, from Grand Coulee east to Spokane, and from Portland south to Eugene. These interties made possible a region-wide power-sale program with everyone paying the same rate for the federal power—one of the key objectives of the Bonneville Project Act.[30]

Proving the prescience of J. D. Ross, the demand for Bonneville power expanded as rapidly as did BPA's transmission infrastructure. In fact, by the end of 1940, before Grand Coulee power was even available, BPA had signed contracts or pending negotiations totaling a commitment of 750,000 kilowatts of generation. Bonneville's four installed generators at that time could provide only a third of that amount of power. Clearly Grand Coulee needed to be brought online quickly and integrated into BPA's transmission grid to meet demand. Both public and commercial utilities as well as aluminum manufacturers signed contracts in these first two years. Not surprisingly, BPA asked for more funding to accelerate generator installations. No longer just a prediction, the rapidly expanding market for Bonneville power meant rapidly expanding revenues for the U.S. Treasury. On June 30, 1939, before the Vancouver transmission line was completed, BPA had only two customers, producing a total income for the preceding fiscal year of $50,007. Exactly one year later, BPA had signed nineteen multi-year power-sale contracts worth $16.5 million in revenue over the life of the contracts. Its income for the fiscal year ending June 1941 was $1.9 million, and it expected more than three times that amount for the next fiscal year. To put this in perspective, the entire cost of construction for Bonneville Dam was $53 million. This made it much easier to convince President Roosevelt and Congress to accelerate funding for generation and transmission.[31]

This exploding market for federal power in the Northwest mostly silenced the skeptics who had called the two dams boondoggles. Commercial utilities' desire for a piece of the pie moderated their traditional hostility toward government intervention. Raver, more than Ross, understood the institutional and public relations value of bringing investor-owned utilities into the fold; he signed contracts with PGE and other large privately owned utilities and aggressively courted industrial customers, especially the Aluminum Company of America (ALCOA). The growing military conflicts in Europe and Asia made the crucial difference, though. By 1940 the United States was overtly mobilizing for war, ramping up military production to sell to allies and to bolster U.S. military capacity. This profoundly influenced BPA marketing. BPA's 1940 annual report crows about increasing its power-sale contracts to public agencies to 20,300 kilowatts. But this amount was dwarfed by power sales to ALCOA, totaling 162,500 kilowatts, at the same time. Raver's report notes that the boom in the industrial market resulted from "tremendous stimulation from the nation's preparedness program," and that BPA's three large contracts to ALCOA responded to "the national defense need for aluminum metal in the aircraft industry."[32] The Pacific Northwest and the world were about to change, and BPA would be a key change agent. In a section titled "New Power Plays Defense Role," Raver's 1940 report anticipates that Bonneville and Grand Coulee dams "would immediately fill a vital need in providing power for the defense industries of the nation. Nowhere in the United States was there so great an immediate or potential source of the dependable electric power that is so essential to many of the industries engaged in the manufacture of defense materials."[33] This impressive early demand for industrial power in 1939–1940 was quickly exceeded by the almost unimaginably large industrial power purchases after the United States entered the war in December 1941 (see next chapter).

As intended by the Bonneville Project Act and required by BPA's power sales contracts, the utilities that purchased Bonneville power quickly and dramatically reduced their retail electric rates. After BPA's first customer, the town of Cascade Locks, received Bonneville power, its primary residential rates decreased by 29 percent. Another early BPA customer, the Oregon town of Forest Grove, reduced its rates for 100 kwh by 31 percent. For higher monthly loads of 750 kwh or more, Forest Grove's rates dropped 60 percent. Both towns had municipally owned utilities. For its 1940 annual report, BPA compiled data from the Public Utility Commissions of Washington and Oregon, documenting commercial utility rate reductions in the two states. In 1939 and again in 1940, electric rate reductions in each state exceeded the total of reductions seen in the four previous years combined. Some of these rate reductions were due to the utilities contracting for Bonneville power, and some resulted from the commercial utilities' attempt to mollify customers and stymie efforts to create public power systems, which were experiencing unprecedented growth in the late 1930s with support from BPA.[34]

THE ASCENSION OF PUBLIC POWER

BPA historian Gene Tollefson aptly calls the 1930s the "high-water mark" of the public power movement in the Northwest. As discussed in the previous chapter, both Washington and Oregon passed legislation in 1930 authorizing the creation of public utility districts (people's utility districts in Oregon), and Roosevelt established the Rural Electrification Administration in 1935 to promote rural co-ops. But neither type of system gathered much momentum until the completion of Bonneville Dam and the creation of the BPA. Only a few PUDs formed in Washington and Oregon before 1936, and none of them had advanced very far in their development. But with Bonneville Dam nearing completion and J. D. Ross at the helm of the new BPA, momentum quickly built. The 1936 ballot in Washington state had a flood of PUD proposals. Voters approved fifteen new PUDs during that election—two out of every three on the ballot. In the 1938 election, voters approved seven more PUDs in Washington. In contrast, the Idaho legislature refused to provide enabling legislation for PUDs in the 1930s. The pro-public-power governor of Idaho, Barzilla Clark, sent a PUD-enabling law to the legislature for approval in 1937, but the Idaho Senate narrowly defeated it. Consequently, municipal-owned utilities and REA co-ops were Idaho's only public-power options that decade. Without any new, grandiose federal dams in that state in the 1930s, public-power systems had little assistance for expansion.[35]

Oregon had three PUDs in 1936 and three more by 1938, a small number in comparison to Washington, but not for lack of effort. PUD advocates in Oregon had a more difficult approval process mandated in the 1930 enabling legislation. Inspired by Washington's success and increasingly optimistic after Bonneville Dam came online, Oregon public-power supporters put a proposal on the ballot in 1938 for a seven-county PUD. BPA staff had told them it would simplify power sales and financing if they created a multi-county PUD. But to their great disappointment, the measure failed by a wide margin, losing in five of the seven affected counties. Commercial utilities put up a good fight because they stood to lose much of their business and assets. PUDs had the authority to purchase existing utilities within their districts and absorb them into their nonprofit system. But voters, too, had several reasons to be skeptical in Oregon. For one, the Oregon enabling legislation did not provide authority for local city and county governments to tax the revenues or property of the PUDs or provide for the PUDs to make payments in lieu of taxes. That meant that if a PUD acquired a privately owned utility, the local governments would no longer receive tax revenue from the utility. Many of these rural counties were chronically short of revenue and hard-pressed to provide expected social services (roads, schools, police, fire, water, sewer, etc.). The commercial utilities hammered home the message that local services would suffer if PUDs took over commercial utilities. The argument worked effectively. Washington state did provide taxing authority or payments in lieu of taxes, so that argument did not work as well there. After the 1938 defeat at the Oregon polls, J. D. Ross tried to help the situation by stipulating that PUDs buying Bonneville power pay 2

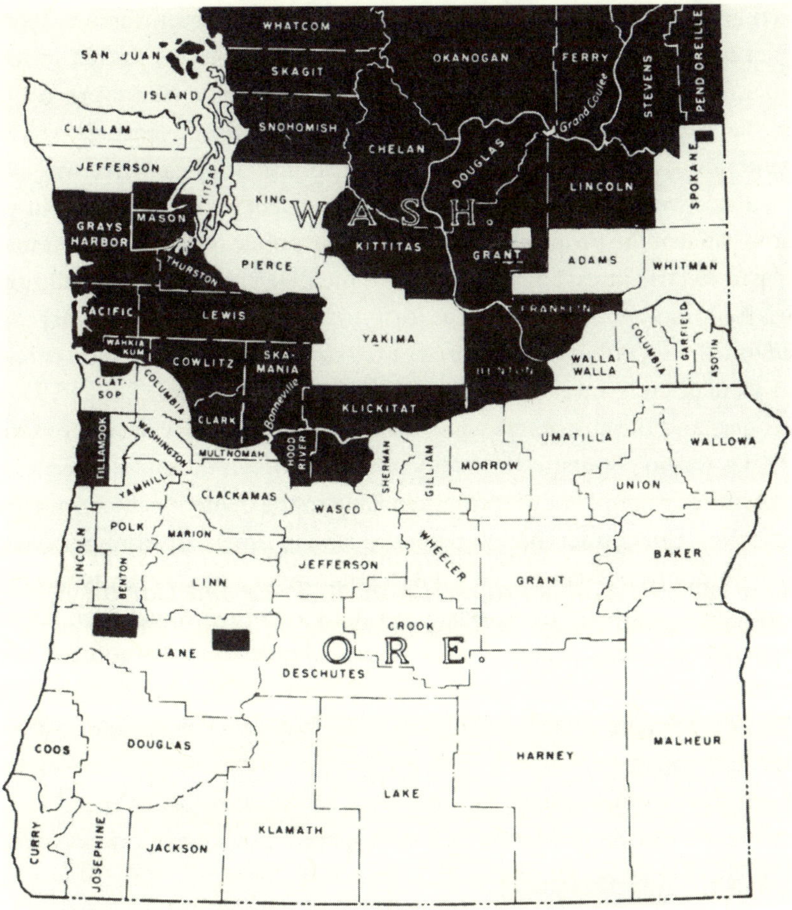

Figure 10.5. Public Utility Districts in Washington and Oregon, 1938. Source: Gene Tollefson, *BPA and the Struggle for Power at Cost* (Portland: BPA, 1987), 134.

percent of revenue in lieu of taxes to local governments. But Oregon remained a tough sell, and the high tide for PUDs was ebbing by 1939. Only six PUDs had formed in Oregon during the decade, and only four of them became active. Still, a total of thirty-four PUDs had been authorized in the two states by 1939, an impressive accomplishment in one decade.[36] Figure 10.5 shows in black the Washington and Oregon counties that approved public utility districts in the 1930s. Keep in mind that the PUDs did not own or control all the electrical facilities and services in those counties. They simply had authority to build and administer a nonprofit electric utility there.

The rapid establishment of PUDs in 1936 and 1938 not surprisingly stimulated a vigorous response from the commercial sector. While the public-owned and investor-owned utilities eventually came to a mutual acceptance in the post-WWII era, in the 1930s the two sides remained locked in mortal combat. Before Bonneville power and BPA transmission lines were available, most PUDs had few options for buying affordable power and few resources for generating their own. The best option from their perspective was to take over an existing for-profit utility—all the infrastructure was already in place, and the value of the property confirmed by the public utility commissions that regulated electric rates. Unfortunately for the PUDs, most private utility companies had no interest in selling, at least not to a public power agency. When asked to sell, they either flatly refused or raised the price to make it unattractive. Years were often spent haggling over the sale price, frustrating the PUD commissioners and the customers who had voted for change. A second option for the PUDs was to condemn the utility property (which the enabling legislation permitted) and simply pay fair market value to the owners as determined by a court of law. But condemnation procedures were time-consuming, expensive, and highly controversial. Nevertheless, with Bonneville power available in 1938 and transmission lines under construction, PUD commissioners felt a sense of urgency and launched a large number of condemnation proceedings. The top target for acquisition was Puget Sound Power & Light. For the next three years, the contest between public and private power rose to a crescendo, coming to a head in the 1940 election.[37]

Both sides in the contest had mixed success in 1940 ballot measures. The commercial industry defeated an effort by public-power proponents to form a PUD in the crucial Portland area and acquire the for-profit utilities serving the city. This was the election that administrator Raver mentioned in his speech at Reed College. The measure met resounding defeat. Philip Funigiello identified three reasons for the defeat: (1) the public-power proponents themselves were divided over whether to support a PUD or create a municipally owned utility; (2) Portland General Electric made two strategic rate reductions just prior to the elections to undercut critics and mollify its customers; and (3) the commercial industry sponsored a barrage of anti-public-power publicity, including standard charges about lost tax revenue and socialism. The disappointing results of both the 1938 and 1940 elections in Oregon more or less permanently doused Oregonians' enthusiasm for PUD initiatives.[38]

In contrast, public-power advocates won a significant victory in Washington in 1940. To stymie the pro-PUD tidal wave, commercial utilities covertly sponsored ballot initiative 139, which was ostensibly led by a "citizens" organization called "Let the People Vote League." The measure would have made it more difficult for PUDs to purchase privately owned electrical distribution systems. It, too, went down to resounding defeat, thus bolstering public-power interests. Despite the defeat, though, public-power supporters complained to their congressional representatives that the commercial utilities were up to their old anti-public-power propaganda tricks. This resulted in a Federal Power Commission investigation reminiscent of the muckraking Federal Trade Commission investigations of the 1920s and early 1930s. The FPC used court orders to acquire utility company financial records and reported in 1941 that several of the large commercial utilities in Washington state had spent over a million dollars ($15.5 million in 2011 dollars[39]) during the previous five years "in political activities opposing publicly owned and operated utility districts and public ownership movements." The commercial utilities donated a good chunk of that million dollars to the "Washington Taxpayers Association," which then quietly passed it on to the Let the People Vote League, obscuring the source of the funds. Even worse, the FPC found that much of that money as well as utility company staff time spent on the anti-public-power campaigns had been "charged to ratepayers" as utility operating expenses. In 1941 FPC Commissioner John Scott angrily appended a letter to the FPC's report, excoriating the utilities for engaging in "economic warfare" against democratic processes:

> The record discloses that by subterfuge these companies sought to pollute the political process of free choice at public elections. Funds obtained from the consuming public have been lavishly expended to prevent the people from obtaining electric energy through publicly or cooperatively controlled organizations.[40]

Clearly, antagonism remained high between public and private power interests.

The formidable investigative powers of the federal government helped level the playing field in the contest between local nonprofit public-power utilities and the much larger nationally financed privately owned utilities. But exposing the lobbying expenditures of the commercial sector did not really address their substantive arguments. So the FPC helped the public-power movement rebut the specific charges of their opponents. A scholarly study in 1940 by Ernest Abrams, criticizing public power and defending private power, laid out the basic arguments that investor-owned utilities proffered in that decade. To blunt the advance of public power, private industry supporters asserted that state regulation of utilities had been effective. Indeed, utility executives ironically and prominently defended state regulation during these battles. Abrams made the case by showing how far the industry had come since state regulation first appeared in 1907, focusing on three key statistics: the spread of electrification, the rise in consumption rates, and the decline in prices. First, he noted that only 1 in 38 households had electricity in 1907 while 2 out of every 3 had electrical service by 1939. Second, the average household consumed 25 kilowatt-hours per month in 1907, while the average in 1939 was triple that. Third, the average cost of residential service per kilowatt-hour fell from 10.5 cents to 4 cents in those years. While impressive, these statistics reflect technological progress as much as or more than they reveal anything specific about the effectiveness of state regulation. Moreover, the data conflate private- and public-owned utilities, so a comparison cannot be made between the two. Certainly, state regulation improved industry performance overall, as I argue in chapter 4, but judgments about whether state regulation was effective varied widely over time and place. It was, in large part, a matter of opinion.[41]

Electric rates provided the main quantitative method for judging the effectiveness of state regulation, although they are a blunt instrument for judging since rates are affected by so many factors. Nevertheless, rates were what mattered most to consumers, state utility commissions, and public-power advocates. Commercial utilities, however, were disadvantaged in this arena as publicly owned utilities paid lower interest rates on debt and did not have to provide profits to investors. Many studies over decades repeatedly confirmed that publicly owned utilities charged lower rates than their rivals. The FPC reported on rate differentials in its Nineteenth Annual Report for fiscal year 1939: "The average typical bills charged in 1937 by privately owned utilities were, with but one exception, higher for all quantities of electricity in all communities of 2,500 or more population." Specifically, the FPC noted that rates charged by the commercial utilities "were from 8 to 44 percent higher for commercial light, 11 to 62 percent higher for commercial power and 2 to 37 percent higher for industrial service than the rates charged by publicly owned utilities for similar services."[42]

Saddled with this kind of unequal record, the commercial utilities constructed an argument to divert public attention from the obvious cause for the higher rates—profits for investors—promulgating instead an "unequal tax burden" argument as the source of the rate difference. Quoting Abrams:

> According to Federal census data for 1937, the privately owned electric utilities of the country contributed slightly more than fourteen cents of each dollar of operating revenues received in that year to the support of Federal, State and local governments, while the combined tax bill of our publicly owned electric utilities was equivalent to but one and a half cents of each revenue dollar.

Abrams does not provide a citation for these data but goes on to argue that this differential tax burden "forced" the commercial utilities to charge higher rates and that the consumers of public power, moreover, did not in fact save any money because they had to "pay a part of the true cost of their electricity in the form of increased taxes."[43] Once again, he offered no citation to any studies or statistics quantifying the added tax burden for communities served by public utilities. But the argument's superficial logic made it useful anyway for public relations battles. This particular assertion had the added power of portraying the playing field as "unfair": the beleaguered electric industry unfairly disadvantaged by government agencies bent on weakening or destroying private enterprise. It played well in the court of conservative public opinion. But the argument was misleading and deeply flawed. The tax figure for publicly owned utilities was very low because, as nonprofit units of local or state government, these utilities were generally not taxable by local or state government, although Washington did permit taxing of municipal utilities. This is the case for all nonprofits (churches, schools, etc.) and a wide array of local, state, and federal agencies. Because of this situation, federal, state, and local laws often require public agencies to make "payments in lieu of taxes" (PILT) to local jurisdictions to compensate for lost tax revenue. The federal government has been doing this since early in the twentieth century.[44] Many publicly owned electric utilities similarly made cash payments to local governments based on gross or net revenues, but the payments were not called taxes. To rebut arguments like Abrams's, the FPC in 1939 calculated both taxes and tax equivalents paid by commercial and public utilities and concluded that publicly owned utilities "paid 18 percent of their revenue in taxes and net cash contributions . . . as against 14.4 percent for privately owned utilities." In addition, the FPC compared the amount of "free service" provided by privately owned versus publicly owned utilities. All utilities, including utilities in Canada, reported the amount of uncompensated service they provided. This involved services such as public street lighting. Not surprisingly, public agencies provided vastly more free service than their for-profit counterparts. The FPC calculated that the value of these services equaled another 8.5 percent of gross revenues for public-power systems but only 0.0024 percent of gross revenue for commercial utilities. Of course, one should not expect a for-profit company to give away its products and services, except for promotional purposes. Nevertheless, the tax burden argument comes up hollow when everything is taken into account. Publicly owned utilities paid out even more in taxes and tax equivalents than investor-owned utilities and provided a large amount of free electrical service and charged lower rates. It is hard to see how taxpayers and electric consumers were harmed by public ownership.[45]

Public power advanced remarkably during the first few years of BPA's ex-

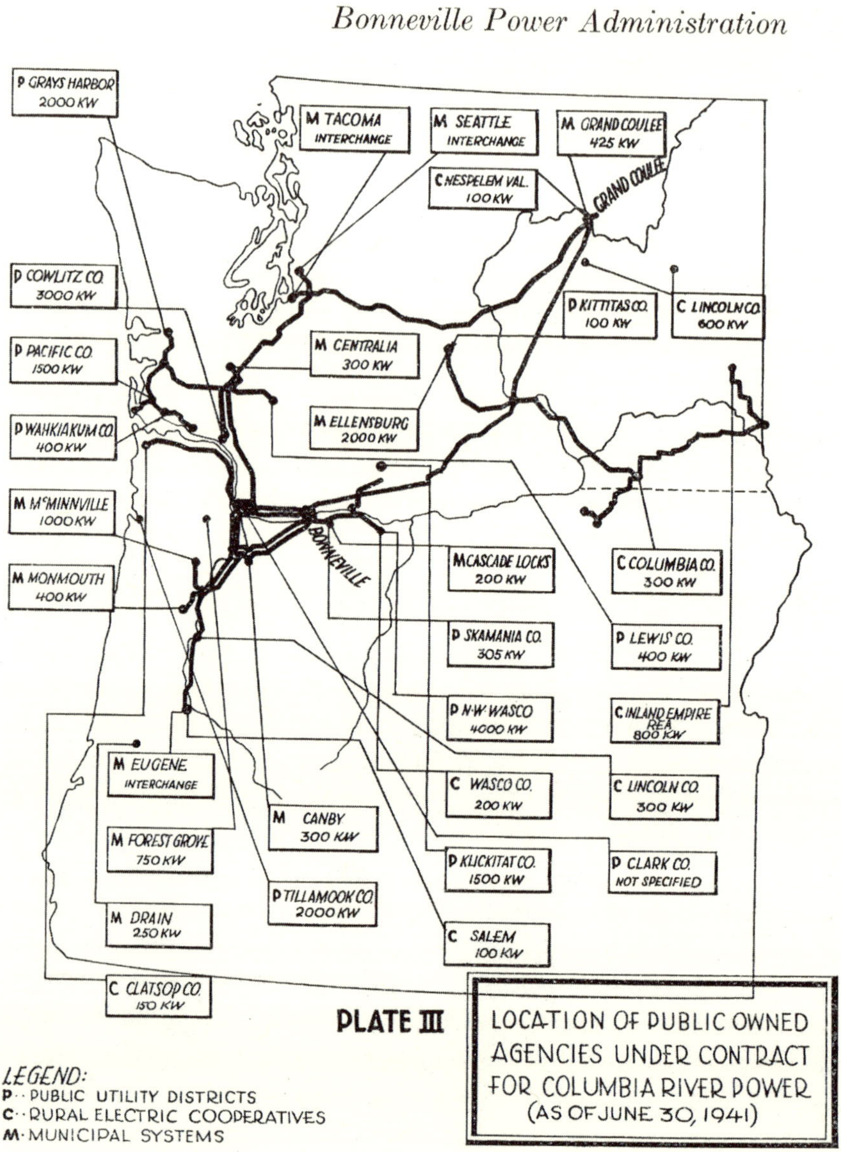

Figure 10.6. Publicly Owned Utilities under Contract with BPA, FY 1941. Source: *Fourth Annual Report of the Administrator of the Bonneville Power Administration* (Portland: BPA, 1941), 67.

istence. Stimulated by federal preference policies, BPA encouragement, and cheap Bonneville power, by summer 1941, ninety municipal utilities and rural co-ops flourished in Washington and Oregon, plus forty-four PUDs. Most of the PUDs were so new that only eleven were actually operating in 1941. Because municipals had been in existence for so long and because co-ops were easy to launch, eighty out of the ninety of them were in operation in 1941. Of the total operating public-power systems, BPA had already signed power-sale contracts with a third of them by the end of fiscal year 1941.[46] (See Figure 10.6.)

PUDs had the most trouble getting going mainly due to the difficulty of acquiring the privately owned systems within their districts. Most of the private utilities by this time owned extensive power systems that crossed county and state boundaries, so acquiring the facilities within just one county-bounded PUD meant severing it from its larger system. This was expensive and problematic. Raver and his staff tried to solve the problem by negotiating joint purchases of several entire private utilities. BPA would buy the utility's backbone transmission lines to integrate into its own grid, while the PUDs would buy the power production facilities and the distribution lines that served customers. This was also complicated and required greater cooperation between neighboring PUDs, but by June 30, 1941, BPA had successfully negotiated two of these purchases,[47] with more negotiated over the next ten years. Thus, even though few new public-power authorities were created after 1940, public-power systems continued to expand their operations despite persistent resistance from the private utilities.

BPA AND THE COLUMBIA RIVER AS REGIONAL INTEGRATOR

Geographer Eve Vogel convincingly argues that BPA and its hydropower system initiated with Bonneville and Grand Coulee dams became the key integrating and region-defining institution of the U.S. Northwest after 1937. Its transmission system literally wired the Columbia River region together, starting with Oregon and Washington in the late 1930s and continuing with Idaho and Montana in the 1940s. As President Roosevelt admonished in his 1937 dedication address at Bonneville Dam, "it is increasingly important that we think of [the Columbia River] region as a unit." The BPA's 1941 annual report announced that it had achieved the first stage of this vision as its vastly increased, broadly distributed power deliveries that year had raised the agency to "major status as a utility enterprise and as a regional institution." By the 1960s,

BPA connected its grid with the Pacific Southwest and British Columbia, but its regional transmission grid and its evolving regional preference policy (in contrast to public-power preference) constructed a de facto interdependence among the states of the Columbia River watershed, with lasting economic, political, and social ramifications.[48]

BPA's transmission line construction program expanded like gangbusters in 1940–1941, establishing the foundation for this regionalization. Of a total 1,177 miles of transmission line in operation by summer 1941, 85 percent of it had gone into service during that fiscal year, along with seventeen new substations. Twenty-five individual transmission lines BPA energized in FY 1941 integrated over two dozen cities and several defense manufacturing facilities into BPA's grid. Administrator Raver proudly announced that this expanded grid allowed it "to render service to nearly every major power load center in the states of Oregon and Washington."[49] Importantly, this integration was not just geographic. BPA interconnected public and private utilities and coordinated interstate economic planning and industrial development. Of the forty-four power-sale contracts that BPA had executed by 1941, thirty-one were small contracts for public-power systems. Most of the remaining contracts served direct industrial customers, including ALCOA (three contracts), Reynolds Aluminum (two contracts), Electro Metallurgical Company (two contracts), and three other industrial manufacturers in both Washington and Oregon. BPA also began selling power to investor-owned utilities that year, with short-term contracts to Portland General Electric (PGE) and Pacific Power & Light (PP&L).[50]

Power sales to all three types of customers expanded steadily in the coming years, making BPA a potent integrative force among formerly distinct and hostile sectors of the industry. Because everyone paid the same price for BPA power and because BPA's electrical generation capacity grew dramatically every year, there were few conflicts over allocations and sales. But that did not mean the allocation was equitable. Because the big municipal systems already had their own sources of power and the emerging PUDs and rural co-ops had fairly low demands, the thirty-one power-sale contracts to public agencies amounted to an energy pittance: only 30,000 kilowatts of the 321,880 kilowatts contracted that year. The contracts with the commercial utilities were likewise for small amounts of power in 1941, though that would change in the next decade. Industrial loads stimulated by defense mobilization quickly became the tail wagging the dog. This volume of sales to electric-intensive manufacturing businesses might have been controversial if not for the nation's steady march toward war. The 1941 BPA annual report dedicated fifteen pages, more attention than any other single topic, to a discussion of national defense and its implications for BPA and power generation. As the next chapter shows, the war accelerated the integration of the Northwest electric power system and region-wide hydropower-based economic development.[51]

Looking back on this period, BPA's 1970s-era historian Gus Norwood observes that while public power continued to expand in the postwar era, public support for the establishment of new PUDs "had run its course by 1940." Most of the purchases of commercial utilities by PUDs occurred during the 1940s, he notes, except for the important acquisition of Puget Sound Power & Light's Seattle properties in 1951 by Seattle City Light. The formation of rural co-ops stalled during the war because of federal rationing of copper as a critical war material, but co-ops had a second wind after the war and expanded significantly in the region through 1960. But their growth came almost entirely in the areas not yet served by electricity, so it remained relatively uncontroversial. Norwood concludes, "The equilibrium between investor-owned and consumer-owned systems has changed little since 1961, giving the region its present pluralistic complexion and a general reconciliation to a live-and-let-live cooperative climate since about 1962."[52] The integrating force of BPA and its Columbia River Hydropower System served as the primary facilitator of this regional integration and eventual détente between public and private power systems.

THE NEW DEAL'S LEGACY

During the New Deal, the federal government constructed hydropower dams and transmission lines in the Pacific Northwest that many intelligent and neutral observers thought were unnecessary and undesirable. So when the government completed Bonneville Dam in 1937 and made its power available to customers, the BPA desperately sought markets for the electricity to prove the skeptics wrong. It consequently engaged in a widespread public relations campaign to encourage people to consume more electricity: it offered rock-bottom, dirt-cheap wholesale prices for power; required retail rate reductions by utilities that bought Bonneville power; helped establish PUDs and co-ops; subsidized rural lines to serve more people and farms; and, most especially, recruited power-consuming industries to the region. This did much to stimulate the economy. It met many of the goals of the administration, but it had the unfortunate result of encouraging profligate consumption that led to a vicious cycle of needing more dams to feed more consumption, leading to more demand and more dams. It established a "maximization" ethos and affirmed the Army Corps of Engineer's view of the Columbia River as an abstract, quantifiable resource of acre-feet, pounds per inch, and kilowatt capacity. B.C. premier Richard McBride's goal in 1914 was to "let every inch of water do its duty"; the U.S. government, of like mind, dedicated its considerable resources from 1937 onward to maximizing the utility value of the Columbia River and its tributaries. And in that dedication to full utilization, electric power took the lion's share of attention. After 1937 cheap federal power became a "vast hydroelectric servant" of regional development and the BPA a virtual regional chamber of commerce.[53]

The messy, complicated process of authorizing the BPA and implementing New Deal policies in the U.S. Northwest reveals not just the continuing strength of New Deal opponents but also the New Deal coalition's own political and ideological fragmentation. As political scientist Philip Funigiello argued back in 1973, New Dealers advocated a suite of occasionally inconsistent policies. New Dealers wanted both regulated monopoly and regulated competition in the utility industry; they wanted to "bust the trusts," but also to work closely with big corporations to facilitate comprehensive economic planning. Decentralizers in Roosevelt's administration vied with the centralizers, while avid public-ownership advocates vied with those who felt government-business cooperation would best facilitate New Deal objectives. In this fragmented policy environment, no single national energy policy could emerge. The heads of various agencies often jockeyed to undercut or overreach their counterparts in other agencies. Interior Secretary Harold Ickes in particular engaged in bureaucratic squabbles and policy power plays to advance the agendas he favored, sometimes to the embarrassment and disfavor of the president. Funigiello summarizes the situation nicely:

> The competition among those New Dealers who favored a rational, government-sponsored business community, those who hoped to restore a freely competitive system, and those who envisioned a form of democratic centralism in which the monopoly power of the utilities was transferred to the state, unfortunately, precipitated fierce and continuing struggles for control of the power program.[54]

Roosevelt himself, a political pragmatist more than an ideologue, wavered among these positions from situation to situation, although, as historian Clayton Koppes argued a decade after Funigiello, FDR's administration leaned more toward "commonwealth liberalism": a political orientation that favored government-led comprehensive planning and integrated resource management to empower the many as opposed to the few and distribute the wealth of the nation more equitably. With the advent of World War II and especially after Roosevelt's death, commonwealth liberals gave increasing ground to another New Deal political faction Koppes calls the "corporate liberals," who sought a stronger federal role in the economy but as an ally and partner of capitalists rather than as a director or antagonist. The latter faction supported policies and incentives that strengthened America's biggest businesses in part because it needed those businesses to accomplish many policy goals and in part because corporate liberals generally felt that a rising tide would lift all boats. Redistributive politics is always much more difficult and contentious than distributive politics. As Koppes argues convincingly and as the next chapter shows, the trend toward corporate liberalism started as soon as the nation began mobilizing for war and shifted even more decisively in that direction during Truman's administration.[55]

In light of these divisions among the New Dealers, it is remarkable that FDR's administration accomplished so much during the 1930s with strained budgets, internal dissent, and vigorous opposition from without. Federal dams on the Columbia River and BPA's transmission system that wired together Northwest utilities were developed quickly and ambitiously because, as historian Richard White pointed out in 1995, the one thing all New Dealers as well as their conservative opponents agreed on was the desire for economic growth. Federal dams on the Columbia River and their low-cost hydropower unquestionably stimulated regional development.[56] Once the United States entered World War II at the end of 1941, that development accelerated at a pace far beyond anyone's expectations, most of it supported by BPA's new integrated regional power grid.

CHAPTER 11

War-Accelerated Transitions II

In every province and in every city men and women were laying aside their individual dreams for the common good. Hands that had done little more than hold a pen or push a lawn mower were making shells and ships. . . . the great machine of [BC Electric] was harnessing itself with gathering momentum to the war effort of the Dominion.

—Cecil Maiden, 1948[1]

Under the impact of war the Bonneville Power Administration has telescoped ten years of normal growth into a brief five years. It has . . . become one of the three largest power systems in the country.

—Bonneville Power Administration, 1944[2]

The effects of the two world wars were quite similar in many ways, and similarly affected the United States and Canada. Both wars greatly stimulated gross domestic production in general and electricity consumption in particular, and both accelerated trends in electric utility evolution toward greater economies of scale, more widespread utility interconnections, and increased government involvement in developing and regulating the electric utility system. As with World War I, the Second World War created a transformationally large demand for food and war materials in the United States and Canada, especially near the coasts, where military personnel and supplies could be quickly transported to regions of conflict in Europe, Africa, and Asia.

Even more so than the first war two decades earlier, war production during World War II relied very heavily on electric power in manufacturing. Air power played a much more significant role in warfare in the 1940s than previously; consequently, the production of aluminum for airplanes (a dominant sector of the "light metals" industry) became a major war material. Aluminum is processed electrolytically and thus requires a tremendous amount of power. Most aluminum manufacturing globally at that time concentrated near large hydropower facilities for that reason. The North American aluminum industry (ALCOA and ALCAN) and the U.S. and Canadian governments understood that the abundant low-cost hydropower in the Pacific

Northwest offered an excellent prospect for creating a Pacific coast aluminum-manufacturing industry, and they assisted in developing it. The construction of several new aluminum plants in the Columbia River Basin vastly increased electricity consumption and supported the region's expanded airplane and shipbuilding facilities—especially in the greater Portland/Vancouver, Puget Sound, and Spokane urban regions. Though British Columbia did not successfully recruit any new aluminum-manufacturing facilities during the war, it nevertheless hosted many important electrochemical facilities and munitions, ship, and plane manufacturing plants in the greater Vancouver and Kootenay regions. These industries all consumed prodigious amounts of electricity. COMINCO, the mining company in the Kootenay region, got a contract from the Canadian government to transform a fertilizer plant in the aptly named Warfield, British Columbia, into a munitions factory. COMINCO also ramped up its zinc production as it had in World War I and received a subsidy from Ottawa in 1942 to quickly build a new power plant on the lower Kootenay to keep up with war-related electrical demand.[3]

Much as the earlier world war had, World War II resurrected tight government controls of production and consumption of all things war related. In 1940, Canada created a Wartime Industries Control Board and a Department of Munitions and Supply to direct materials production and consumption. Likewise, in the United States about a month after the Japanese bombed Pearl Harbor, President Roosevelt established the War Production Board, similar to the War Industries Board established during World War I. The two national governments ordered factories to produce certain materials to support the war effort and compensated them handsomely for it. Critical war materials like copper, petroleum, and certain foods were rationed. Both governments assiduously worked to ensure that adequate electric power would be available to support war manufacturing. The U.S. government designated the BPA a War Production Agency in March 1942 and accelerated the financing of additional generators at Bonneville and Grand Coulee dams to meet power demand, including heavy new demands from both privately owned and government-owned aluminum-manufacturing plants being built in Washington and Oregon.

Government economic intervention and regulation greatly accelerated at all levels during the war in both countries. Americans had become more inured to such activity during the New Deal than Canadians, whose central government played a more restrained role during the Depression. As historian Jeremy Mouat explains, "Wartime brought with it a much greater degree of government regulation for Canadians," and as during World War I, most people accepted it as a "necessary consequence of war." Included in that increased regulatory attention, Mouat says, was "far more public scrutiny and accountability for utility companies."[4]

As explained in chapter 6, World War I advanced the nascent development of integrated regional electric systems, the concept referred to as "Superpower" by proponents of investor-owned utilities in the United States and as "Giant Power" by the proponents of public-power systems. This early stage of integration advanced quickly during World War I because the United States and Canadian governments ordered certain utilities to interconnect their power grids to improve electrical reliability and facilitate power exchanges. After the war, both governments stopped ordering interconnections, but the integration trend continued, albeit at a slower pace. During the Depression and New Deal, as noted in chapter 9, the U.S. government again became active in ordering interconnections through new authority and mandates given to the Federal Power Commission (FPC) in 1935, but this time under economic duress rather than as a war measure. In both cases, the goal was to overcome utility competition and improve system efficiency. The FPC used its power to order interconnections only gingerly, however. During World War II, all restraints dissolved as the U.S. and Canadian governments both aggressively ordered interconnections and power sharing wherever and whenever they felt it was necessary to support war production.[5]

Interconnecting utility transmission grids was but one part of a broader trend toward cooperation and integration that both wars stimulated. In World War I, cooperation and integration in the U.S. Northwest occurred mostly between privately owned utilities, since the only public utilities at the time were municipally owned systems serving single towns and therefore offering little opportunity for regional interconnections. But during World War II, the public-power sector in the U.S. Northwest included rural electric co-ops, county-wide public utility districts, and the federal power system managed by BPA. Together, these three new types of public-power systems, along with the municipal utilities, offered ample opportunities for regional cooperation within the public-power sector and larger transmission grid integration between both publicly owned and privately owned utilities. The advent of World War II greatly accelerated this integrative agenda.

Extending the practice of ordering interconnections, the U.S. government during World War II ordered the creation of a region-wide "Northwest Power Pool," composed of all types of utility systems and facilitated by the BPA. This kind of direct federal integration of both public and private utilities in a federally owned transmission system did not occur during World War I. Also for the first time, British Columbia interconnected with BPA during the war years and was formally and permanently integrated into the Northwest Power Pool in 1949.

In British Columbia, region-wide utility system integration during and after World War I took the form of utility consolidation in the investor-owned sector, leaving just two main companies serving the great majority of the province's population. No new public-power authorities arose in British Columbia during the 1920s and 1930s, so the situation on the eve of World War II was substantially the same as at the end of World War I. The coming of the Second World War, however, initiated a trend toward provincial utility regulation and also toward rural electrification, which had been largely neglected in the province during the 1930s.

Just one year before Canada formally entered the war, the B.C. provincial legislature established a public utility commission, British Columbia being one of the last provinces in Canada to do so, but the PUC played a low-key role in its first few years (see chapter 10). During the war, however, the B.C. utility commission asserted itself much more vigorously, completing valuation studies of the province's biggest utility and ordering rate reductions. The provincial legislature also authorized a government power authority during the war to coordinate rural electrification and consolidate small, scattered systems into a single, more efficient, government-run electric utility. This British Columbia Power Commission began consolidating power generation and distribution for the small towns and rural areas not served by BC Electric in the west or West Kootenay Power and Light in the east. The B.C. Power Commission unintentionally laid the groundwork for an eventual takeover of BC Electric and the formation of today's provincially owned BC Hydro.

The government retrenchment seen after World War I did not occur to the same degree after World War II. During World War I, many of the federal agencies and their assertive economic interventions dissolved once the war ended. Wartime economic controls were lifted, war production contracts cancelled, and business regulations lessened. In the United States, conservatives regained power in Washington, D.C., in 1920 and concertedly retrenched the interventionist and regulatory activities of the national government for the next twelve years. Canada did not turn as hard to the right in the 1920s, but its dominant Liberal Party nevertheless eschewed the economic interventionism of the war years. The situation was quite different after World War II. Certainly, Canada's Wartime Industries Control Board and America's War Production Board cancelled many military contracts, war agencies disbanded, and rationing lifted. But unlike the generally peaceful 1920s, the post-WWII

Cold War kept military production booming. Moreover, Keynesian economic policy had gained more acceptance in the United States and Canada by World War II. Both governments, having learned valuable lessons in the postwar economic slump following World War I, invested significant public resources into postwar economic conversion and into jobs and education programs for returning soldiers. In the United States, liberal Democrat Harry Truman succeeded Roosevelt in the White House, extending the commitment to an activist government and many New Deal programs until the early 1950s. In British Columbia, the Liberal-Conservative coalition formed in 1941 continued after the war and pursued a vigorous public works program in the province with treasury funds made flush by war-stimulated industrial production.

WAR MOBILIZATION AND ELECTRIC POWER

In the divided, politically contentious environments of the U.S. Northwest and British Columbia at the start of World War II, nearly everyone agreed on the need for centralized economic planning. Once again, a "war emergency" justified all manner of authoritarian government controls supported equally by liberals, conservatives, progressives, and socialists. As political scientist Martin Robin aptly states, "If there was any consensus in British Columbia at the turn of the decade it revolved around Canada's sacred war mission, which implied a national patriotism and a willingness to sacrifice local claims to the larger national interest."[6] While many local, state, and provincial projects were put on hold during the war, anything related to war production got widespread support, including electric power development, which was so critical to war production. Like the 1930s, though, government backing for hydropower development was stronger in the United States than in Canada. For example, while Congress approved virtually all BPA's accelerated power-generation requests during the war, the Canadian government twice turned down BC Electric's pleas to fast-track construction of its Bridge River project, resulting in a power shortage in British Columbia in 1943 that required the province to import electricity from Puget Sound Power & Light.[7]

Enthusiastic support for war mobilization and the development of war industries finally lifted both the U.S. and Canadian economies out of the Great Depression. A massive Keynesian infusion of government spending (fueled by debt) accomplished this feat. With hundreds of thousands of men and women joining the armed forces and even larger numbers flocking to the defense industries for well-paying jobs, unemployment plummeted, wages rose, and income and profits soared. By 1940, British Columbia's chronically tight provincial budget was in the black for the first time since 1926, and the province was making "record breaking disbursements" for public works and social programs. Both the U.S. Northwest states and British Columbia had significant shipbuilding and airplane-manufacturing capacity, as well as electrochemical and electrometallurgical industries, which war demands greatly expanded.[8] In Vancouver, British Columbia, military bases sprang up, manufacturing industries boomed, and workers flocked to the new jobs, packed like sardines into the electric trolleys to get to their jobs. The BC Electric Company's modest power surplus evaporated by 1942, and the company consequently lobbied the Canadian Department of Munitions and Supply to allocate it critical war materials like copper and hydroelectric generators so it could construct a power plant on Bridge River. But, as noted above, the government turned down the request, underestimating the region's war-related power needs and presuming the Bridge River facility would not be completed before the end of the war. As BC Electric predicted, it could not meet power demand the next year, despite electricity rationing and regular planned "brown-outs."[9]

The situation in Vancouver significantly contrasted with the experience south of the border, where aggressive development of additional power and transmission kept supply just ahead of demand throughout the war. Electrical generation by all sources combined in Washington, Oregon, and Idaho more than doubled from 1940 to 1945,[10] although the vast majority of that growth came from new power generated at Bonneville and Grand Coulee dams. In the states of Washington and Oregon in particular, electric power played a central role in war mobilization, with the two federal dams on the Columbia serving as the central nervous system of this electricity-driven development. BPA made an especially effective instrument for the federal government to facilitate war production in the region. BPA's 1942 Annual Report appropriately characterized Columbia River power as an "aggressive war weapon."[11] President Roosevelt used that weapon to significant advantage. Shortly after his declaration of war, FDR called for the production of *60,000 new military aircraft per year* to fight the war. The scale of that demand, coming after more than a decade of depressed business activity, was truly astonishing. The Boeing Airplane Company, founded in 1916 in Seattle, was well positioned to provide military aircraft. It had prospered during World War I, supplying airplanes for that conflict, but Boeing's transformation during World War II was even more dramatic. During the peak years of war production, Boeing's Seattle-area plants employed 40,000–50,000 workers on nonstop 24-hour production lines capable of finishing a mind-boggling sixteen B-17 "Flying Fortresses" every 24

Figure 11.1. "Striking Power," Bonneville Power Administration War Production Poster. Source: Lloyd Hoff war poster, Bonneville Power Administration Research Library.

hours. Boeing's workers also produced a B-29 aircraft every five days at each of several different assembly lines near Renton, Washington. By the end of the war, the company had provided 7,000 B-17s and 3,000 B-29s to the U.S. government for the war effort.[12] BPA provided most of the power for these plants (see Figure 11.1).

Aluminum was the best material for building planes, and BPA had already recruited four large aluminum-manufacturing plants to the region before Roosevelt declared war—one in 1939 and three more in 1941. During the war, two additional plants were built in the Northwest. Nationwide in 1939, aluminum-processing capacity in the United States totaled 435,000 tons. By 1943, processing capacity had ballooned to 2.4 million tons, and 25 percent of that was produced in the Pacific Northwest states (more on this below).[13]

Industrialist Henry Kaiser won government contracts to build three new shipyards in the Portland-Vancouver area during the war. During 1942–1944, the three peak years of war production, 24-hour-a-day assembly lines at these three shipyards finished over 700 *Liberty* ships, aircraft carriers, and troop transport and cargo ships. By 1943, the Vancouver shipyard broke all production records, completing one aircraft carrier every week. New techniques for electric welding greatly sped up fabrication time (see Figure 11.2).

The Kaiser shipyards employed up to 92,000 workers at its peak (many of them women because so many men served in the armed forces), used over two million tons of steel, and paid out a billion dollars in wages. Another 23,000 laborers built or repaired ships in non-Kaiser-owned plants. The flood of workers moving to the cities with war industries caused chronic housing shortages. A new housing development, Vanport, near the shipyards on the Columbia River, quickly built from scratch by Kaiser and the U.S. government, became Oregon's second largest population center by 1944. Cheaply constructed out of particle-board, with wood blocks as foundations, the houses in this community were completely wiped out in the Columbia River flood of May 1948.

Puget Sound's shipyards grew by leaps and bounds, too. The Bremerton shipyards in 1942 repaired three of the battleships damaged by the Japanese bombing of Pearl Harbor in Hawaii. To energize the Portland-Vancouver shipyards, BPA built three transmission lines to them directly from Bonneville Dam. Getting power to Bremerton was more complicated, as it was much farther from Bonneville than Kaiser's shipyards were. The War Production Board determined that it needed 40,000 kilowatts of power at the Bremerton shipyards, and it ordered the nearby Tacoma municipal utility to build a transmission line to supply the shipyard. BPA provided the power by interconnecting with the Tacoma system. Congress enabled BPA to build these and

Figure 11.2. "Bonneville Fights Time," BPA War Production Poster. Source: Lloyd Hoff war poster, Bonneville Power Administration Research Library.

other high-voltage lines quickly by doubling its transmission-line construction budget in 1942. In addition to these aircraft- and shipbuilding and aluminum plants, power was needed for munitions, worker housing, and military bases. By June 30, 1942, war-related production consumed 92 percent of BPA power contracts and commitments.[14]

BPA historian Gene Tollefson characterizes BPA's war record as prodigious: transmission-line mileage increased from a few hundred miles in 1940 to over 2,000 miles in 1945; BPA's industrial and utility customers increased from five to eighty; BPA supplied 20 billion kilowatt-hours to war industries; and power-sale revenues went from a few hundred thousand dollars to $23 million. BPA administrator Paul Raver noted that the taxes paid by the war industries in the Northwest during these years exceeded the entire original cost of Bonneville and Grand Coulee dams.[15]

Most of the electricity demand for the war industries in the Northwest fed the needs of the aluminum-reduction factories, an industry BPA concertedly recruited to the region to build a market for its power. This posed an odd challenge to the agency; its founding policies directed it to encourage "the widest possible use" of electric power, emphasizing low rates for residential consumers, with priority given to public agencies and co-ops. BPA unquestioningly accomplished this, but the residential and rural markets consumed only a tiny fraction of BPA's potential power supply. To prove the critics of Bonneville and Grand Coulee wrong, BPA needed to "build load," and that required finding much bigger markets. The aluminum industry fit the bill. If not for the war, BPA might have struggled to justify selling so much of "the people's power" to ALCOA and Reynolds Aluminum. BPA did not just obscure an industrial sales agenda behind a mask of patriotism. Instead, it recruited the light metals industry to the region under terms consistent with New Deal social policies. The degree to which it succeeded is debatable, but its effort was genuine. Historian Harry Stein has convincingly argued that more than simply building load, the BPA "wanted to create a vertically integrated, competitive, permanent, and prospering aluminum industry in the Pacific Northwest."[16]

Drawing on New Deal employment policies and regional planning principles, BPA tried to establish a permanent start-to-finish light metals industry that integrated aluminum ore reduction with ingot production, sheet-metal rolling mills, and end-product manufacturing (e.g., airplane building). The agency's 1942 Annual Report states, "Every electro industry established during the past 4 years in the Pacific Northwest is part of a pattern of industrial development providing for the complete processing of Northwest resources from raw materials to consumer goods." BPA aimed to secure this industry for the long term. BPA wanted a competitive industry, too, so it sought to break ALCOA's aluminum-manufacturing monopoly by limiting its control of Northwest factories and recruiting competitors such as Reynolds and Olin. ALCOA had been fighting a U.S. antitrust suit since 1938, and as late as 1944, it still controlled 91 percent of U.S. aluminum capacity. Consistent with the Bonneville Project Act's requirement to "prevent monopolization . . . by limited groups," BPA supported the government's antitrust efforts against ALCOA. (The government succeeded in breaking up ALCOA in 1945.) The federal government and BPA also purposely distributed aluminum manufacturing across several urban nodes hundreds of miles apart in areas near Vancouver (Washington), Tacoma, and Spokane. The U.S. government's Defense Plant Corporation, a war agency established in 1940, facilitated this by building four of the six new Northwest aluminum-manufacturing plants. Despite protests from ALCOA, BPA and the Defense Plant Corporation argued that for the good of the region, the facilities should be dispersed. In short, BPA sought to diversify the Northwest economy, provide decentralized employment, and create a stable foundation for a postwar manufacturing economy—all central goals of New Deal policy. Indicatively, BPA's administrator Paul Raver claimed the agency in those years was "part of a great experiment improving social and economic conditions for the region." BPA's aluminum policy director, Samuel Moment, fondly recalls, "we were part of the Roosevelt spirit."[17]

The attention BPA lavished on aluminum production in these war years is most graphically illustrated in the agency's energy delivery statistics. (See Table 11.1.) At the peak year of war production in 1944, the aluminum industry alone consumed 63 percent of BPA's electrical generation. By then, Grand Coulee Dam was the largest single power-producing facility in the nation, and BPA managed the third largest electric system in the nation, producing half of all the power generated in the five Columbia Basin states of Washington, Oregon, Idaho, Montana, and Utah.[18] It is truly remarkable that two-thirds of all BPA's power went to one relatively new industry and that nearly all of its power supported the war effort.

One additional war-related industry significantly boosted the economy in a remote location of the Northwest, although few people at the time knew much about it. Under a shroud of secrecy, the U.S. government's Manhattan Project began work in 1942 on a nuclear weapon in Hanford, the low-desert region along the mid–Columbia River. Scientists, engineers, and skilled laborers flocked to the top-secret facility. According to Gene Tollefson, of the million or so people who migrated to the Northwest in search of jobs during the war, one in ten went to Hanford. The nearby farming village of Richland,

Table 11.1. BPA Electric Sales (Megawatt-Hours), Fiscal Years 1940–1944

Class of Customer	FY 1940 mwh	FY 1942 mwh	FY 1944 mwh
Aluminum Industry	0	1,845,249	5,453,893
Other Industry	21	76,580	934,588
Military Establishments	0	2,575	87,889
Publicly Owned Utilities	3,101	142,491	727,642
Privately Owned Utilities	188,806	357,704	1,467,304
Total	191,928	2,424,599	8,671,316

Source: Bonneville Power Administration, *Report on the Columbia River Power System, Fiscal Year 1945*, Table II, 7. A megawatt-hour is 1,000 kilowatt-hours.

with a population of 247, grew in a matter of months into a U.S. government–owned municipality of 11,000 people. Additional tent camps and barracks for thousands more workers sprang up in the sagebrush flats near the Columbia River. The government built eight mess halls for the workers, serving fifty tons of food per meal. Under contract to the federal government, Hanford workers and the DuPont Corporation produced the plutonium used in the world's first atomic bomb. The process required a large block of power, which BPA supplied, although for years the agency did not know what the power was for. From a rather modest draw of 473 kilowatts in 1943, the Hanford "mystery load" grew to 55,000 kilowatts in 1945.[19]

While BPA was the dominant energy provider in the U.S. Northwest by World War II, it was not the whole story. Virtually all utilities in the region experienced significant growth and change during the war years. Their primary challenge was to support war mobilization and get adequate electricity where it was needed for war production. While BPA directly supplied most of the power needed for war production, the public and private utilities also contributed in a variety of ways. Because of the Depression and federal construction of the Columbia River dams, investor-owned utilities put new plant construction on hold throughout the 1930s. When electric demand suddenly spiked in 1940, BPA was ready to serve it, so the private sector continued to delay new construction. Washington Water Power (WWP), for example, had not built any new power-generation facilities since 1928 and did not initiate a new hydropower dam until the early 1950s, when it undertook construction of Cabinet Gorge Dam.[20] Instead, privately owned utilities either bought wholesale power from BPA, which was quicker and cheaper than building new plants, or sought to meet war-related energy demands by redirecting power supplies, improving efficiencies, interconnecting their systems, and encouraging less consumption by homeowners and other "nonessential" customers. Annual reports for the war years by WWP and Puget Sound Power & Light (PSP&L) mention how they curtailed appliance sales to residential customers as a conservation measure (as BC Electric also did during the war), cut back on rural line extensions to conserve copper, dimmed streetlights, made modifications to their power plants to increase efficiency, and cooperated with government-sponsored energy-rationing programs, redirecting the saved power to uses deemed important to the war effort. PSP&L creatively expanded its grid's power supply by connecting to twenty lumber mills with their own steam-powered electric plants fired by wood waste. Many of the mills had excess power capacity, allowing PSP&L to bring up to 25,000 more kilowatts into its grid.[21]

Despite having access to cheap Bonneville power on a preferential basis, Seattle City Light pushed its hydropower program forward aggressively in the 1940s, with encouragement from the federal government. City Light had just finished construction of Ross Dam on the Skagit River in 1940, although it was not yet producing any power. Its other two dams on the Skagit River still provided the utility a handy surplus of power that contributed substantially to war needs in the harbor city. The U.S. government built 3,568 housing units in Seattle in 1942 for war-industry workers and military families, and City Light provided their electrical service. So many people flocked to the city for jobs that residential consumption of electricity increased by 27 percent from 1941 to 1942, even though the utility curtailed sales of electric ranges and water heaters in mid-1942. Ridership on the urban trolley system owned and managed by City Light jumped from 91 million to nearly 127 million in these two years. As a result, City Light petitioned the War Production Board to allow it to increase the height of Ross Dam by 185 feet in order to add hydropower capacity. The War Production Board gave City Light approval to "proceed . . . with all possible speed," and federal financial assistance helped launch the project in early 1943. A $9 million investment increased Ross Dam's generation capacity by a third, adding 300 million kwh of annual capacity when the expansion was completed in 1949. Ross Dam produced no power, however, until the 1950s.[22]

Canada experienced a war boom, too, in both production and electrical generation, but not as dramatic as in the United States. For the nation as a whole and for British Columbia and the Yukon specifically, electrical generation increased by just one third from 1940 to 1945. In comparison, electrical generation more than doubled in the U.S. Northwest in these years.[23] All of Canada's added power supply and more went to war mobilization. Canada added two million horsepower of hydroelectric generation during the war years, most of it in Ontario and Québec, and the Dominion Power Controller directed all of it to war facilities and also diverted a great deal of civilian electric consumption to war needs. Cecil Maiden's history of BC Electric mentions electricity rationing and brown-outs as common in the Vancouver and Victoria areas, similar to practices in the United States. It seems, however, that western British Columbia had to meet even more of its power needs by rationing and conservation than did Washington and Oregon, which benefited from more aggressive construction of additional generation capacity by public-power agencies like BPA and Seattle City Light.[24]

In southeastern British Columbia, the situation was quite different. The Canadian minister of munitions and supply, C. D. Howe, deemed COMINCO critical to the war effort. The mining company's affiliate and power supplier West Kootenay Power and Light (WKP&L) did not have enough generation capacity to support full production of COMINCO's metals- and chemicals-processing plants, which included the production of "heavy water," which is essential to nuclear weapons production. So in 1941, COMINCO asked the minister to approve construction of a new hydropower facility at Brilliant, which had been proposed a decade earlier but shelved due to the Depression. The government granted permission along with a tax write-off incentive to facilitate private financing and rapid construction. Near the confluence of the Kootenay and Columbia rivers, the Brilliant facility captured the last increment of "fall" from the Kootenay River below Kootenay Lake. With five dams, the lower Kootenay was fully developed when the Brilliant project came online in summer 1944. By then, of course, the war was winding down, although nuclear weapons production was ramping up. Seeking sources of heavy water for its Hanford facility, the U.S. government arranged to build a heavy-water manufacturing plant near Trail, British Columbia, in 1942 and turned it over to COMINCO to operate. COMINCO produced and sold heavy water back to the U.S. government. Its production was so energy-intensive that the facility eventually consumed a third of WKP&L's electric load.[25]

The aluminum industry in Canada, dominated by ALCAN, prospered, too, though not directly in British Columbia. ALCAN had once been a subsidiary of ALCOA but became independent in 1928, though its independence was more nominal than actual. The Canadian company boomed after 1937 due to increased war-production needs, accentuated by additional demand from the United States. Hoping to reduce ALCOA's monopoly, the U.S. government contracted for a very large amount of aluminum from ALCAN, which expanded its Québec facilities fivefold from 1940 to 1943. ALCAN's gross sales went from $49 million in 1937 to $259 million in 1944. But all of this war-production growth was concentrated in eastern provinces, even though B.C. officials tried hard to entice ALCAN to the West Coast. The best the province could get was mild interest, several studies of the possibilities, and several "show-me" trips after the war. BC Electric lacked adequate surplus capacity to support a large new aluminum plant, and British Columbia's best remaining undeveloped hydropower sites were too remote to attract the company. ALCAN needed a huge amount of power, a deep-water shipping port, and urban infrastructure to support a population of tens of thousands. Not until the 1950s would ALCAN build a facility in British Columbia, selecting a Nechako River location west of Prince George. It was an unprecedentedly large project; as historian Matthew Evenden notes, "no industrial development in the province's recent past could match it for size or importance."[26]

If a B.C. site had been selected by ALCAN during the war, British Columbia's wartime development and power consumption might have more closely matched that of the U.S. Northwest. Lack of support for BC Electric's Bridge River hydropower proposal from the Canadian government likely hampered the province's ability to attract more industry during the war years. In contrast, the development occurring south of British Columbia's border was undeniably facilitated by national policy and national investments. Regional boosters on both sides of the border understood the lessons of the war and applied them liberally after hostilities ended, lobbying during the Cold War for their national governments to sponsor defense-related industries as an instrument of regional economic development.

Also of note at this time was the precedent set for electric energy conservation. Oddly, for all the genuine dedication to conservation, there was no talk whatsoever about conserving electric power during the 1930s among New Dealers in the Northwest or liberals in British Columbia, or anyone else in the region, for that matter. All the conversations, lobbying, advertisements, editorials, and policy pronouncements consistently focused on promoting more consumption, reducing rates, and encouraging the widest possible use of electricity. Even more ironically, government agencies and conservationists spoke constantly in the 1930s about the importance of conserving all the other major sources of energy: wood, coal, oil, and natural gas. The inconsistent attitude toward electric power seems not to have troubled anyone, at least not in the Pacific Northwest. In fact, policy makers and electric power boosters in the region, including the BPA, actually argued that people could help to conserve energy resources by consuming *more* electricity. This is perhaps understandable as the region's power supply was about 90 percent derived from "renewable" hydropower, which appeared exempt from fears of exhaustion at that time. But such attitudes were naive at best, delusional at worst. Rivers offer a limited capacity for power production; good sites for building dams are even more limited; and the economic, social, and environmental costs of developing rivers for hydropower mount quickly the more a river is dammed. Prior to World War II, Americans and Canadians in the Pacific Northwest enjoyed few restraints on hydropower development and never concerned themselves with conserving electric power because there was so much opportunity for developing additional supplies. This is typical of the politics of abundance.[27] But when war-production needs caused a supply bottleneck during World War II, suddenly everyone was talking about conserving power (except the war industries, of course). As soon as the war ended, conservation of this again abundant resource disappeared from the political lexicon. But northwesterners would enjoy only a few more decades of blissful avoidance of the term and its institutional and behavioral implications for the electric power system. Conservation of all forms of energy would return and remain a permanent and increasingly important fixture of federal and regional power policy by the 1970s.

NORTHWEST POWER POOL

The most important measure the U.S. government took to increase the power supply in the Northwest during World War II was to add more generators to Bonneville and Grand Coulee dams. The second most important measure for increasing the power supply was the creation of the Northwest Power Pool. As noted above, interconnecting utility systems offered a relatively quick and affordable means for making increments of additional power available by transferring power from areas with surpluses to areas with deficits. All utilities had enough power capacity to meet their peak loads, plus a little extra, so each had significant surpluses during nonpeak periods. Moreover, every utility had a unique set of power demands with varied timing for peak loads. Utilities also had varied mixes of hydropower and thermal-generated power, and each served a different region with different rain and snow and river flow patterns affecting their hydropower-generation capacities. The more utilities that combined in a power pool and the larger the region covered by that pool, the more opportunities there were to even out individual surpluses and deficits and wring greater efficiencies from the system. That is why the U.S. and Canadian governments ordered interconnections during World War I and why they did so again during World War II. It is the same reason the holding companies often interconnected their subsidiary operating companies during the 1920s and 1930s.[28]

The Northwest Power Pool, formally created in 1942 following a mandate from the War Production Board, was particularly expansive, much more so than the Northwest interconnections accomplished or contemplated during World War I. The particular character of the power pool derived from two main sources: the BPA, with its focus on the entire Columbia River watershed, and the private utilities currently or formerly connected with the Electric Bond and Share (EBASCO) holding company empire. Steve Blewett's history of WWP claims that EBASCO affiliates pioneered the power pool in 1941 when WWP, PP&L, Northwestern Electric, Montana Power, Idaho Power, and Utah Power & Light all interconnected their systems.[29] This pool reached across five states and included the majority of the power supply in Idaho, Montana, and Utah. But it did not interconnect with other private utilities or any of the big public utilities in western Washington or Oregon, and it did not pool resources with BPA. Completing the job, the War Production Board in May 1942 ordered additional interconnections to facilitate a full regional power pool that included all the major public and private utilities in five northwestern states that are part of the Columbia River Basin (Washington, Oregon, Idaho, Montana, Utah). The power pool began its coordinated operations on August 1, 1942, and even expanded to include British Columbia a few years later, which contains the Columbia River headwaters and about a third of the river itself. BC Electric had experienced a power shortage in 1943 and had to buy surplus electricity from PSP&L, its neighbor south of the border. Seeing the benefits of interties and not yet able to begin construction of its planned Bridge River facility, BC Electric lobbied American and Canadian authorities to authorize an intertie with the BPA grid. By the end of the war, BPA had completed a transmission line through the Puget Sound region to the U.S.-Canada border. BC Electric built a transmission line from Vancouver to the border to connect with it and started buying BPA power in 1946. The same year, with wartime rationing ended, BC Electric began construction of its Bridge River facility, which came online in 1948. Other interties across the border in the interior Northwest followed in subsequent years; eventually, even Alberta joined the Northwest Power Pool.[30]

Not surprisingly, BPA immediately became the most important member of the power pool. Climate stress launched this role for the agency. A drought in the winter of 1941–1942 resulted in "the lowest streamflow conditions in 54 years of recorded runoff history on all rivers but the Columbia," according to BPA. As a result, the hydropower facilities of the region's utilities (built on tributary rivers) could not produce enough power to meet their needs in the summer and fall of 1942. The Columbia River's flow was not affected as much since its headwaters are in the deep ice fields of the northern Rocky Mountains in British Columbia, so BPA made up the shortfall by pushing its installed turbines at Bonneville and Grand Coulee dams to the maximum—beyond their design capacity. BPA ended up providing 50 percent of the entire power load of Washington and Oregon in the fiscal year ending June 1943. The following year, with even more turbines installed at the two dams, BPA generated 50 percent of the power produced by *all* the power pool utilities in *all five Northwest states*. Besides direct sales to aluminum plants and other war manufacturers and military bases, BPA supplied almost 2 billion kilowatt-hours to other utilities in the power pool, helping them to meet their loads in fiscal year 1944. Almost 20 percent of BPA's generation that year supplied other utilities. The two largest recipients were both commercial utilities: 800 million kwh to Portland General Electric (PGE) and 340 million kwh to PSP&L.[31] Not only did the federal government facilitate creation of the Northwest Power Pool, but a federal agency, the BPA, provided the backbone power for the pool and the backbone transmission lines to connect its power to the members of the pool. Federal Columbia River hydropower literally united a formerly fractious collection of public and private utilities into a cooperative regional association. This kind of development advanced even further in the coming decades. Environmental law scholar Michael Blumm claimed in the 1990s that the interconnections facilitated by the Northwest Power Pool "signaled the beginning of an era in which public and private utilities would operate their projects as essentially one utility, which remains the chief characteristic of the system operations today."[32]

PUBLIC/PRIVATE UTILITY RELATIONS AND RURAL ELECTRIFICATION

The power pool put a premium on cooperation. In the grip of war, coordination and unity became the dominant leitmotif of corporate and government public relations. Reports, editorials, newspaper ads, and public speeches of leaders in both the private and public sector all called for cooperative effort toward the national defense in both the U.S. and Canada. But beneath the wartime rhetoric of cooperation, the contest between public and private power in the United States continued during the war, though in a more muted fashion. About a dozen PUDs continued their condemnation proceedings to acquire some of the facilities of the private utilities, congressional proponents of public-power introduced several more (unsuccessful) bills during and after the war to create a TVA-like power authority in the Northwest, and public-power referenda continued to appear on ballots in Washington and Oregon. In response to this, investor-owned utility (IOU) executives continued their spirited defense of the commercial sector of the industry. They disputed the superiority of public power, charged the PUDs and municipal systems with unfair competition, and complained that the BPA was attempting to transform the entire Northwest into a wholly public-power system. In British Columbia the question of public power started out muted, but as the war dragged on, voices calling for rural electrification and a public-power authority grew in volume and stridency until they finally brought results in 1945.

At the start of the war, in the United States both the publicly owned utilities and the privately owned utilities feared the worst regarding the possible effects of the war on their fortunes. Both jockeyed to take advantage of the changed political conditions while publicly endorsing calls for unity. The private sector worried that the Roosevelt administration, which clearly favored public power, would use the war emergency to strengthen public ownership in the Northwest through the BPA—which it did, in fact. On the other hand, public ownership advocates, who had momentum on their side, feared that the investor-owned sector would use the war emergency to strengthen private power by acquiring large war contracts and dampening the acquisition efforts of the PUDs in the name of unity and truce.[33]

As described in the previous chapter, over thirty PUDs were authorized by public initiative in Washington and Oregon before the war, but very few of them were actually up and running. Seven PUDs started providing electrical service during the war, but PUD service still represented a small fraction of the whole. Most were attempting to acquire the power plants and distribution systems of private utilities within their district, often with the assistance of the BPA. Because the private utilities in most cases did not want to sell, the PUDs had to initiate condemnation proceedings to get a court-determined value for the properties and then sell bonds to acquire them. It all took time. During the war nearly a dozen of these condemnation proceedings moved forward, to the consternation of the investor-owned utilities, which argued that these piecemeal county-sized acquisitions by individual PUDs would break up a coherent system and harm the objectives of the power pool. Of course, the power pool would remain intact, and the PUDs would be connected to it, so the argument made only rhetorical sense, but IOU executives often employed it to their advantage. They liberally used patriotic rhetoric to try to force a truce in the battle between themselves and the public ownership movement.[34]

In a speech to the Edison Electric Institute Annual Meeting in 1940, WWP president Kinsey Robinson said that if the United States entered the war, "the first thing to be achieved is a renewal of our national spirit and the elimination from our social and economic structure of class consciousness and divisionalism. . . . United we stand, Divided we fall." Puget Sound Power & Light was the utility most beleaguered by expanding PUDs and municipals, so it generated the most vigorous media campaign in defense of patriotic unity. The company's *Report to Stockholders* during the war years reprinted a large number of full-page advertisements published in area newspapers, all signed by PSP&L president Frank McLaughlin. One published just after the bombing of Pearl Harbor contained the bold title "UNITY WITHOUT MALICE—AND SHOULDERS TO THE WHEEL." It quoted President Abraham Lincoln calling for "malice toward none" and President Franklin Roosevelt calling on Americans to lay aside their political divisions during the war emergency. The next month, another PSP&L ad appeared in area newspapers, admonishing that "With United Effort We Must Forge Ahead Toward the Common Goal of Total Victory and a Higher Future Destiny." Both Robinson's and McLaughlin's calls for unity and cooperation were a preamble to a defense of for-profit utilities and "free enterprise." PSP&L's "Unity Without Malice" ad included a paragraph saying, "there is no one who regrets more than we that we have to fight for the very life of our business at the same time we must fight for the welfare of our country. . . . The survival of our business and the welfare of our country are both vital to us."[35] There was indeed a kind of truce during the war, but it was more superficial than substantive, and it masked continuing antagonisms and skirmishes between the two sides.

More lasting than the war emergency in moving public and private power toward détente was the influence of the BPA as a central source of low-cost power available to all utilities and as a facilitator of a unified electric grid in the Northwest Power Pool. In Washington, the Clark County PUD (near Vancouver, Washington) launched operations in 1942 even before it acquired the properties of Northwestern Electric and Portland General Electric (both IOUs) by outbidding the private utilities to serve the Vancouver shipyard with BPA power. According to BPA historian Gene Tollefson, the PUD's low bid saved the U.S. government $1.5 million over the next five years. Both PGE and Northwestern Electric fought condemnation suits over the next several years, but federal courts finally determined the fair market value of their properties in Clark County in 1944 and 1945. The PUD issued bonds and successfully acquired those properties in 1946 and 1948. The Grant County PUD (in the Columbia Basin, central Washington) got fair market value determinations from the courts in 1944 and 1945 and successfully purchased PSP&L and WWP properties within its district boundaries. WWP also lost its facilities in Douglas and Okanagan counties in 1945 to PUDs. The Douglas County PUD financed its acquisition with a loan from the REA.[36]

The BPA strongly supported these acquisitions and even intervened to try to negotiate a purchase of the entire PSP&L properties on behalf of several PUDs and Seattle City Light in 1944, but PSP&L turned down the $90 million offer. It did not argue that the price was unfair, but rather that the BPA lacked statutory authority to finance and acquire the property. Technically the PSP&L lawyers were correct, but it was a stalling tactic. Several pro-public-power legislators had introduced bills in Congress to grant BPA the authority to purchase private utility systems, but they ran afoul of other efforts to create a Columbia Valley Authority. In 1942, Washington senator Homer T. Bone had sponsored a bill to create an independent Columbia Valley Authority (CVA), while Washington representative Knute Hill sponsored a competing proposal to create a Columbia River Power Authority (CPA), similar to BPA, housed within the Department of Interior. Secretary Ickes opposed Bone's proposal for an independent agency, so Senator Bone adopted Hill's CPA proposal, adding language and funding authorization to permit the new power authority to acquire private utility systems. As FDR's administration lined up behind the compromise bill, Roosevelt, Ickes, and BPA administrator Raver publicly argued that complete public ownership of the Northwest power system would significantly advance the war effort. The Oregon press quoted Raver as saying, "Acquisition of all private power supplies in the Northwest by the Federal Government in the next six months is necessary for national defense and the region's industrial growth."[37] This understandably roused a most vigorous response from the commercial utility executives in the Northwest and in Washington, D.C., stalling the Bone and Hill bills in committee. The lack of action in 1942 left BPA as it was, with no new authorities. In the meantime, Raver pursued an alternative path to purchasing private utilities. He went to New York in 1942 to create a quasi-public organization that would get bank financing to buy commercial utilities in Oregon and Washington. This effort failed, too. Later that year, when the War Production Board ordered the creation of the Northwest Power Pool, the overt effort to buy all commercial utilities in the Northwest disintegrated. But BPA's efforts to assist in individual acquisitions continued. By 1943, the handwriting was on the wall for PSP&L. That year, the Seattle City Council passed a unanimous resolution announcing that Seattle City Light would serve all electric needs within the city after PSP&L's franchise expired in 1952. In other words, after 1952, PSP&L would lose its Seattle market whether it liked it or not. Seattle did in fact acquire those properties (at fair market value) in 1951.[38]

Advocates for the commercial utilities resented BPA's collusion with the public-power movement. A change of administration after the war softened BPA's advocacy, but throughout the New Deal and World War II era, BPA's promotion of publicly owned utilities rankled IOU executives. Pacific Power & Light's historian, John Dierdorff, disapprovingly lists the public-power leaders working for BPA: Seattle City Light director J. D. Ross, who became BPA's first administrator; BPA's director of public information Stephen Kahn, who cofounded the People's Power League in Oregon and produced several classic promotional films for the agency; BPA "consultant" Carl Thompson, who headed the Public Ownership League of America; Bonneville field representatives Fred Chamberlin of the Washington State Grange and Morton Tompkins of the Oregon State Grange; and many more. Reflecting the pervasive disgruntlement of IOU executives regarding BPA's blatant bias toward public power, Dierdorff prefaced his list of BPA's public-power men with the following passage:

> A great financial shot-in-the-arm for the public power movement of the period came through the early appointment of many of the faithful to the Bonneville payroll when the federal agency was being set up. Federal paychecks provided the meal tickets, government cars provided the transportation, government printing presses ground out literature to be spread along the way, and government photographers were busy making movies of the glamorous river developments.[39]

The public-power advocates kept up the pressure. After the Bone and Hill bills stalled in 1942, public-power advocates in Washington state organized behind a referendum for the 1944 ballot to expand the authority of PUDs to band together to purchase entire commercial utilities instead of merely those pieces contained within one utility district. An extensive media campaign by the private utilities helped defeat the measure by a significant margin. Interestingly, even some pro-labor newspapers in the region parted company with the state granges and editorialized against the measure, adopting the red-baiting language commonly employed by the business sector. A July 1943 editorial in the *Bellingham Labor News* stated,

> Due to the war, nearly one-fifth of the productive capacity of this country is now owned outright by the government. In plain words, a fifth of industry is already socialized. When the war is over, this era of socialism will either expand or be gradually replaced by private ownership. If it is not replaced by private ownership, the American workman will eventually lose his freedom.

A January 1944 editorial in the *Washington Teamster* warns, "Let us not fool ourselves! How far do we want to go toward government ownership of all industries? Do we want America Sovietized? The Teamsters do not want our American system destroyed. We are definitely opposed to a substitution of socialism for free enterprise."[40]

Shortly before Washington voters decided the fate of Referendum 25, Senator Bone accepted an appointment to the Ninth Circuit Court of Appeals and left the legislative arena. His replacement, Senator Hugh Mitchell, collaborated with then-Congressman Henry "Scoop" Jackson of Washington state in 1944 to reintroduce a third incarnation of a CVA bill. They were likewise unable to move it past the opposition. A fourth attempt in 1949, backed by President Harry Truman, also failed to gain approval. Consequently, the postwar BPA remained substantially as authorized in 1937 and 1940, and the mix of public and private power finally stabilized around the end of the decade. BPA historian Gene Tollefson concludes that "the die was cast by the 1942 interconnection decision." By the end of the war, he says, "the time when an all-public system of generation and distribution seemed possible had passed."[41]

The 1940s "cold war" between public and private ownership of electric systems resulted in a compromise. Each side lost some and preserved some of their goals and interests during World War II. Overall, however, public power gained more during the war than did its competitor. New Deal principles and initiatives continued to advance, while investor-owned utilities lost ground to PUD condemnation suits and became increasingly dependent on BPA power. U.S. Bureau of the Census data indicate that 66 percent of the power generated in the states of Washington, Oregon, and Idaho in 1945 was publicly owned. Interestingly, this is a dramatic reversal of the situation in 1940, when investor-owned utilities supplied 74 percent of the power in the three Northwest states combined. Another related development during the war was that the private utilities in Washington and Oregon bought increasingly larger percentages of their power load from BPA. In 1945, private utilities bought more than twice as much power from BPA as public utilities did. Even more remarkable, BPA supplied PSP&L with 30 percent of its total power and PGE with 68 percent of its power load. This reliance on BPA to supply a significant percentage of IOU power loads was an absolutely novel trend, launched during World War II, that persisted, in varying degrees, after the war and up to the present. World War II essentially forced a shotgun marriage between public and private power in the U.S. Northwest.[42]

In British Columbia, investor-owned utilities continued to completely dominate the province's power system throughout the war years, at least where its population and industry were concentrated. Rural areas were mostly without electricity or served by small, independent systems. In 1943, British Columbia and the Yukon together had only nine publicly owned power plants, generating only 17 million kilowatt-hours, serving only 21,449 customers. In contrast, the two provinces had sixty-two privately owned power plants, generating 2.6 billion kwh, serving 190,578 customers. In other words, the privately owned utilities generated over 99 percent of the power and served 8 of every 9 electric customers.[43] The rural areas of British Columbia were poorly served and electricity expensive and often unreliable, leading to growing complaints from customers and small municipalities. "Wartime dissatisfaction drove change," according to B.C. historian Matthew Evenden. In response to these problems and complaints, the B.C. government grew more interested in public power and rural electrification, and the B.C. Public Utility Commission grew more vigorous in regulating the investor-owned utilities. The rising fortunes of the socialist Cooperative Commonwealth Federation Party (CCF), which had won a plurality in the B.C. election of 1941, can be substantially credited for bringing public power and rural electrification to the forefront of political debate. The CCF party platforms in 1943 and 1944 urged public ownership of public utilities and provincially organized rural electrification, along with comprehensive social security insurance, a graduated income tax, collective farming and crop marketing, and more. During a provincial legislature budget debate in 1943, CCF leader W. W. Lefeaux called for the government to expropriate BC Electric and transform it into a provincially owned utility. This was too radical a proposal for the time in British Columbia, but it clearly started the ball rolling. The legislature that session established a committee to investigate the opportunities for greater rural electrification and provincial hydropower development.[44]

These events in British Columbia did not happen in a vacuum. The CCF, making significant headway in many provinces, pushed rural electrification and public ownership of utilities across the nation. The Liberal Party tended to favor public power, too, though more as a reform effort than a move toward state socialism. Six Canadian provinces had already established power commissions between 1906 and 1937: Ontario, Nova Scotia, Manitoba, New Brunswick, Saskatchewan, and Québec. Ontario passed a law to promote and subsidize rural electrification in 1930, three years before Franklin Roosevelt created the Rural Electrification Administration. Nova Scotia came next, passing a Rural Electrification Act in 1937. The provinces with the highest percentage of public power had the highest per capita consumption of electricity and the lowest electric rates. British Columbia's Rural Electrification Committee reported in 1944 that residential consumers of electricity in British Columbia used half as much power per capita and paid twice as much for it as their counterparts in Ontario. British Columbia had not yet taken the public-power route, but during the war it turned decisively in that direction. Québec may have served as an impetus, too. In 1944, the province's electric authority had accomplished little. But a new provincial law that year directed the new Québec Hydro-Electric Commission to take over several large privately owned utilities: Montreal Light, Heat, and Power; Montreal Island Power Company; and the Beauharnois Light, Heat, and Power Company. This expropriation included over 900,000 horsepower of electrical generation capacity and 365 miles of transmission lines, serving 1,500,000 people.[45]

Québec's bold move to provincialize its large urban utilities in the Montreal region served as an object lesson, worrying British Columbia's investor-owned utilities while encouraging the province's public-power advocates. Of even more concern to BC Electric were two municipal conferences held in British Columbia in 1944 to discuss public power. The first conference of twenty-nine towns and cities served by BC Electric voted unanimously in favor of a resolution asking the provincial government to expropriate the BC Electric Company and its subsidiaries. A second conference of nine municipalities in the Kootenay Valley voted two-to-one in favor of a similar resolution. (Two of the towns voting against it already had municipally owned utilities.) In fact, the growing support for some kind of provincially owned power authority was fairly widespread in the province by the 1940s, about as widespread as the brewing dissatisfaction with the province's commercial utilities that had done little to promote widespread electrification as envisioned and as implemented south of the border. Besides municipalities, other groups favoring positive provincial action to advance public power and rural electrification in 1944 included chambers of commerce, unions, service organizations, women's groups, and more. Not wanting to abandon the field to the CCF, the ruling Liberal-Conservative coalition warmed to the idea too. Of course, the Liberal Party under Premier Pattullo had advocated for public power and rural electrification all through the 1930s without success. Whether the war was coincidental or causal, the time for public power had finally arrived.[46]

Within a year, the B.C. legislature approved the Electric Power Act of March 28, 1945, creating the B.C. Power Commission (BCPC). Titled "An Act to provide for improving the Availability and Supply of Electrical Power," the law authorized the BCPC to develop hydropower sites, generate and distribute electricity, and purchase or lease in whole or in part other electric utilities. Though more conservative than the Québec law, this was nevertheless a significant departure from precedent in the province. The law protected the property and interests of the province's two largest utilities, BC Electric and WKP&L, by authorizing the BCPC to work only in areas not already served by those two companies and their subsidiaries. In this way the enabling act was similar to the law establishing the Rural Electrification Administration in the United States. Outside the territories served by the two big utilities, the enabling law encouraged the BCPC to actively acquire and consolidate smaller private and public utilities and to seek wider access, increased consumption, and lower rates through economies of scale. The law even permitted the lieutenant-governor in council to authorize the BCPC "to take by expropriation" any electric power property "without the consent of the owner," after paying fair market value. Capping off this accomplishment, in 1945 the legislature voted to appropriate $10 million to the BCPC for rural electrification. This was just one of many public works projects being funded during this time of flush budgets, projects former premier Pattullo longed to see approved during the Depression, when the province was, unfortunately, broke.[47]

In response to these developments, BC Electric announced in late 1944 a $50 million investment upgrade of its facilities, spread over the next five years. The following year it announced a five-year $1 million rural electrification program.[48] Many observers felt it was too little too late, while others were satisfied that the threat of expropriation had turned the company onto a more socially responsive path. As it turned out, the B.C. Power Commission and BC Electric actually cooperated for mutual benefit in the early postwar years. The Power Commission offered to take over some of BC Electric's more scattered, less profitable service areas, the kinds of places a government-run nonprofit utility can serve more effectively while offering lower rates to customers. BC Electric was apparently happy to oblige. As the company historian Cecil Maiden explains, "The end of 1946 saw the further simplification of operations (and a very wise confining of them to the basic areas) when . . . the outlying electrical properties at Kamloops and of the company's interests in the central portion of Vancouver Island were disposed of to the B.C. Power Commission." Maiden adds, ironically and perhaps wishfully, "More and more was the Company now becoming identified with all that was new and progressive in the story of British Columbia."[49] Little did he suspect when he finished his narrative of the company in 1948 that thirteen years later BC Electric would cease to exist, expropriated by the province under the Social Credit government of former Conservative W.A.C. Bennett to form the BC Hydro and Power Authority.

While the provincial legislature was creating the B.C. Power Commission, the B.C. Public Utility Commission grew a backbone and started ordering rate reductions and more customer-friendly policies regarding electric-line extensions during 1943 to 1945. Such a change of heart in a public utility commission usually comes after a change of leadership on the commission or a post-election change in governing policies. Neither explains why the B.C. utility commission suddenly started tightening the screws. The exact same three commission members served throughout the war years, and the same ruling coalition was in power in Victoria from 1941 to 1944 and reelected in 1945. Nevertheless, the commission shifted away from policies and decisions that preserved the commercial utility-dominated status quo (see chapter 9) toward more consumer-oriented regulatory activism. As in the United States, the B.C. utilities commission examined the books of for-profit utilities to determine the company's value. Combined with a commission-established "fair rate of return" on investment, the valuation determined the allowable rates the company could charge customers. Higher property valuations permitted higher electric rates. The commission had launched a valuation study of the BC Electric Company and its subsidiaries in 1939. After two years, it published a draft valuation report and held public hearings through early 1943 in various towns and cities served by the company. Not surprisingly, BC Electric's executives roundly criticized the conclusions reached by the commission's chief engineer and its accounting consultant, George A. Touche & Co. In one of the public hearings, BC Electric's vicepresident, E. H. Adams, excoriated the valuation report as "irrelevant, incomplete, inaccurate, one-sided, and inadequate." Nevertheless, the valuation stuck. In 1943 the PUC formally determined the company's rate base as $110 million and allowed it a 5.3 percent annual rate of return. The company was disappointed because this valuation and rate of return justified a commission order providing "immediate relief" to customers in the form of a "free billing period" of one to two months that year to most of the company's customers.[50]

This shift in direction by the PUC caught BC Electric and the other big commercial utility company, West Kootenay Power and Light, by surprise. They scrambled to adjust to the new regulatory environment and recover their public image as "public service" corporations. BC Electric's historian in the 1940s, Cecil Maiden, was so confused by the turn of events that he completely misinterpreted the commission's ruling. Maiden erroneously claimed that the report by the commission "found the Company's rates to be much less than sufficient to provide a fair return on the rate base recommended." The utility commission reports confirm the exact opposite conclusion. Not only did the utility commission require the company to provide rebates in 1943, but it arm-twisted the company to provide additional rate reductions the next year. As the 1944 PUC annual report says, "Subsequent discussions between the Commission and the Company officials . . . led the company to voluntarily reduce both domestic and commercial rates in the Vancouver Island South area and the Alberni District served by the company." The PUC next turned its attention to West Kootenay Power and Light, initiating valuation proceedings in 1944. Seeing the handwriting on the wall, WKP&L executives opted for an informal agreement with the commission in April 1945 to lower its commercial and domestic rates by 20 percent.[51]

In addition to lowering rates, the B.C. utilities commission took a different tack on rural line extensions in 1944 and 1945 than it had earlier. In 1940 and 1941 the commission nearly always sided with the companies in settling complaints by customers about the unwillingness of the commercial utilities to bring new electrical service to homes and farms lacking it or about the high cost of line extensions that companies did make. In regard to those complaints, the commission commonly concluded, "revenue requirements of Company not unreasonable or discriminatory."[52] With the agitation for rural electrification becoming mainstream by 1944, the PUC pressured the utilities to accommodate more extension requests and make them more affordable. As reported in the PUC's 1944 annual report, of seven customer requests for extensions, the commission granted four of them and denied two, while another was under "arrangement."[53] The next year, to facilitate more extensions, the commission completely revised utility policies regarding service extension financing. It convinced BC Electric to "construct up to 300 feet of pole-line per customer without making any special requirement as to revenue." For longer extensions the company would require only a modest $2/month revenue guarantee from the customer, who would also pay 30 cents for each additional 100 feet of electric line beyond the initial 300 feet provided free by the utility. WKP&L similarly agreed that year to revise its extension policies, promising to construct distribution lines without charge to customers wherever and whenever there were at least ten customers per mile. This was an improvement over previous extension policies, although the REA in the United States, by comparison, required only three customers per mile for a rural co-op seeking a federal loan to build distribution lines.[54]

Reflecting its conservative roots, the B.C. utilities commission preferred to let the utilities do the right thing "voluntarily," if possible. It saved time, money, and grief. Many utility executives, too, preferred this route; it saved face and allowed the company to claim it was lowering rates or liberalizing its extension policy on its own volition as a public service to its customers. But just as in the United States, significant action to limit excess profits and protect the interests of the consumer usually required either the assertive hand of a regulator or at least the threat of action by a suitably empowered government agency.

CONCLUSION

With its protected deep-water harbors, access to Asia, and abundant natural resources, the north Pacific coast region was a prime location for war-mobilization efforts. Both world wars greatly accelerated generation and consumption of electricity, increased economies of scale in electrical generation, expanded government regulation of the industry, and advanced electric system interconnections. Mobilization for the Second World War was much more extensively fueled by cheap, abundant hydropower than the war boom two decades earlier. The prewar presence of a federal power authority in the U.S. Northwest, the BPA, greatly facilitated electricity-driven industrial development. Because of the extraordinary boon of surplus federal power from the Bonneville and Grand Coulee dams, the war industries in the U.S. Northwest expanded much faster and much more extensively than in British Columbia. The lack of a provincial power authority in British Columbia and the Canadian government's focus on war-production facilities in Ontario and Québec left British Columbia somewhat marginalized compared to eastern provinces and British Columbia's neighboring states south of the border. Nevertheless, the war significantly increased the presence of and public acceptance of government controls in both Canada and the United States. Historian Matthew Evenden observes that in British Columbia the war brought "a new role for the state in economic development and collective affairs."[55] The Depression and New Deal had accomplished that same change in political culture and political institutions in the United States a few years earlier.

As in World War I, both governments during World War II actively worked to ensure that adequate power would be available when and where it was needed. They built or subsidized new generation, increased power system efficiencies through utility interties, and redirected electricity from low-priority uses to high-priority uses. In the Northwest states, the U.S. government pursued the first two options so aggressively that little rationing of electric consumption by households and businesses was needed during the war. In contrast, other than the Brilliant power plant built by mining giant COMINCO on the lower Kootenay River, British Columbia added hardly any new generation capacity during the war. Consequently, wartime electricity-supply challenges in British Columbia had to be solved almost exclusively by rationing and redirecting consumption. As a result, British Columbians experienced a great deal more electricity restrictions, prohibitions, and brown-outs than in the U.S. Northwest. This likely explains British Columbia's popular dissatisfaction with the investor-owned utilities after 1941 and the calls for provincial expropriation that led to the regulatory reforms of 1943–1945 described in this chapter. To be fair, BC Electric had twice lobbied the Canadian government during the war for permission to use rationed war materials to build a new hydropower facility on Bridge River, and the dominion government twice turned the company down. The political situation in British Columbia regarding electric utilities might have been different if Ottawa had approved the request.

At war's end, a new and different electric system existed in the Northwest, from which there would be no backtracking. On both sides of the border, public power came to play a much larger role in the energy mix. This trend started in the U.S. Northwest in the 1930s, and the war accelerated it. From 1940 to 1945, the ratio of public power to private power in the U.S. Northwest fully reversed, with public power making strong gains over the commercial sector. In British Columbia, the trend toward public power started during the war but would not mature until well afterward. But in both cases, a unique blend of public and private power solidified and became a permanent fixture of the region. In British Columbia, according to Evenden, "By the end of the war, a distinctive, mixed system of private power in the cities and public power in the hinterland had emerged."[56] In the U.S. Northwest, the geographic distribution of public and private utilities was more complex. Rural co-ops dominated the most remote areas; municipal-owned utilities were scattered throughout the region in both large and small cities; and county-wide PUDs dotted the Northwest, competing with preestablished private utilities serving extensive interstate territories.

Surrounding and profoundly influencing the public-power movement was the BPA, a federal power agency that some thought would be only temporary but that became the most influential single organization in the entire Northwest power system for the remainder of the twentieth century. In its first decade, 1937–1947, the BPA strongly promoted and advantaged publicly owned utilities. But as it grew increasingly essential to the entire integrated, public-private Northwest power system, it softened its partisanship. In fact, even while BPA was steeped unapologetically in its pro-public-power policies of the war years, publicly owned utilities received only a very small percentage of BPA's electrical output. Right from the start, BPA served as an integrating force among the contentious and formerly insular nonprofit and for-profit sectors of the industry. The exigencies of war started this trend, which would continue after the war. By 1952, according to historian Gene Tollefson, the "Northwest Power War" had concluded, with "little change" in the mix of public and private power since then.[57] With those two wars out of the way, a mighty, concerted, unified effort ensued on both sides of the border to build on wartime precedents and develop more power, more industry, and more wealth by fully controlling the region's rivers to maximize hydropower production.

CHAPTER 12

"Endless Limitless Development": A Telescoped View of the Postwar Era

When the Liberal-Conservative coalition government in British Columbia geared up for the elections of 1945, the two coalition leaders, John Hart (Liberal) and Royal Maitland (Conservative), proffered on behalf of their parties a forty-three-point governing manifesto, the theme of which, according to political scientist Martin Robin, "was development, endless limitless development; the old boom song of frontier conquest."[1] With an overflowing treasury and an end to wartime restrictions on materials, the B.C. government launched a massive postwar public works program building bridges, highways, and hydropower facilities, while assiduously recruiting industry to the province. The B.C. populace, after the privations of the Depression and the rationing and sacrifices of the war, jumped at the chance to consume freely, bolstered by high employment, good wages, and a general culture of victory and economic optimism. The creation of the B.C. Power Commission in 1945 and the agreement to divide the province into cooperative public and private power sectors "unleashed an unprecedented hydroelectric building program in BC," according to B.C. historian Matthew Evenden. Immediately after the war, BC Electric started construction of its long-delayed Bridge River facility and interconnected with the BPA grid at the international border. British Columbia's Power Commission acquired numerous small hinterland utilities, initiated an improved transmission system to eventually integrate the scattered towns it would serve, and started construction on two hydropower facilities, one to serve central and northern Vancouver Island and another in the Okanagan Valley.[2] Development defined the next generation in British Columbia.

The exact same forces and inclinations animated the states south of British Columbia. The U.S. Congress went on a veritable water development binge during the generation following World War II. In the 1945 River and Harbor Act, Congress authorized 257 major construction projects for the U.S. Army Corps of Engineers, including twenty-four flood-control, navigation, and hydropower projects in Washington, Oregon, and Idaho. That came only a year after Congress had authorized six Corps of Engineers projects in the Northwest under the Flood Control Act of 1944. Among these immediate postwar projects were a series of dams on the lower Snake River to provide slackwater navigation all the way to Lewiston, Idaho; a series of projects on the Willamette River above Portland; and the huge Umatilla (McNary) Dam on the Columbia River, finished in 1953. The U.S. Bureau of Reclamation had a new portfolio of projects, too, including extensive irrigation development in the Columbia Basin made possible by its dam at Grand Coulee and a new massive storage reservoir in Montana called Hungry Horse, authorized in 1944. Congress gave the Bonneville Power Administration authority to market the power from all these additional federal dams, as well as other federal projects launched in subsequent years, including these gigantic Columbia River main stem dams: Chief Joseph, finished in 1955; The Dalles, finished in 1957; and John Day Dam, finished in 1968.[3] The PUDs got into the act, too, building Columbia River main stem dams at Priest Rapids (1957), Rocky Reach (1961), Wanapum Dam (1963), and Wells Dam (1967). Besides these Columbia main stem dams, the federal government, PUDs, municipals, and private utilities all built hydropower facilities on the region's tributary rivers, too. Congress authorized the Corps of Engineers to build Libby Dam, creating the gigantic Koocanusa storage reservoir on Montana's reach of the Kootenai River in 1950 (though it was not completed until 1975) and Albeni Falls Dam on the Pend Oreille River in 1951; The Bureau of Reclamation built the Chandler (1956) and Roza (1958) dams on the Yakima River in Washington; WWP built Cabinet Gorge Dam (1952) and Noxon Rapids Dam (1960) on the Clark Fork River in Montana; Seattle City Light expanded Ross Dam (1949) and built the Boundary Dam on the Washington segment of the Pend Oreille River in 1967; Idaho Power Company built Brownlee Dam (1958), Oxbow Dam (1961), and Hells Canyon Dam (1967) on the Snake River. And there were more.[4]

A new and different crisis in the Northwest helped precipitate much of this development. Just as the Depression and the war stimulated government programs to address those crises, disastrous region-wide floods in the spring of 1948 spurred a rash of flood-control and comprehensive river basin development schemes on both sides of the border. The second largest flood ever recorded on the Columbia River wrought havoc on floodplain communities throughout the watershed in late May and early June 1948. The flood completely wiped out the hastily built World War II worker community of Vanport, which for a time was the largest suburb in Oregon. Local resident Manly Maben describes the scene:

As the surging waves first moved in, they quickly hit the many sloughs. One report described showers of spray 50 feet high upon impact with the slough water. Then, for 35 to 40 minutes a creeping inundation occurred as the sloughs absorbed much of the water. After they were filled and a sheet of water had spread over Vanport, the waves began to roll again; cars were sent careening, houses wrenched apart.[5]

Crises often precipitate institutional change, either accelerating existing trends, as in the two wars, or stimulating reform and innovation, as in the Depression. The 1948 flood served as an accelerator. In this postwar Keynesian political economy, Americans and Canadians embraced river development more enthusiastically than ever. Monumental ambitions for reengineering nature defined the postwar generation, spurred on by a desire to prevent any recurrence of disasters like the 1948 flood. Flood control in the postwar era emphasized dams and storage reservoirs, and every flood-control dam was outfitted with hydropower turbines. This entwined the power system with flood-control and river development regimes even more tightly than before the war. In the United States, the Albeni Falls, Libby, Priest Rapids, John Day, The Dalles, and Chief Joseph dams were all authorized or reauthorized as flood-control measures in the 1950 River and Harbor and Flood Control Act.[6]

British Columbia suffered from epic flooding in the spring of 1948, too, with the same political response as in the United States. The second largest Fraser River flood in recorded history raged through greater Vancouver at the end of May, inundating thousands of acres of farmland and destroying dikes, bridges, roads, rail lines, power lines, and 2,300 homes. Immediately afterward, the media and policy makers searched for solutions, many calling for a comprehensive watershed approach to the control of floods. Typical of the era, the preferred approach involved extensive reengineering of the Fraser River and its tributaries rather than changing how people occupied the river valleys: flood *control* was the operative word. A leading provincial newspaper, *The Province*, wasted no time editorializing in the midst of the flood about its policy implications:

> The place to control the Fraser is up above, where the water comes from. The way to control it is to control the tributaries. For this we need a Fraser River Authority—a board with authority enough and resources enough to take the river in hand, control its transports, coordinate its industries and make it work consistently for the benefit of the province as a whole.

As Evenden notes, most engineers and policy makers believed the way to control the tributaries was to dam them: "Just as on the swollen Columbia in 1948, the excuse of a flood was quickly deployed to promote dam construction." That fall the province and dominion established a joint board to devise a plan to prevent future floods. It eventually proposed construction of ten dams in the upper Fraser Basin. And that was only one of five development schemes proposed for the Fraser Basin in the 1950s. Later in the decade, the B.C. Social Credit government scrapped Fraser River development over concerns about the salmon fishery and instead moved forward aggressively with development of the province's other two biggest rivers, the Columbia and Peace.[7]

All this river development in British Columbia and the Northwest states profoundly influenced the evolution of the regional power system from the late 1940s into the 1960s. BPA secured its position in the region as power czar and regional chamber of commerce. The investor-owned utilities in the United States held their ground against the acquisition efforts of the PUDs after 1951, expanding their facilities and regaining their clout in the next two decades. The private utility sector in British Columbia, in contrast, grew comparatively weaker. Public power in British Columbia advanced dramatically in the 1950s and especially the 1960s in terms of power supplies and territory served. It did so by building new facilities as well as acquiring privately owned facilities. In the United States, the public-power utilities also dramatically increased their power supplies, but by building their own dams rather than by taking over private systems. Cooperation between private-owned and public-owned utilities in the United States advanced in these years, as did cooperation between both of them and the BPA. In contrast, cooperation between public and private utilities in British Columbia eventually broke down in this era.

Grid integration advanced both within the region and beyond, as BPA sponsored interties in the 1960s from its Northwest grid south to California and east to Alberta and Montana. Joint U.S.-Canadian development of the transboundary Columbia River in the 1960s greatly enhanced regional cooperation and integration. This epic building boom virtually destroyed the multimillion-dollar annual salmon fishery in the United States, while British Columbia succeeded in preserving a large portion of its fishery even while developing more hydropower than the province needed.

Technological progress continued to make power generation more efficient and electric rates cheaper during the postwar years until the 1970s, when economies of scale finally reached a peak and ceased to reduce the cost of power. The constant increase in supply and decrease in cost through the 1960s kept per capita power consumption steeply rising and supported great inefficiencies in the end-use of electricity. This, too, came to a stressful end in the 1970s.

INVESTOR-OWNED UTILITIES IN THE POSTWAR YEARS

Investor-owned utilities (IOUs) experienced some losses and some gains in the postwar generation—more gains in the U.S. Northwest and more losses in British Columbia. Perhaps most significantly, the IOUs in the Northwest states gained their independence from the utility holding companies that had hung on from the prewar decades. Under pressure from the SEC, which enforced the Public Utility Holding Company Act of 1935 (PUHCA), the big holding companies like Stone & Webster and Electric Bond and Share (EBASCO) incrementally lost control over their subsidiary properties in the Northwest and elsewhere in the United States throughout the 1940s and 1950s. Stone & Webster offered to sell its holdings of PSP&L as early as 1943, though negotiations dragged on for a decade. EBASCO sold its Idaho Power holdings in 1944. Then in 1946 the U.S. Supreme Court upheld an SEC order directing EBASCO and two other holding companies to dispose of their ownership in five other Northwest utilities, including WWP, Montana Power, PP&L, and Northwestern Electric. The divestitures still took some years to accomplish but by the early 1950s all these IOUs had become independent of their former holding companies. Historian Gene Tollefson notes that, nationwide, 759 utility operating companies gained independence from holding companies between 1935 and 1950, and the total number of holding companies dwindled from 216 to 18 due mainly to enforcement of PUHCA.[8]

The year 1951 marked a milestone for the private utilities in the Northwest states. That year, PSP&L sold its Seattle facilities to the city of Seattle to be absorbed by the municipal utility Seattle City Light. During the next five years, City Light reduced rates to PSP&L's former customers by a total of $2.5 million. This was the largest single purchase of an investor-owned utility property by a public-owned utility in the region's history, and it marked the end of large public-power acquisitions of private utilities. On the bright side for the region's IOUs, 1951 also heralded the start of several large new hydropower dams, stimulated by a rising defense-related power market due to the Korean War and generous federal tax incentives arranged by the new Defense Electric Power Administration. WWP started Cabinet Gorge Dam on Idaho's Clark Fork River, its first new hydropower facility since 1928, and finished it in a remarkably quick twenty-one months. The dam could generate up to 200,000 kilowatts, doubling WWP's generation capacity. The same year, Pacific Power & Light started Yale Dam on Washington's Lewis River, completed two years later with a generation capacity of 108,000 kw. Shortly after, PP&L built Swift Dam on the Lewis River, completed in 1958 with a capacity of 268,000 kw.[9]

Also in 1951, Portland General Electric dusted off and expanded an older proposal to build Pelton Dam on the lower Deschutes River of Oregon. This turned into a bitter seven-year battle over fisheries, which foreshadowed dozens more similar battles to come over the next three decades. It involved state, federal, and tribal agencies and pitted fisheries interests against power interests and the federal government against state government. A defunct Columbia Valley Power Company had received a license from the FPC back in 1924 to develop the Pelton dam site but never followed through. The FPC consequently cancelled the license in 1936. Twelve years later, a consortium composed of PGE, PP&L, and WWP resurrected the original plan and applied for an FPC license. By then, the devastating effects of large power dams on salmon and other anadromous fisheries were well-known, and Northwest fisheries interests were better organized to defend their interests. Oregon legislators introduced two bills in 1948 to designate the Deschutes and Metolius rivers off-limits to power production to preserve their valuable fisheries. Neither bill passed, but the opening volley had been fired. In the aftermath, two contending sides emerged: on one side were arrayed the three utilities backed by the Warm Springs Tribal Council, the FPC, and the federal government's National Security Resources Board, which supported additional power generation for the Korean War effort. On the other side were arrayed the Oregon Fish Commission, the Oregon Hydropower Commission, the Chinook Indian Nation, and the U.S. Fish and Wildlife Service. The proposed dam site and its reservoir bordered the Warm Springs reservation and federal lands. The controversy delayed approval of the license request for three years, causing PP&L and WWP to back out of the project in 1951. PGE then revised and resubmitted a new proposal in 1951, promising a $1 million investment in fisheries-mitigation measures, but the fisheries interests were not placated.[10]

In an interesting development, the U.S. government's FPC granted PGE the license for developing Pelton Dam while Oregon state's Hydropower Commission denied the license; both sides went to court. The Oregon Supreme Court sided with the state agency against PGE, while the U.S. Supreme Court sided with the FPC against the state of Oregon. Once PGE got the go-ahead from the U.S. Supreme Court, it started construction. That prompted Oregon's senators, Richard Neuberger and Wayne Morse, to introduce legislation in the U.S. Senate to revoke FPC's permit in order to undermine the Supreme Court ruling.

In the meantime, the Chinook Tribe, concerned about the effects of the dam on the entire lower Columbia River fishery, which was declining yearly due to dam development, publicly lobbied the Warm Springs tribes to deny PGE permission to use tribal land for the dam and reservoir. The Warm Springs tribes had already signed an agreement with PGE, entailing annual payments that the tribal council considered adequate. Dueling letters to the editor in regional newspapers carried on the fight in 1956. A letter to the editor of the *Oregonian* in March 1956 by J. Grant Elliott, chairman of the Chinook Nation, stated in part: "It is our opinion that Portland General Electric . . . has misled the Warm Springs Indians into accepting an amount for a dam site on their land that is equivalent to about one fortieth of the annual value of the salmon and steelhead runs that spawn in the Deschutes River system and may be destroyed by it." None of the efforts stopped the dam. PGE broke ground in 1956 and dedicated the dam's completion in 1958. The dam included a three-mile-long fish ladder, at that time the longest in the world, but the fishery nevertheless suffered. Adding insult to injury, the FPC also granted PGE a license in 1952 to build another dam nine miles above the Pelton Dam at Round Butte, backing slackwater up the Deschutes, Metolius, and Crooked rivers. PGE began construction on the 300,000 kw Round Butte hydropower dam in 1961, finishing it in 1964. The Warm Springs tribe now owns a third of it. As happened at many other dams in the region, the fish-passage facilities failed. Some adult salmon made their way up the fish ladders, but the young salmon could not find their way to the downstream passage facility at Round Butte, so the fishery collapsed, and PGE closed the dam's fish-passage facilities in 1968.[11]

Conflicts between hydropower advocates and salmon advocates erupted all over the region on both sides of the international border. In one of the most infamous fish-passage failure incidents, an entire run of Chinook salmon perished in the fall of 1958 at the base of Oxbow Dam, under construction by the Idaho Power Company in Hells Canyon on the Snake River. Embarrassed by the publicized tragedy, Idaho Power tried again and again in the next few years to make its fish-passage facilities work effectively but finally admitted defeat in 1960. This Oxbow incident galvanized fishery advocates in Idaho and in part led to the political decision to keep dams off Idaho's Salmon River, a major tributary of the Snake River downstream from Idaho Power's complex of three dams in Hells Canyon.

Despite this high-profile failure and growing clashes with fisheries interests, Idaho Power's developments in Hells Canyon reveal the rising fortunes of the investor-owned utility sector after the election of Republican president Dwight Eisenhower. The IOUs substantially benefited from the change of administration. Eisenhower announced what he called his "partnership" policy, a euphemism for reversing the activist tendencies of the federal government since the New Deal. His preference for business-government partnerships meant, in practice, relying more on corporations and market forces to guide economic planning and development. In the field of electric power, the administration and its congressional allies wanted to end or reverse the advances of public power and allow the commercial utilities to determine future development. This policy shift played out dramatically in Hells Canyon. Starting in 1948, an epic struggle erupted between Idaho Power and the federal government over who should develop hydropower sites in Hells Canyon, the nation's deepest gorge. While Truman had been in office, public-power advocates had something of an upper hand, with support from BPA and FPC for the Bureau of Reclamation to build a "high" dam in Hells Canyon to bolster the BPA power grid. But controversy stalled a resolution of the question for years, until the election of 1952 put a Republican in the White House. Eisenhower's partnership policy immediately translated into his controversial "no new starts" policy regarding federal hydropower dams. With the Reclamation Bureau thus stymied and new political appointees to head the FPC and the BPA, Idaho Power gained the upper hand. In 1955, the FPC finally approved Idaho Power's application to build three dams in Hells Canyon. After a brief legal challenge, both the U.S. Supreme Court and Congress ratified the FPC's decision in 1957. Idaho Power subsequently built the three dams.[12]

In British Columbia, private corporations in the electric utility industry flourished from the end of the war through the 1950s and then suffered a dramatic setback. Voters elected a new right-of-center government in 1952, led by the ambitious W.A.C. Bennett of the newly victorious Social Credit Party. His party held power for the next twenty years, but their electric power policies took a surprising turn in 1961. A flurry of river development proposals surfaced in the 1950s under the Social Credit government; they were focused on three different river basins: the Fraser, the Columbia, and the Peace. The privately owned BC Electric Company's main service territory was in the lower Fraser Basin, so that is where it concentrated its attention. After completing its powerhouse at Bridge River in 1948, the company built several more upstream storage reservoirs in the next decade to divert water through tunnels to the power plant and added three more generators, bringing its capacity to 180,000 kw. It also started a new 60,000 kw hydropower plant on Wahleach Lake near Chilliwack in 1950 and more than doubled the generation capacity at its Buntzen Lake hydroelectric facility in 1951 by replacing seven small turbines with one larger, more efficient turbine. Where hydropower facilities were not available, BC Electric added a few diesel-fired electric plants to supplement its growing power needs.

In addition to these smaller construction projects, the company got on board for several ambitious plans hatched in the 1950s by wealthy investors and provincial and national political leaders to dam the Fraser River and its tributaries. Some of these proposed projects, like the gigantic Moran Dam proposed for the main stem of the Fraser, rivaled Grand Coulee. A coalition of interests including the provincial government and the influential General Andrew McNaughton, chair of the Canadian Section of the International Joint Commission on Boundary Waters, pushed comprehensive Fraser River Basin development. All the plans called for the private sector to undertake financing and construction of the projects with government support and incentives. But McNaughton's plan was so politically complicated and costly that government approval and private-sector financing remained elusive, continuously delaying development of the Fraser. A major factor complicating hydropower development there was the world-class salmon fishery, of which more will be said below.[13]

Meanwhile, in the Columbia River watershed, a population boom in the Kootenay Valley and general industrial prosperity led COMINCO and West Kootenay Power and Light to seek additional power supplies. In the early 1950s, COMINCO got a green light to build a large dam on British Columbia's portion of the Pend Oreille River near its confluence with the Columbia River. WKP&L meanwhile met some of its load growth in scattered small towns by adding diesel generators. But the big story in the region was a proposal for comprehensive development of the Columbia River. The United States had been pressuring Canada since 1944 to collaborate on developing the Canadian portion of the Columbia to improve flood control and hydropower system performance in the U.S. reach of the river. Basically, the United States wanted Canada to build a series of large storage reservoirs to capture snowmelt in the spring and then release it in the low-water season of late summer and fall. The flood of 1948 added urgency to the lobbying by the United States. But British Columbia's government in the 1940s and early 1950s was more interested in developing the Fraser, which was closer to its main population center, so the joint development dreams of U.S. policy makers languished unfulfilled. Finally, America's wealthy industrialist Henry Kaiser stepped in and asked the B.C. government to allow him to develop a hydropower dam and storage reservoir in the Arrow Lakes region of the Columbia watershed. Kaiser promised the B.C. government 20 percent of the power generated at the plant in exchange for the privilege of building the facility and won tentative approval in 1954 from Premier Bennett. But the Canadian government scuttled the deal in 1955 because it was sponsored by an American company primarily to serve American interests downstream. McNaughton had a hand in its veto, too. He wanted to develop Canadian waters for Canadian benefit and had proposed to divert a portion of the Columbia River near Mica Lake through a huge pipeline over the mountains into the Fraser watershed where it would be used to supplement the flow of the Fraser, which he hoped would be developed with ten dams. American-promoted development of the Columbia in Canada threatened his plans.[14]

To make matters even more complicated, Premier Bennett latched on to what was either a mad or brilliant scheme to develop the remote but mighty Peace River in northeast British Columbia. It troubled him that development of the province was so uneven, and he longed to bring the presumed blessings of electric power and industry to the more remote central and northern regions of the province. Once again hoping to attract private investment capital to accomplish this dream with government incentives, Bennett solicited proposals from the private sector. The person who emerged with a plan as grandiose as Bennett's ambition was the Swedish millionaire arms dealer Axel Wenner-Gren. He proposed in 1956 a plan for a gigantic dam on the Peace River whose abundant hydropower would support new mining, manufacturing, and pulp and paper mills, along with a collection of new working-class towns that would spring up around those industries. Bennett was thrilled and announced to the province in 1957 that the Peace River and northern British Columbia finally would experience major development. But complications stalled the project. Many politicians in British Columbia did not trust the Wenner-Gren Corporation and resented that a foreign capitalist was going to develop B.C. resources. Others felt the plan was unrealistically ambitious. Financing, once again, posed the biggest problem, however. Wenner-Gren could not convince investors to put up the money for the project without having firm contracts from utilities to buy power from the dam. The hydropower revenue stream was the essential lynchpin for the whole development scheme. Unfortunately, the Peace lay so far distant from the centers of population and industry that no ready power market was available. Hoping to save the project, Premier Bennett stepped in and asked BC Electric to contract for a large block of power from the proposed Peace River project, but the company flatly refused. It believed the transmission distance was too great, the costs would be too high, and the project too speculative. So the Peace River development collapsed, at least for the time being.[15]

With his more grandiose development ambitions frustrated at every turn, Premier Bennett took a bold and surprising path around the stumbling blocks. In August of 1961, just as the Canadian government was thick in final negotiations with the United States over joint development of the Columbia River,

Bennett's administration submitted to the B.C. legislature a Power Development Act providing for provincial expropriation of both the Peace River Power Development Company (the successor to the Wenner-Gren Corporation) and BC Electric. Unexpectedly, it passed easily and quickly, in part with the assistance of the Cooperative Commonwealth Federation opposition party that had advocated for expropriation of the province's privately owned utilities for at least two decades (see chapter 11). The law stipulated that both companies be paid fair market value for their properties and investments. But in fact, when all was said and done in 1963, the B.C. government paid about 15 percent more than assessed value for BC Electric. The legislature passed a related law in 1962 merging these two expropriated companies together with the provincially owned B.C. Power Commission to form the BC Hydro and Power Authority, today known as BC Hydro. The expropriation left COMINCO and WKP&L alone, but BC Hydro eventually absorbed most of the rest of the power systems in the province into its unified provincial public-power system.[16]

Bennett, a quasi-conservative admirer of capitalism, pushed for this development, not to stifle private enterprise, as his critics charged, but to promote it. Though publicly excoriated in much of the pro-business press for this maneuver, Bennett was only interested in promoting development on the scale seen in the U.S. Northwest. He and his allies in government saw the private sector as timid and obstructionist. So they decided in this case to do it themselves. A contemporary report from the newly established B.C. Energy Board had determined that a nonprofit provincial power authority could not only promote jobs and industry in northern British Columbia but also supply power at a competitive price to the urban and industrial regions of the southern B.C. mainland. Provincial acquisition of the Peace River Power Development Company guaranteed that the Peace would be developed while acquisition of BC Electric guaranteed a market to make the project financially feasible. Moreover, that left the Columbia River available to serve both B.C. and U.S. interests and took the Fraser River off the table entirely, much to the relief of fishery interests. This outcome, in which the Peace and Columbia rivers would both be developed, the former to serve British Columbia exclusively and the latter to serve international interests, was the fulfillment of Bennett's Two Rivers policy that he had pursued for the previous five years. It led to the kind of hydroelectric-based, cheap energy–driven economic boom that the Northwest states had enjoyed since 1938. Colossal hydroelectric development projects so powerfully influenced events for the rest of the decade that Martin Robin titled his chapter on late 1960s B.C. political history "Big Dam Government."[17]

POSTWAR PUBLIC POWER IN THE U.S. NORTHWEST

After World War II, the PUDs in Washington and Oregon continued their piecemeal acquisitions of private utilities within their county boundaries, some by purchase agreement but most by condemnation and court proceedings. Between 1945 and 1951, eleven PUDs acquired significant investor-owned properties in their districts and began or expanded their nonprofit service.[18] Nearly two dozen PUDs were in operation in Washington and Oregon by 1947, plus over two dozen municipal utilities in Washington, Oregon, and Idaho. Reflecting the maturation of the public-power sector, the Northwest Public Power Association (NWPPA), initially formed in 1940, hired its first executive secretary in 1947, Gus Norwood. Norwood was a Harvard- and Annapolis-trained World War II naval officer who transformed the NWPPA into a major advocate of public power in the Northwest. He later went to work for BPA and wrote one of three major histories of the federal agency, cited often in this study.[19] NWPPA spoke for both PUDs and municipals early in its organizational life, then Norwood brought on board the rural electric co-ops in 1947. In addition to NWPPA, the Washington Public Utility Districts' Association in Seattle, the Oregon State R.E.A. Association, and other similar organizations all advocated for public power in the region. Often these state and regional associations were offshoots of national organizations that provided additional resources and support, especially the American Public Power Association (APPA) and the National Rural Electric Cooperative Association (NRECA), which remain active today. Shortly after joining NWPPA, Norwood launched the "One-Cent Club," a promotion for utilities whose residential retail power rates were at or below 1 cent per kwh. The average retail residential rate in the region in 1947 was 1.7 cents/kwh. The municipal utilities in Salem and Tacoma, along with several PUDs in southwest Washington, were the first to join the club. Most PUDs and municipals met the criteria for membership in the next few years, with the exception of rural co-ops, which never had enough customers to get the economies of scale needed for rates that low. The One-Cent Club remained in force until NWPPA disbanded it in 1968 in the face of rising electric rates.[20]

As noted in chapter 10, expansion of the rural electric co-ops slowed significantly during World War II but boomed again after the war, when critical war materials rationing ended and the REA began making significant loans again. Congress reauthorized the REA in 1944 (the Pace Act) and altered its loan terms, dropping the interest rate to a set 2 percent and extending the repayment term to thirty-five years. That allowed the co-ops to reach even less densely populated areas, no longer needing 3 customers per mile of distribution line to be financially viable. In 1946 the REA extended $3.5 million in loans to Washington co-ops and $2.6 million in loans to Oregon co-ops. These loans supported 3,400 miles of new distribution lines, serving 5,900 new customers by 1947. In that year, the total electrification rate reached 87 percent in Washington and 89 percent in Oregon. After another fifteen years of rural electrification, 99 percent of farms in Washington, 98 percent of farms in Oregon, and 97.5 percent of farms in Idaho had electrical service. Existing co-ops accomplished most of the construction of needed rural lines. In fact, in the years after World War II, only nine new co-ops formed in the Northwest (six in Oregon, two in western Montana, and one in Idaho), making a total of thirty-four Northwest co-ops. Most of them bought BPA power as soon as they possibly could in order to save money and provide lower rates for their customers. For example, the Flathead rural co-op in western Montana had to wait until October 1952 to switch to BPA power due to a pre-existing contract with its supplier, the Flathead Irrigation District. The month before the Flathead co-op switched to BPA power, its electricity bill from the irrigation district was $5,943. The next month, its bill from BPA was $2,132.[21]

BPA historian Gene Tollefson argues that 1951 marked the end of an era. By that year over 90 percent of Northwest businesses, homes, and farms were electrified, and publicly owned utilities served nearly half of those customers. Much had been accomplished to extend electrification, reduce rates, rein in corporate profiteering, and promote greater per capita consumption of electricity. Public power had carved out for itself a permanent, prominent place in the electric utility industry of the Northwest. All of these accomplishments had been promoted and enabled by reformist New Deal–era government policies and agencies. By 1951, according to Tollefson, "Twenty tumultuous years of reform were at an end," followed by "a 30-year period of increasing cooperation and consensus . . . between public and private power." Creation of the Pacific Northwest Utilities Conference Committee (PNUCC) in 1946 symbolized this transition. A voluntary association of public and private utilities, PNUCC cooperatively assessed regional power needs and supported congressional appropriations for more federal development of hydropower. It remains an active organization today, with forty-three publicly owned utility members and six investor-owned utility members, plus ALCOA. While increased cooperation and coordinated regional power became the dominant trend after the war, antagonism between the two sectors of the power industry did not disappear. In fact, in his book *People, Politics, and Public Power*, Washington PUD lobbyist and activist Ken Billington calls the period from 1951 to 1956 an era of "Face-off, Fights, and Failures." Not until the 1969–1974 period does he acknowledge a "Peace Pipe and Joint Efforts with Private Power."[22] Nevertheless, as the mix of public and private power finally stabilized in the 1950s, the trend toward cooperation and integration became the new era's leitmotif, especially after the election of Dwight Eisenhower in 1952. Gus Norwood characterized the eight years of the Eisenhower administration as an evolving "power partnership" among public power, private power, and the BPA. To be sure, BPA formed the center of the Northwest power grid. In its capacity as primary supplier of energy to the Northwest Power Pool, it helped to craft a cooperative, integrated Northwest electric system in the 1950s and 1960s.[23]

BPA'S POSTWAR EVOLUTION

The power generated by BPA continued its impressive upward trajectory for decades after the war. The Bureau of Reclamation continued adding generators to Grand Coulee Dam, three new ones in 1950 for a total of thirteen that year, and a combined Bonneville and Grand Coulee generation capacity of 2.1 million kilowatts. Demand for BPA power had reached a wartime peak of 9.2 billion kwh in 1944, then fell to a low of 6.2 billion kwh in 1946, then climbed quickly and steadily back up thereafter. By 1948 demand for BPA power exceeded the wartime peak by more than 15 percent. In 1950 BPA generated over 14 billion kwh, half again as much power as the wartime peak. Clearly, the postwar slump had been brief and recovery rapid.

More federal dams came online in the 1950s, greatly boosting BPA's power supply. By 1960, BPA was marketing power from nine Corps of Engineers dams and four Bureau of Reclamation dams with a total generation capacity of six million kilowatts—triple its capacity of 1950. To connect to all these new federal dams, BPA continued its transmission construction program after the war, reaching 4,040 circuit miles of high-voltage transmission lines by 1950, connecting Washington, Oregon, Idaho, and Montana in what was then the largest single-operator electric grid in the nation. It didn't stop there. By 1960, BPA administered over 8,000 circuit miles of transmission lines, 201 substations, and 424 separate points of contact with other utilities and industries. It had indeed become the region's central nervous system.[24] Figure 12.1 shows the BPA transmission system as of 2008.

Aware that the cost of federal river development projects elicited controversy and criticism by opponents of public power, the BPA diligently tracked and annually reported its revenues and repayments to the U.S. Treasury. Re-

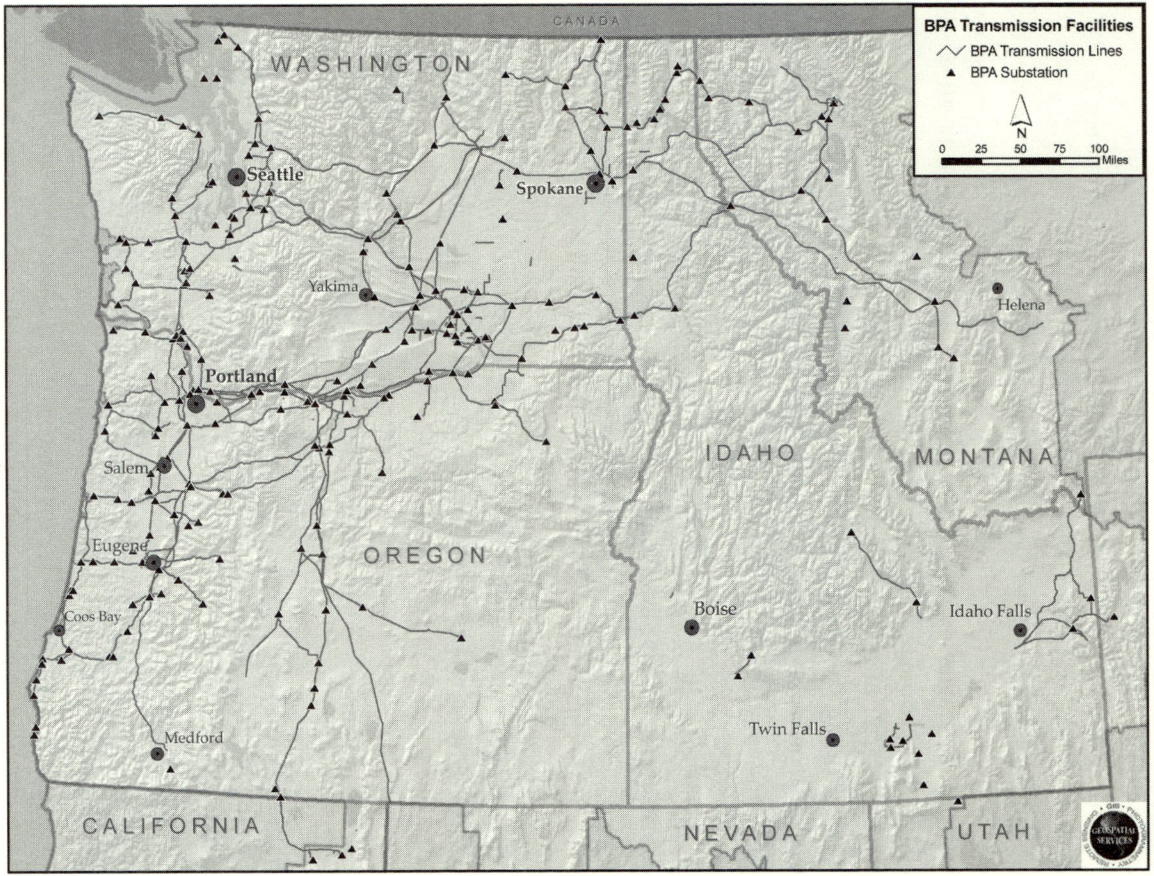

Figure 12.1. BPA Transmission Facilities, 2008. Source: Bonneville Power Administration, Geographic Information Services website, http://transmission .bpa .gov/lancom/geographic_ information_services/maps.cfm.

vealing how lucrative a cash machine federal power sales were, BPA's 1950 annual report states that it had already repaid about 25 percent of the cost of Bonneville Dam (the costs allocated by the FPC to power generation in 1938), had repaid 13 percent of the "repayable commercial power construction costs" of Grand Coulee Dam, and repaid about 20 percent of the cost of the transmission system. Way ahead of its required repayment schedule, BPA's payments to the U.S. Treasury were three times more than the amount required for Grand Coulee and 75–80 percent more than required for Bonneville Dam and the transmission system.[25]

During the war, BPA catered to three main classes of customers: publicly owned utilities, privately owned utilities, and industries. These same three customer classes dominated BPA's market in the decades after the war, but the proportion of BPA's power going to each shifted significantly. During the war, industrial sales, mostly to the aluminum industry, exceeded all other sales combined by more than four to one. By 1950, industrial sales were only a little more than half of BPA's load. This resulted from a postwar explosion in the residential and commercial power market. All of the Northwest aluminum facilities cut back on production after the war, and several closed down. Not until 1949 did the power demand from aluminum industries again reach its 1944 wartime peak. Power demand from other industrial customers did not reach the wartime peak again until 1950. In contrast, BPA's power sales to public and private utilities sank below the 1945 wartime peak for only one year, then shot back up quickly. Rapid population growth during and after the war, coupled with rapid increases in per capita consumption of electricity, sent residential and commercial power consumption through the roof. The Northwest states experienced a 44 percent increase in population between 1940 and 1949, compared to a 13 percent increase for the United States as a whole. To serve this growing demand from 1945 to 1950, the private utilities increased their purchases of BPA power by more than 50 percent, while public utilities more than tripled their purchases. The great imbalance in BPA's World War II load quickly shifted to less overwhelming dominance by industrial customers and more of a balance between public and private utility purchases, which served mainly commercial and residential customers. In the 1950s, BPA sales to publicly owned utilities grew the fastest. By 1960, that class of customers was purchasing the largest load from the agency (see Table 12.1).[26]

The Northwest Power Pool, created in 1942, continued to function after the war, serving as a force of integration among the various public and private utilities as well as a regional integrator across five states and British Columbia. Some utilities put more power into the shared grid than they took from

Table 12.1. BPA Energy Sales by Class of Customers for Fiscal Years 1950 and 1960

Class of Customers	Percentage of Energy Sales FY 1950	Percentage of Energy Sales FY 1960
Publicly owned utilities	22	35.7
Privately owned utilities	25	18.8
Aluminum industry	45	30.1
Other industries & federal government	8	15.4
TOTAL	100.0	100.0

Sources: Bonneville Power Administration, *Report on the Columbia River Power System, 1950*, 18; *1960 Annual Report on the U.S. Columbia River Power System*, 16.

it, while others were net purchasers of power from the pool. BPA, of course, was the main contributor, providing more than half of all the energy requirements of the region in 1950. Three privately owned utilities had surpluses that they contributed to the Power Pool that year: Montana Power, Idaho Power, and BC Electric. BC Electric formally joined the Power Pool in 1949, making a net contribution of 200 million kwh to the grid in 1950. Seven large public and private utilities had deficits and drew excess power from the pool to meet their needs: Pacific Power & Light, Seattle City Light, Puget Sound Power & Light, Washington Water Power, Utah Power & Light, Tacoma City Light, and Portland General Electric. PGE, in fact, bought 68 percent of its total electric load from the Power Pool surpluses that year. The other deficit utilities drew 15–40 percent of their power from the pool. Since BPA supplied 84 percent of the surplus power available for purchase in the Power Pool, it was in effect still the main source of electricity for meeting these utility deficits (see Figure 12.2). Ten years later, the same seven utilities plus BC Electric were buying power from the pool, while BPA, Idaho Power, and Montana Power continued providing surpluses for sale. About 72 percent of the surplus came from BPA.[27]

RIVERS, SALMON, AND SOCIAL JUSTICE

The qualities that made Northwest rivers excellent hydropower resources also made them excellent habitat for anadromous fisheries, particularly salmon and steelhead. The rapid development of the region's rivers for hydropower after the war led directly to an equally rapid decline of the fisheries. Before hydro-

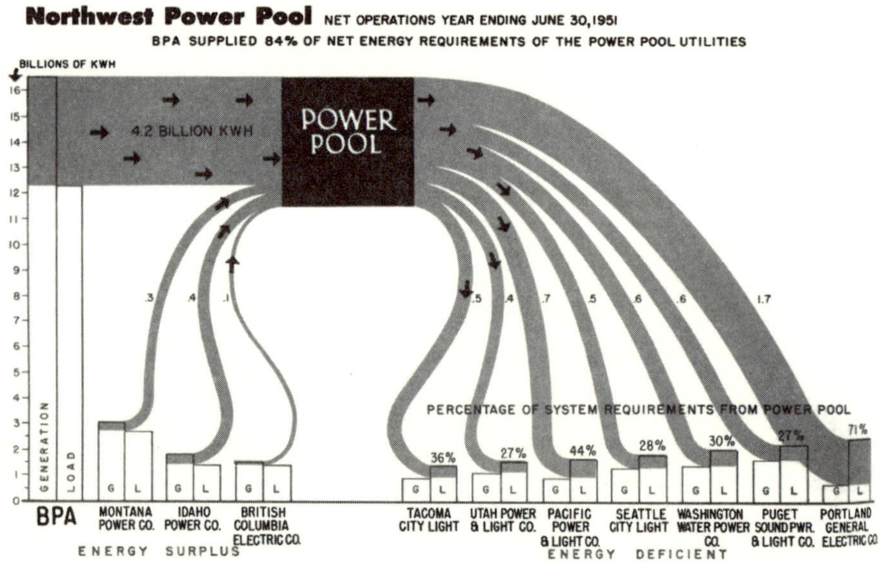

Figure 12.2. Northwest Power Pool in 1951. Source: Bonneville Power Administration, *Report on the Columbia River Power System 1951*, 28.

power development, an estimated 12–15 million adult wild salmon annually returned from the ocean to the Columbia and its tributaries to spawn. The number now is closer to 1 million, and most of those fish are hatchery-reared at public expense. Federal studies in the 1980s determined that about 80 percent of the loss of the salmon fishery in the U.S. Northwest is due to hydropower development.[28] When Puget Sound Power & Light (financed by Stone & Webster) built the very first dam on the Columbia River main stem at Rock Island in 1933, they installed fish ladders to allow adult salmon to migrate upstream, but the ladders were poorly designed and largely failed. The year after the gates closed at the dam, Colville Indians upstream suffered an 88 percent decline in their salmon catch. Later, in 1944, the U.S. government rebuilt the fish ladders to improve upstream migration, but additional dams made the effort to save the Columbia River fishery moot. When the Army Corps of Engineers started construction of Bonneville Dam on the lower Columbia in 1937, it made a much more concerted effort to build functional fish ladders. The ladders worked well enough to get adult salmon up the river, but the dam posed a difficult obstacle to the young salmon smolts headed downstream to the ocean for the mid-portion of their life cycle, so the fishery declined further. In addition, the reservoir behind the dam buried thirty-five Indian fishing sites, including Cascade Falls, the third most important indigenous fishing site on the river.[29]

The next dam on the Columbia River, Grand Coulee, was so large that effective fish ladders could not be built. The U.S. government had scuttled alternative plans in the 1930s for a series of smaller dams in the upper tributaries with functional fish ladders, in favor of the big dam at Grand Coulee. When completed in 1941, Grand Coulee Dam permanently blocked the entire upper third of the Columbia River salmon spawning grounds and buried Kettle Falls, the second most important indigenous salmon-fishing site on the river, along with numerous towns, burial sites, and most of the Colville Indians' farmland. Native Americans from around the Pacific Northwest held a three-day "Ceremony of Tears" at Kettle Falls in June 1940 as the gates to the dam closed. Compensation for these losses was meager and long delayed.[30]

The U.S. government then built a dam in 1955 almost as big as Grand Coulee but downstream from the colossus. Likewise too tall for fish ladders, it also blocked all upstream migration. As ironic and useless compensation, the government named the dam after Nez Perce chief Joseph. Two years later the Dalles Dam provided the next insult to the Indians and the fishery. Just upstream from Bonneville Dam, The Dalles, finished in 1957, buried the single most important Indian fishing site on the Columbia River: Celilo Falls. It also forced relocation of the Celilo Indian Village and burial grounds. Finally, in the 1950s, the Army Corps of Engineers began building four dams on the lower Snake River, completing them over the next two decades. These dams soon put the salmon runs in that watershed on the endangered species list. Indigenous advocates and fishery interests fought all of these developments to no avail.[31]

In the face of all this river development, efforts by Indian tribes in the U.S. Northwest to preserve functional fishing rights guaranteed to them by bedrock treaties in the 1850s met with little success. In fact, throughout most of the twentieth century, state fish and game authorities regularly restricted Indian fishing in favor of non-Indian commercial and sports-fishing interests, increasingly so as the salmon runs diminished. The turning point came with an act of civil disobedience by Indian fisherman David Sohappy in the late 1960s—fishing for salmon in defiance of state regulations and in defense of treaty rights—which led to the case of *Sohappy v. Smith* (302 F. Supp. 899, D. Oregon 1969). The upshot of this case and two related U.S. District Court decisions in 1974 and 1979, referred to as the "Boldt Decisions" after the presiding judge, was to allocate to Northwest tribes *half* the annual allowable salmon harvest in the Northwest. It led to a revolutionary reordering of salmon politics and significant empowerment of the tribes. Unfortunately, most salmon runs in the U.S. Northwest by then were either endangered or extirpated, so rights to half the catch gave Indians half of very little.[32]

A different outcome unfolded in British Columbia. Fishery interests there focused most of their attention on the Fraser River, the mainstay of the salmon-fishing industry and critically important to First Nations peoples living in the watershed. The Fraser is the largest river in British Columbia and carries half again as much water as the Canadian reach of the Columbia River. All five species of Pacific salmon spawn in the Fraser watershed, with its sockeye run the largest in the world. As discussed in chapter 5, huge landslides in Hells Gate canyon from 1911 to 1914, from Canadian Northern Railway construction, blocked salmon passage and a great deal of upstream spawning habitat for several decades, drastically reducing the salmon runs. The Fraser sockeye were important to American salmon fishers in the Puget Sound region, too, since the salmon passed through international waters on their way from the open ocean to the Fraser to spawn. In fact, studies in the 1930s and 1940s indicated that American fishermen took as much as two thirds of the sockeye run as it passed through the Strait of Juan de Fuca and Puget Sound. Consequently, both Canadian and American fishing interests collaborated to protect the Fraser River runs. In 1937 the two nations signed a convention establishing the International Pacific Salmon Fisheries Commission (IPSFC), charged with recovering the Fraser sockeye run. But the authorizing treaty required the commission to study the problem for five years before taking direct regulatory and restoration actions. Consequently, not until 1943 did the IPSFC finally get to work clearing the obstruction enough to facilitate greater upstream passage and begin restoring the once world-class runs.[33]

Just as the Fraser sockeye fishery rebounded in the late 1940s and early 1950s, the Columbia River salmon fishery crashed. Canadians watched these two contrasting events with great interest. As noted earlier in this chapter, when the B.C. government under Premier Bennett announced its support in the early 1950s for ten large dams to be built in the Fraser Basin, B.C. fisheries interests united in opposition, supported by Puget Sound fishermen's organizations in the United States, who saw the Fraser runs as their last hope for a sustainable fishery. While the United States remained committed to artificial hatcheries to compensate for salmon losses, B.C. fishery authorities had given up that ghost two decades earlier in favor of a policy of keeping dams off important salmon rivers (see chapter 5). Concern over the fishery stalled plans to develop the Fraser in the 1950s, but the Columbia and Peace rivers were spared that pressure since the United States had already blocked salmon passage on the Columbia, while the Peace flowed northeast away from British Columbia's main population and its coastal fishing industry. By 1958 Premier Bennett's Two Rivers policy turned development efforts toward the Columbia and Peace, leaving the main stem Fraser undammed. That precedent has held to the present, and the Fraser today still supports the most productive sockeye fishery in the world and is one of the longest undammed rivers in the world.[34]

In the postwar U.S. Northwest, salmon were considered expendable, despite the fact that the United States had treaty obligations to Indians, spent millions of dollars trying to sustain the fishery, and passed numerous laws promoting fish and wildlife conservation. Most policy makers, fish culturists, economists, and dam engineers believed that technology would solve the fishery problem and allow both maximum hydropower development and acceptable salmon survival. Dissenting voices—and there were many—could not overcome the technological optimists. When it became clear for a particular river or dam proposal that hydropower and salmon were incompatible, the hydropower advocates argued successfully that cheap electricity provided greater social benefits. If a "minority" were harmed, it was for the greater good. This vision of social welfare certainly provided an array of benefits to northwesterners, but those benefits were not equitably distributed. Native Americans, for example, lost much more than they gained—in land, fish, autonomy, security, and culture. When they got electricity or irrigation water from these dams, it usually came later and less reliably than it did for whites, cities, and industry. This profound bias toward the majority white population and toward hydropower in river decision-making caused losses that northwesterners now regret and inequities they struggle to redress. But all was not lost. Tribes formed the Columbia River Inter-Tribal Fish Commission (CRITFC) in 1977 to give them a more united voice, and they gained a powerful seat at the fisheries and hydropower decision-making table due to mandates in the 1980 Northwest Power and Conservation Act. More recently, obsolete dams on once-productive salmon rivers like the Elwha on the Olympic Peninsula are being removed to restore salmon runs. There are even serious proposals to remove the lower Snake River dams to restore fisheries in that basin. River-restoration proposals will likely be a more common occurrence in the future.[35]

THE COLUMBIA RIVER TREATY

Despite the objections of Native peoples and fishing interests, the predominant postwar goal of policy makers in both nations regarding the Northwest's rivers was maximally efficient development, with an almost obsessive focus on hydropower generation. Since the river's flow fluctuated seasonally, with more water than needed in the spring and less water than needed in the fall, engineers and hydropower planners in both nations proposed massive dams upstream in the watershed to store spring snowmelt and release it in the summer and fall to improve the efficiency and reliability of the hydropower system. This strategy reduced the Columbia River to little more than numeric abstractions: cubic feet per second of flow, acre-feet of storage, and kilowatt-hours of generation capacity. As noted above, the floods of 1948 added complementary flood-control considerations to this hydropower calculus. Fortunately for the advocates of comprehensive maximum development, large storage reservoirs served both purposes simultaneously, so they enhanced each other's prospects. B.C. engineers and policy makers proposed building "upstream storage" reservoirs on the Fraser River, while American engineers and policy makers proposed upstream storage on the Columbia. But potential sites for storage reservoirs in the U.S. portion of the Columbia watershed, such as Hungry Horse and Libby, were limited and already approved for development. British Columbia contained the Columbia River headwaters, along with several good sites for large reservoirs. As a consequence, more storage on the Columbia River for the United States meant convincing Canada in general and British Columbia in particular to dam their portion of the watershed. While the United States had dammed or started damming the main stems of nearly all its biggest rivers in the Northwest by the early 1950s,[36] British Columbia had not yet developed the main stems of *any* of its three biggest rivers—the Fraser, Columbia, or Peace. As early as the 1940s, U.S. policy makers and river engineers saw British Columbia as a land of opportunity and understood clearly that flood control and the efficient management of the Columbia River hydropower system in the United States required coordinated development and management between the United States and Canada.

Earlier in the century, the United States and Canada formalized their intent to negotiate over shared water resources in the 1909 U.S.-Canada Boundary Waters Treaty, which created the International Joint Commission (IJC) to oversee dispute resolution. By midcentury the IJC also became deeply involved in promoting cooperative development of water resources. In 1944 both the U.S. and Canadian governments directed the IJC to investigate possibilities for joint development of the Columbia River, mostly at the urging of the United States. While the IJC pursued its studies, U.S. river development agencies refined their own reports regarding cooperative development opportunities that would enhance river management goals in the United States. The Bureau of Reclamation, for example, published during the Truman administration a number of extensive regional watershed reports under the series title "Our Rivers: Total Use for Greater Wealth." Several volumes on the Columbia River

Figure 12.3. Major Dams of the Columbia River Basin. Source: Bonneville Power Administration Research Library.

were published from 1947 to 1950. While focusing on the U.S. side of the border, the bureau acknowledged the importance of international cooperation:

> The Canadian part of the basin has natural resources, especially water resources, which are large and important. Despite the presence of the International Boundary, the unity of the basin creates a broad mutuality of interest between the Canadian and United

States portions. . . . [T]he desirability is recognized for appropriate coordination of plans for resource development through international agreement.[37]

But leaders in British Columbia in the late 1940s and early 1950s were much more interested in developing the Fraser and Peace rivers than they were the Columbia, so negotiations dragged on with complex disputes over funding, timing, and the sharing of costs and benefits from Canadian development of the Columbia. Eventually, in the face of vigorous opposition from fisheries interests to Fraser River development, Premier Bennett opted for the aforementioned Two Rivers policy aimed at the Columbia and Peace, which clinched the negotiations between the United States and Canada over joint development of the Columbia River.[38] The two nations' executives signed a treaty in 1961. After several revisions, both nations finally ratified it in 1964. Canada agreed to build three large dams in the upper Columbia watershed (Keenleyside Dam, finished in 1968; Mica Dam, in 1973; and Revelstoke Dam, in 1984). It also allowed the United States to build a dam south of the border (Libby Dam) that backed water up into Canada. The United States agreed to pay Canada for half the downstream flood-control and power-generation benefits it derived from Canadian water storage. Importantly, the United States agreed to pay Canada a lump sum in advance for the first thirty years of hydropower benefits.[39] As a result, the Columbia River became fully developed for efficient hydropower generation, a triumph of the heirs to the New Deal who believed the public welfare was best served by maximizing the availability of inexpensive electricity.

The Two Rivers policy and the Columbia River Treaty marked the pinnacle of an era dedicated to endless, limitless development or, as the Bureau of Reclamation called it, "total use for greater wealth." Dramatic changes in social values and federal policies occurred in the two decades following the treaty, however, altering the context and direction of river and hydropower development in the U.S. Northwest and British Columbia. The same year the U.S. Congress ratified the Columbia River Treaty, it also passed the National Wilderness Preservation Act, marking the start of a flood of environmental legislation reforming federal environmental policy and natural resource development. The National Environmental Policy Act of 1969, the Clean Water Act of 1972, and the Endangered Species Act of 1973 are just three of more than a dozen environmental laws of this decade that steered the American ship of state in new directions. By the late 1970s, the U.S. government was out of the business of building big dams. Canada experienced a similar reform of environmental policies in subsequent years and got out of the business of big dam construction in the 1980s after completing its three treaty dams on the Columbia River and two dams on the Peace River—at least for a couple of decades, as it is now proposing another project on the Peace River. The "environmental era" of the 1970s enshrined policies requiring more caution regarding development, greater government transparency and disclosure requirements, greater acknowledgment of ecological complexity and social justice issues, a more diverse representation of values, and many opportunities for stakeholders and the general public to participate in decision-making. The consensus for endless, limitless development of hydropower slowly gave ground to a more complex focus on meeting electric demand through a combination of smaller-scale developments, efficiency improvements, and conservation or "demand-side management." One era had ended, and another began. The transition, however, would be long and occasionally tumultuous.[40]

EPILOGUE: OUT WITH THE OLD, IN WITH THE NEW

As historian Richard Hirsh convincingly demonstrates in his 1999 book *Power Loss*, the institutional structure of the electric power system established during the first three quarters of the twentieth century "shattered" in the 1970s due to three separate but mutually reinforcing stresses. What Hirsh calls the "utility consensus," a system of state-regulated monopolies enjoying guaranteed profits in exchange for state oversight of rates and services, broke down in the 1970s as technological stasis, the energy crisis of 1973, and the rise of the environmental movement undermined the foundation for that consensus. Technology, the first stressor, played a key role in this transition. Since the beginning of the twentieth century, technological progress had provided a reliable basis for the growth of the industry, the success of regulation, and the satisfaction of customers. Constantly increasing efficiency in the production and distribution of electric power through ever expanding economies of scale had simultaneously secured both profits for commercial utilities and lower rates for consumers. But economies of scale cannot expand infinitely. Due to technological and economic limitations, power-generation economies of scale reached a peak in the 1970s, despite many hopes and efforts to the contrary. From 1902 through the early 1960s, the "thermal efficiencies" of steam-powered electric turbines in North America steadily rose from less than 5 percent to over 30 percent, some units even reaching 40 percent efficiency in converting fuel energy into electric power. But engineers achieved these thermodynamic efficiencies by increasing steam temperatures and pressures, which stressed the equipment and required increasingly expensive parts and materials. After eighty years of steady increases in thermodynamic efficiencies, technological progress in steam generation based on temperature and pressure finally reached a practical limit in the 1960s. Similarly, the maximum capacity of individual thermal generating units steadily increased from a few thousand kilowatts in the 1910s to over a thousand *megawatts* in the 1970s, enhancing economies of scale for utility companies. But practical limits again ended that trend as numerous equipment failures and maintenance problems erased the economic advantages of the largest units. By the 1990s, utilities installing new power plants had largely abandoned the steam-turbine units over 1,000 megawatts capacity and reverted to smaller turbines that proved more reliable and profitable. As increased economies of scale came to an end, so did the advantages provided to producers and consumers from the ever declining cost of production. In other words, electric rates began to climb rather than continue to fall.[41]

This technological stasis, as Hirsh terms it, in thermal-generated power would not have affected the U.S. Northwest much because of its reliance on hydropower, except that the big dam-building era came to an end at the same time that increased economies of scale in thermal generation ended. In the mid-1960s, sixteen major generating facilities were under construction in the bi-national Pacific Northwest—eight federal dams, three Canadian dams on the Columbia River, four commercial (nonfederal) dams in the Northwest states, plus a nuclear power plant on the Hanford Reserve. According to the BPA, this represented "the greatest construction program in the history of the electric industry of the Northwest."[42] But it was a last hurrah for dam-building south of the international border. By the 1970s, U.S. Northwest utilities turned increasingly to thermal power plants to meet load growth. By 2010, hydropower supplied just a little over 44 percent of the "firm" power in the Northwest states, with coal-fired plants providing 19 percent of the load, natural gas–fired combustion turbines providing almost 12 percent, and a variety of other sources making up the rest.[43] Electric power from these plants cost more to generate than hydropower, and those costs escalated as fuel prices climbed after the 1973 OPEC oil embargo and as nuclear power plant construction costs skyrocketed in that decade. Nationwide, as well as in the Northwest, electricity prices climbed in the 1970s for the first time in history, both in absolute terms and as adjusted for inflation, which challenged the industry, soured consumers, bolstered conservation programs, and led, not surprisingly, to significant reductions in per capita electric consumption in the following years that lowered power demand projections and challenged cost recovery for existing power plants.[44]

Nuclear power in particular turned out to be far, far more expensive than originally predicted, leading to price increases in the late 1970s and early

1980s from cost overruns at five nuclear power plants under construction in the Northwest. A consortium of publicly owned utilities in Washington state formed the Washington Public Power Supply System (WPPSS) in 1957 to pool their financial resources to build joint power supplies. They turned to nuclear power in the 1960s in anticipation of the day, soon to come, when all the good dam sites on the region's rivers were fully developed. Over the next two decades, a series of fateful decisions that would prove disastrous resulted in the undertaking of construction of five nuclear power plants. Early postwar advocates of nuclear power were almost utopian in their enthusiasm for this power source, which boosters claimed would be so cheap that utilities would not even bother to meter its consumption. But this glossy veneer soon tarnished. Environmental concerns, coupled with the high cost and risk of nuclear power, slowed and eventually reversed the momentum for nuclear power across North America. In the Northwest, the Washington Environmental Council sued to force Seattle City Light, a member of the WPPSS consortium, to do an Environmental Impact Statement (EIS) for the WPPSS nuclear power plants, which would have significantly delayed construction. It withdrew its lawsuit only when City Light agreed to appoint a citizens' committee including several environmental representatives to oversee energy planning for Seattle. Under pressure from this citizens' committee, the Seattle City Council withdrew from projects four and five of the WPPSS nuclear program in the 1970s and adopted conservation measures to help meet future projected load growth, including a cleverly named "kill-a-watt" energy-savings program adopted in 1974. Delays, design problems, high interest rates, construction incompetence, costly safety features, and widespread mismanagement led to extraordinary cost overruns of three to four times original projections for the five power plants, which ultimately led WPPSS to halt construction of four of the five partially built nuclear power plants and default on its bonds in 1982. At the time, this was the largest municipal bond default in American history. Adding insult to injury, the WPPSS acronym became widely referred to as "whoops!" The consortium finished only one plant, brought online in 1984: Hanford 2. No new nuclear power plants have been built in the Northwest since then.[45] The only other nuclear plant in the region was Portland General Electric's Trojan power plant in Rainier, Oregon, started in 1968 and completed in 1975. It, too, was besieged by criticism, including multiple on-site mass protests and arrests in 1977 and 1978. Design flaws led to numerous maintenance problems, shutdowns, and repairs until PGE, under significant environmental and public pressure, closed the plant in 1992, twenty years ahead of its designed end of lifespan.[46]

British Columbia avoided these problems by never adopting nuclear power or even much thermal power of any kind. In the 1960s, as the U.S. Northwest embarked on its final push to develop all its remaining big dam sites in the Columbia River watershed, British Columbia was just beginning to develop the main stems of the Columbia and Peace rivers within the province. Not until the 1980s did it complete these five Columbia and Peace River megadams, and they provided a significant power surplus for more than a decade. By the time British Columbia needed to think seriously about additional sources of energy, enthusiasm and financial backing for nuclear power had substantially dissipated, while public support for conservation and efficiency strategies had substantially increased. British Columbia still relies almost entirely on hydropower. In 2010, over 90 percent of BC Hydro's electricity supply came from hydropower. It owned only one thermal plant (the whole province had only three thermal plants). The B.C. Clean Energy Act of 2010 prohibited nuclear power development in the province and emphasized conservation, efficiency, and clean, renewable sources of power. The following excerpts from B.C.'s 2010 Clean Energy Act illustrate these priorities:

British Columbia's energy objectives:
(a) to achieve electricity self-sufficiency;
(b) to take demand-side measures and to conserve energy, including the objective of the [BC Hydro] authority reducing its expected increase in demand for electricity by the year 2020 by at least 66%;
(c) to generate at least 93% of the electricity in British Columbia from clean or renewable resources and to build the infrastructure necessary to transmit that electricity;
(d) to use and foster the development in British Columbia of innovative technologies that support energy conservation and efficiency and the use of clean or renewable resources; . . .
(g) to reduce BC greenhouse gas emissions . . .
(j) to reduce waste by encouraging the use of waste heat, biogas and biomass . . .
(l) to foster the development of first nation and rural communities through the use and development of clean or renewable resources . . .
(o) to achieve British Columbia's energy objectives without the use of nuclear power . . .[47]

The Northwest Power and Conservation Act of 1980 similarly put BPA on the path toward encouraging conservation and efficiency as a key strategy for meeting future load growth to minimize the need for new generation facilities. It took a while for this policy to get traction, but by the 1990s, with nuclear power dead and energy costs rising, "demand-side" management and efficiency programs rapidly proliferated. By the first decade of the twenty-first century, conservation and efficiency were BPA's primary means for meeting projected future energy demand.[48]

While technological progress in the old power system had stalled, a new frontier of innovation in efficiency, demand management, and distributed, clean, renewable energy production took off. The future institutional structure of the Northwest electric power system remains unclear; the transition from the old arrangements to a new configuration is still in process. However, it is clear that a new form of technological progress is driving change forward, supported by new policies (competition, renewable energy, conservation, and efficiency) driven by changed economic conditions and social values. The evolution of the hybrid public/private electric system continues on its path to provide reliable power at affordable rates, permitting reasonable profits while protecting the public interest. Negotiating among these sometimes competing goals is a messy but fascinating challenge that each generation must engage anew.

CHAPTER 13

Conclusion

For more than a century, citizens of the Pacific Northwest and British Columbia have sought to construct a reliable, efficient, affordable, profitable, and socially responsible electric power system to drive the energy needs of modern society. Every twist and turn on this journey is instructive. Below I reflect on some of the key lessons embedded in the hundred years of history examined here. The Northwest power system is in some ways a unique product of its place, yet it also reflects many common features of power development in North America and throughout the developed world. Residents of the Pacific Northwest may find these conclusions especially relevant to their experience, but the lessons apply far beyond the region.

From the very start, northwesterners marched on the vanguard of electrical development. Despite its great distance from eastern centers of capital and culture, electrical entrepreneurs in the region introduced and experimented with cutting-edge technologies. Portland and Victoria harbors saw some of the world's earliest arc lighting systems. Portland and Seattle installed precedent-setting long-distance transmission lines delivering hydropower from the hinterlands. Northwest cities built some of the first electric streetcar systems; mining and metal companies experimented with the newest electrolytic manufacturing processes; BPA experimented with North America's largest turbines, highest voltages, and biggest capacitors; and the Hanford nuclear facility helped create the technology for the first nuclear power plants. The U.S. Northwest states were among the early adopters of municipal ownership, state regulation, and rural electrification. The region was also on the forefront, along with the Tennessee Valley, of federal hydropower development and coordinated river basin management. Northwesterners embraced these new technologies and institutions early and enthusiastically. Consequently, we can learn from the region's history a great deal about how new technologies and institutions are created, disseminated, adapted, and reformed over time.

Unlike late-developing regions, which can draw from earlier precedents and learn from others' prior successes and failures, the early history of electric power development in the Pacific Northwest exhibits all the messy, gritty trial and error of innovation and market creation in a minimally structured capitalist economy. The first electric light and power systems were inefficient, unsafe, and of limited utility. But they showed promise, and the systems quickly improved. As more customers showed a willingness to purchase electric light and power, the leading electrical manufacturers competed to dominate the market and determine system standards. An era of unregulated cutthroat competition ensued, which turned out to be unacceptably unstable and risky to both producers and consumers.

Within a decade the exciting but chaotic early years of invention and freewheeling entrepreneurialism soon gave way to a more reliable, structured industry. At first, wealthy corporations and eastern financiers provided the organizational structure as it proved increasingly profitable. But their primary interest in profits limited the spread of this technology and led to customer dissatisfaction. Furthermore, cutthroat competition in the early years led to insecurity and instability in the business. Eventually, after decades of contentious trial and error, an institutionalized relationship evolved between industry and government that secured stable profits for investors as well as larger social objectives like greater accessibility, affordability, and safety. From that point forward, the electric power system evolved in a dynamic tension between the profit motive and social goals, leaning alternately in one direction, then the other. Private owners of electric utilities enjoyed the greatest amount of independence and the least pressure from regulators when business was good and customers were satisfied. At other times, users of electricity demanded political interventions to solve problems or improve system performance or achieve goals that the for-profit system failed to advance. This pattern repeats itself with slight variations for most socio-technological systems in the modern world. The Northwest's experience in working through these developmental stages is not so different from what has occurred at other places and times.

Location did create some characteristics unique to the Northwest power system. As described in this study, the region's wet winters, abundant mountain ranges, and high-volume rivers made an ideal geography for hydropower. Hydropower undergirded all of Canada's power systems in the twentieth century, making British Columbia rather typical of the national trend. But the hydropower-based electric supply in the U.S. Northwest was not typical of other American regions. Outside the Pacific coast, northern Rockies, and the Southeast, most of America's regional power systems have depended primarily on fossil fuel–burning power plants, especially coal-fired generators. After the 1960s, thermal-generated power reduced the dominance of hydropower in the U.S. Northwest, too, but never supplanted it. During the period covered in this study, hydropower was virtually always less expensive than thermal power. As a result, from the 1890s through the 1970s, citizens of the region on both sides of the border have enjoyed some of the lowest-cost electricity, the highest per capita consumption rates, and the highest electrification rates in North America.

While hydroelectricity does not require the purchase of fuels, it has other costs associated with its production; namely, the construction of dams, penstocks, and turbines. In the early decades of the industry, the initial capital costs for a hydropower facility were higher than the cost of a simple fuel-fired power station. That meant that larger, better-capitalized corporations thrived in the Northwest. This contributed to the rapid consolidation of Northwest power and light companies into a handful of big utilities associated with a few national holding companies. Once the capital costs of a hydropower facility are paid off, it is essentially a cash cow, so despite the initial high costs, hydropower is a bargain for investors and a boon for consumers.

The dams and reservoirs on the region's rivers entail another kind of cost. A cottage industry of studies on the environmental and social impacts of damming rivers has grown up over the past thirty years. I touch on some of this literature, but it has not been a major component of my study. Clearly, the environmental impacts of the Northwest power system fall most heavily on rivers and fisheries. In other regions, in contrast, coal mining and air pollution are the focus of environmental problem-solving.

While the social and environmental impacts of power production in the Northwest have a particular regional character, the *process* of problem-solving is not so different from that in regions with primarily fuel-based power supplies. The evolutionary stages of problem-solving are fairly uniform. Here is how the stages played out in this case: A simple goal of profits and capital accumulation drove system development in the earliest years. This led to some unacceptable societal repercussions, which drove reforms that integrated more complex social goals to the profit-driven system. Environmental impacts were not yet part of those early stage reforms. In the nineteenth and early twentieth centuries, Americans and Canadians generally accepted environmental impacts as a cost of development. While Progressive Era conservationists brought concerns about waste, pollution, and resource depletion into the policy arena, they accomplished little to resolve the widespread environmental effects of development. Increasingly, from the 1930s through the 1960s, citizens and their governments paid more attention to environmental problems. By the 1970s, a critical mass of public sentiment drove policy makers to further reform the power system to incorporate environmental goals and address environmental problems. In the U.S. Northwest, this meant an end to further dam-building and ambitious efforts to restore the decimated salmon fishery. These reforms, of course, have been limited and contested, and they are constantly evolving. But like the social goals incorporated into power system planning over the decades, these newer environmental goals are likely to remain on the agenda for the foreseeable future. The trend toward greater accommodation of diverse goals and values is not easily reversed.

The prodigious hydropower resources of the Pacific Northwest generated another characteristic of the regional power system with broad implications: the hazards of abundance. As discussed briefly in chapter 11, a certain psychology and set of policies follow from conditions of abundance: optimism, acquisitiveness, and unrestrained consumption. This dynamic is clearly visible in power system development from the 1890s through the 1920s. It becomes prominent again with U.S. federal development of Columbia River dams and with provincial development of the Peace and Columbia rivers. Ironically, the behaviors and policies that thrive in conditions of abundance lead fairly directly to the decline of abundance. With no incentive to conserve or to worry about waste or inefficiency, the windfall is quickly consumed, after which the culture of abundance yields rather uncomfortably to a new politics of restraint.

The decade of the 1970s in the United States and the 1980s in British Columbia marked the most dramatic transition from a politics of abundance to a politics of scarcity. In the end, a region of enormous environmental assets saw its resources consumed with abandon until abundance became just a memory. Nostalgia, regret, and restraint replaced it. This outcome was not inevitable. The region's citizens, businesses, opinion leaders, and politicians could have avoided the big swing from abundance to scarcity simply by modifying their behavior earlier, by adopting some of the concerns and constraints common in times of scarcity. This is largely what the science of sustainability is all about: extending conditions of abundance, postponing conditions of scarcity, and avoiding the wrenching dislocations that attend sudden transitions from one state to the other.

Another interesting characteristic of the Northwest regional power system is its public-power component and the related powerful influence of economic crisis and war on the evolution of the system. The location of the region on the Pacific coast made the Northwest important to the war effort in the 1940s. The two federal governments encouraged war manufacturing, which was significantly supported by electric power development. Particularly in the U.S. Northwest, war mobilization militarized the economy and greatly stimulated public-power development. The BPA, created as a Depression-era relief and recovery agency, became the crucial federal tool for transforming the Northwest economy on a foundation of super-cheap, super-abundant publicly owned hydropower. The Cold War sustained this transformation. As a result, the U.S. Northwest power system became a relatively balanced blend of private and public power, whereas prior to the Depression and World War II the region's power supply had been predominantly privately owned. British Columbia did not experience nearly as much war-related industrial and power development, so federal policies and institutions remained less influential there. As a consequence, for-profit corporations continued to dominate the B.C. power system, electric rates remained at moderate levels, and development proceeded at a slower pace than in the U.S. Northwest, where dirt-cheap federal power provided an economic stimulus somewhat akin to the Federal Reserve System's lowering interest rates. Once the B.C. government expropriated BC Electric in the 1960s, the new provincial power authority became an engine of regional growth, too. When citizens of the region began questioning the ideology of growth for the sake of growth in the 1970s and calling for more conservation and environmental protection, these public agencies, especially the BPA, pioneered new innovations in managing power production and consumption to solve problems and accommodate changing goals and values. Not confined solely to a profit calculus, public-power agencies could more assertively pursue policies that addressed far more than mere revenue concerns, such as salmon habitat restoration and demand-management innovations.

Looking at trends over the hundred years of power system evolution examined in this study (1870s-1970s), several stand out as significant. Many trends are progressive, in the most literal meaning of the term. For those interested in technology, it is undeniably a century of steady progress, of constantly increasing precision, efficiency, scale, complexity, performance, and resilience, enabled by constant innovation. For those interested in business or organizational history, the story is similarly progressive, at least through the 1960s. Entrepreneurial acumen, coupled with government incentives and intelligent regulations, produced an increasingly well-organized, stable, and profitable electrical industry, with the notable exception of the turmoil of the Great Depression. The instability that has come since the energy crisis of the 1970s and deregulation in the 1990s contrasts markedly with the more reliable and predictable era that preceded it.

Another trend in power system evolution during the first century of electrical development is the steady integration of discrete systems into interconnected regional systems. Power companies continuously extended their reach, integrating more and more of the population and the economy into their energy webs. As the outer edge of one electric system met the frontier of another, the utilities interconnected with each other to make local and then regional and then continental power pools wired together by a network of super-high-voltage transmission lines marching hundreds, even thousands, of miles across the landscape. This process of expansion and integration continued unabated from the 1870s to the 1970s and beyond.

Most people considered the consolidation and integration trend a mark of progress, just as the economies of scale provided by ever larger centralized power plants advanced the efficiency and profitability of electrical generation. But in a study published in 1982 titled *Brittle Power*, Amory and L. Hunter Lovins challenged this view. They argued that large centralized power plants, long-distance high-voltage transmission lines, and extensive regional interdependence of power systems all made the power supply vulnerable or "brittle," at least from a national security perspective.[1] At the vanguard of a new wave of critics, the Lovinses pushed for smaller-scale, decentralized, low-risk, and "clean" electric power systems. They were harbingers of a new era questioning the system bequeathed by the earlier generation.

For those interested in social meanings, the first 100 years of electrification is generally seen as a progressive tale of modernization. By the early twentieth century, electric lights and power were so superior to previous forms of harnessed energy, such as waterwheels and oil lamps, that electrification literally defined *progress*. Just owning an electric appliance was a badge of modernity. By the 1920s this new technology became so useful and integral to daily life that access to electricity and electrical appliances transformed into a kind of social entitlement, something that every citizen deserved and should not be denied. At that point the story of electrification merged with narratives of rights, responsibilities, and justice. Increased access, consumption, desire, and dependence on electricity brought increased demands for social governance of electric systems. Electricity accordingly evolved from a privately owned commodity traded in a laissez-faire market to a publicly owned or publicly governed technological system produced and distributed in a managed market. Over the first 100 years of developing this human-centered and socially constructed system we became ever more successful and efficient in manipulating rivers, fuels, turbines, and wires to deliver streams of electrons to cities and farms, factories and trolleys, domiciles and department stores to do work for us—to light our buildings, pump our water, process our metals, run our machines, and turn our wheels.

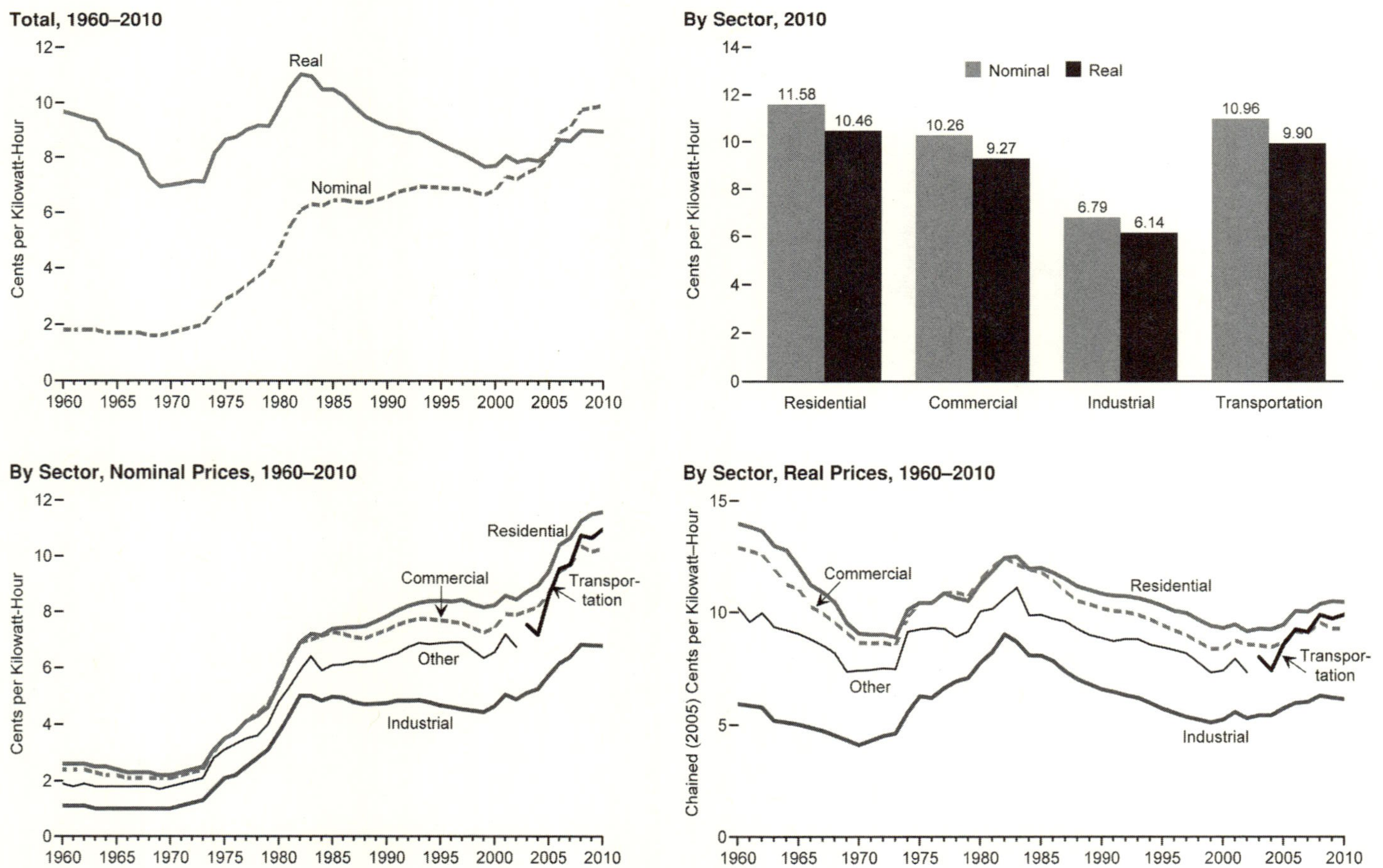

Figure 13.1. Average Retail Prices of Electricity, 1960–2010. Source: U.S. Energy Information Administration, *Annual Energy Review 2010* (Washington, DC: U.S. Department of Energy, 2011), Table 8.10, p. 268.

This is a progressive story, so long as one pays no attention to the environmental and social impacts of damming rivers, extirpating fisheries, mining and burning fossil fuels, polluting the air, depleting groundwater, accumulating nuclear waste, and building energy-inefficient homes and appliances while encouraging ever greater electrical consumption. As I noted above, these problems were mostly ignored or accepted until the 1960s. Ironically, to make progress in solving these problems, concerned citizens, scholars, and policy makers had to challenge the progressive narrative that dominated public understanding of electrification and promote instead the view that the system was flawed and in need of reform. This is the standard recourse when an institution no longer serves or satisfies its clients and constituents.

While environmental concerns played an important role in generating dissatisfaction with the power system, the real trigger, not surprisingly, was economic stress. Electric rates rose (see Figure 13.1), and profits fell in the 1970s and early 1980s.[2]

This sudden reversal of fortunes came after a remarkable century of steadily declining electric rates.[3] As discussed in chapter 12, historian Richard Hirsh convincingly traces the source of this transformational shift in large part to the decline of traditional forms of technological progress (such as economies of scale in electrical generation) that had reliably reduced the cost of producing electricity over time, supporting reduced electric rates *and* enhanced profits. When that trend stalled in the 1970s, its repercussions rippled across the whole system, exacerbated by rising fuel prices caused by the OPEC oil embargo of 1973. Within a decade, a critical mass of utility managers, investors, consumers, regulators, and academics started questioning the power system and re-evaluating the previously unquestioned consensus on what constituted progress and appropriate policy for the utility industry. Just like the Great Depression, the crisis in the 1970s led to significant structural reform, from which we have not yet fully emerged.[4]

As I noted in chapter 1, when the electric power system works well, it is generally out of sight and out of mind. We allow the institutional structure of the system to reproduce and self-regulate. When the system is not performing to most people's satisfaction, institutional arrangements are scrutinized, complaints articulated, and adjustments made to try to solve perceived problems. Problems in the availability, reliability, and affordability of electric power during the first decade of the twentieth century, for example, led to the rise of state regulation in the second decade of the century, which in large part fixed those problems, for a while. State public utility and public service commissions provided incentives, protections, and directives to the power companies that guaranteed their profitability but required a degree of social responsibility in exchange. Concerns about the concentration of wealth and risky speculation in the 1920s led to legislative reforms in the 1930s that broke up holding-company monopolies and restrained some of the worst financial abuses. Complaints about the lack of electrification in rural areas in the 1920s led to government programs in the 1930s and 1940s that virtually accomplished full electrification over a couple of decades. The desire to use electrification as a tool to win wars or to promote dispersed economic development led to new public institutions like the Bonneville Power Administration and the BC Hydro and Power Authority, which promoted electrification for *both* commercial and not-for-profit public purposes.

The more complex a system becomes, however, the more complicated it is to control and to adapt to changing conditions. This is the great challenge in successfully managing large, complex socio-technological systems: to make them both stable and responsive to changing needs. The former goal tends conservatively toward stasis, while the latter tends to undo previously crafted arrangements and upset the status quo. Much like the evolution of power systems, the entire political history of the American and Canadian democracies is marked by a to-and-fro struggle between defenders of the status quo and advocates of change. Problems and especially crises lead to struggles over rights, privileges, costs, and benefits allocated by the institutions that shape our lives. These struggles usually lead to incremental reforms that sustain some existing institutions and values, elevate others, and inhibit still others. Most political change in a democracy is compromised and incremental, so the result usually disappoints both advocates of reform and defenders of the status quo. No wonder democratic politics breeds cynicism.

Increasing complexity is another consistent trend in the century of electrification covered in this book. If anything, since the 1970s the power system has become even more complex, facilitated in part by advances in electronic information technology. Beyond the obvious technological complexity, however, lies an evolving managerial and institutional complexity to the power system, which has been a significant focus of this study. This trend from simplicity to complexity manifested in many ways. In the 1880s, a single dynamo typically sent electricity over a single set of wires to a single end-use, such as downtown streetlights or a saw blade in a lumber mill. Slowly but steadily, electric systems incorporated multiple sources of power: a scattered collection of dams and fuel-burning power plants distributing electricity to many customers on an integrated power grid. The development of power pools extended this complexity further, requiring higher levels of management coordination. Whereas formerly each utility independently operated its own system, by World War II on the U.S. side of the international border, the municipal utilities, PUDs, rural co-ops, Bonneville Power Administration, and investor-owned utilities were all working together to coordinate key aspects of the regional power supply. In the 1960s, the Pacific Northwest–Pacific Southwest Intertie and the U.S.-Canada Columbia River Treaty further expanded the geographic scope and institutional complexity of the power supply. Compared to the systems of the late nineteenth century, the complexity of today's power system boggles the mind.

The "demand" side is similarly complicated. Instead of serving one use, power companies quickly gathered into their customer portfolios a large variety of users distributed widely across an expanding service territory. Each consumer class had a unique "load factor," the character and timing of its power demand. Trolleys consumed lots of power during morning and evening rush hours and little or none late at night. Most lighting demands came after dark; most commercial demands came after sunrise. Most heating demands came in winter; most irrigation demands came in summer. Power companies sought to "balance" these loads to make the most efficient use of their power-generating equipment. As the diversity of demands proliferated, the challenge of efficiently and profitably servicing those demands heightened.

Utility managers and engineers had to coordinate both the supply side and the demand side when administering the technology under their control. As supply and demand grew increasingly complex, so did the challenges of administration. This would be difficult enough in a simple supply-and-demand market environment. But, as I have shown, multiple nodes of authority in planning and management arose in the first half of the twentieth century, with multiple layers of both private and public ownership of power systems. Besides these complex and occasionally overlapping power system authorities, there were provincial, state, and federal agencies involved in permitting dams on rivers, protecting fish and wildlife, regulating commerce and investment securities, and, during times of war, directing production and rationing consumption. The sociological dimension of the power system was easily as complex as its technological dimension, and more difficult to predict and control. Under the circumstances, I find it less surprising that the power system occasionally breaks down than that it has worked so well for so long.

The icing on the cake of complexity, which to me is the most interesting aspect of the power system, is that it is designed to serve multiple and conflicting purposes. At a simplified macro level, the power system appears neatly divided between producers seeking revenue and consumers seeking affordable electricity. To most consumers, the power companies seem to be in charge of the system. But consumers have clout. If they are large customers, like ALCOA or ALCAN, their market power and political influence can significantly shape aspects of the system, such as distribution and rate policies. As residential consumers became an increasingly large customer class starting in the 1920s, they gained clout through democratic channels and a voice through some members of utility regulatory commissions who saw themselves as public-interest advocates.

Ironically, the interests of producers and consumers are both complementary and at odds. Producers want satisfied customers, and both want a reliable, efficient, affordable power supply. At the same time, producers seek enhanced revenue from customers while customers seek maximum value for their dollar from producers. In part because of these conflicting interests, provinces and state legislatures created public utility commissions to arbitrate that relationship between the two. It was the first and most important step toward quasi-democratic governance of the power supply, aimed at achieving a fair balance between the revenue interests of the producer and the value interests of the consumer.

Producers and consumers, however, are not unified classes with common interests among the members in each class. Until the advent of state regulation, producers were at each other's throats, competing for the same customers and service territories. Even after state regulation, the half-century war between publicly owned and privately owned utilities extended these competitive conflicts that exemplified the divided nature of the producer class. Consumers have not had unified interests either. The rock-bottom wholesale electric rates paid by large industrial customers have long rankled residential customers, who routinely pay much higher retail rates.

As the century progressed, the line between producer and consumer blurred, further complicating this division. Starting with utility interconnections in the 1910s and accelerating with the expansion of the Bonneville Power Administration in the 1930s and 1940s, producers themselves became customers of other producers. BPA supplied power directly to industries but also to other public and private utilities, making those utilities customers of BPA. A similar boundary blurring is occurring today with owners of distributed renewable energy systems tied in to the power grid. Homeowners with grid-tied rooftop photovoltaic panels, for example, are both consumers and producers, depending on the time of day. Their utility companies likewise are alternately purchasers and sellers depending on the balance between power generation and use at the solar-powered home. This is a fascinating twist in the evolution of the power system.[5]

A long-term historical view of the power system makes it clear that one must be very cautious when making categorical generalizations about the stakeholders in the power system and their interests and relationships. One must be similarly cautious when making sweeping judgments about any particular institutional arrangement of the power system. For example, while nonprofit publicly owned utilities nearly always provide cheaper service than for-profit commercial utilities, there have been exceptions to the rule. Some privately owned utilities had better equipment and greater economies of scale, allowing them to provide better service and cheaper rates than small municipal utilities. And while publicly owned utilities theoretically better represented their customers (they are literally owned by their customers), these utilities could be just as bureaucratic and unresponsive as many investor-owned companies. Publicly owned utilities in the Northwest also made some spectacularly bad investment decisions that led to unnecessarily high rates, such as the Washington Public Power Supply System's (WPPSS) nuclear power plant debacle, discussed in the previous chapter.

Moreover, while regulation often solves problems, it can also create new problems that require new adjustments. Utility regulatory policies that guaranteed a fair return on investment indirectly encouraged the for-profit sector to invent ingenious methods for padding their expenses in order to increase revenue. Throughout the twentieth century, utilities and regulatory commissions have struggled over what constitutes legitimate expenses and a fair return on investment. The highly successful practice of guaranteeing a fair return on investment early in the century became a highly problematic safety net for economically unwise investments in the 1960s and 1970s. Utilities remained wedded to the bigger-is-better policy even after the bigger plants, especially the nuclear-powered plants, no longer returned positive economies of scale. Utility executives had no incentive to choose smaller, more efficient sources of power (natural gas combustion turbines, for example) because regulatory agencies permitted a fair return on investment, even on massive cost overruns for nuclear-power plants. Consequently, an unusual coalition of conservative economists and environmental activists called for deregulation in the 1980s to increase competition among utility companies and stop rewarding economically inefficient investment decisions. This was one more nail in the coffin of nuclear power.[6]

Beyond the competing goals of profit-seeking producers versus value-seeking consumers, a raft of other objectives animated the architects of power systems over time. As described in chapters 2–5, city leaders, businesses, residential customers, and regulatory agencies all assiduously sought to stabilize the industry and make the power supply more reliable and affordable. In the early twentieth century, community and business leaders saw electrification as a local asset useful for boosting the reputation of the town and attracting new business and capital. They consequently promoted, even subsidized, the establishment and extension of electric systems, creating partnerships between for-profit electrical entrepreneurs and the local political elite.

As more cities got this new technology, intercity competition shifted to the question of who had lower electric rates. At that point, city and regional boosters seeking economic development understood *cheap* power as a crucial tool for economic development, causing many to join the ranks of those calling for restraints on utility company profits and advocating public regulation or public ownership. When this happened, the traditional partnership between the local political elite and utility executives sometimes broke down into political acrimony. City movers and shakers sought to shape the electric system to promote general economic development. Sometimes that meant providing incentives to profit-seeking electric companies, while other times that meant restraining or expropriating those companies to more efficiently direct power development toward broader economic development or other public-interest goals.

Finally, as I described in chapter 9, stimulated by the promise of prosperity in the 1920s and the economic crisis of the 1930s, a number of intellectuals and policy entrepreneurs launched a movement to wield electric power as a tool for social improvement. Beyond mere "modernization," these proto-utopians hoped widespread, cheap electricity would improve working conditions and wages, raise standards of living, reinvigorate depressed rural areas, decentralize industry, reduce the gap between rich and poor, and provide a stable foundation for long-term economic growth. Power policies under Franklin Roosevelt's New Deal and to a lesser extent under Premier Dufferin Patullo's Liberal government in British Columbia in the 1930s advanced many of these social goals. Not surprisingly, though, electrification by itself did not and could not accomplish these lofty aims, although widespread cheap electricity did provide a solid foundation for economic growth during and after the war.

In sum, both internal and external architects of the electric industry in the Pacific Northwest, and other regions of the United States and Canada as well, sought to craft a power system that simultaneously provided revenue and profit, reliable and affordable service, local and regional development, and general social improvement. Such a tall order required cooperation between capitalist and governing institutions, another crucial layer of complexity in the institutional development of the Northwest power system. The development of regulatory institutions is a direct outcome of this creative and constantly evolving effort to achieve social goals and solve problems through the combined tools of market capitalism and government policy.

Both capitalism and democratic governance are complex human institutions reflecting diverse and divided interests and loyalties. So the task of crafting successful rules and relationships is enormously challenging, especially since we constantly adapt our institutions as historical conditions and social values change. Far from an aberration or corruption, regulation is a fundamental and necessary condition of a well-functioning complex system. Authorities, duties, and behaviors in a family are regulated, both formally and informally. The same goes for businesses, schools, clubs, churches, communities, states, nations, and economic systems. Successful capitalism *requires* rules and regulations. Even the biological world has regulatory mechanisms. The body regulates its temperature, digestive systems regulate food processing (bowel movements should be "regular"), hormones regulate sleep patterns and sexual behaviors. The relationships between species in ecosystems, such as that between predators and prey or host and parasite or plant and pollinator, are biologically regulated. None of these regulatory mechanisms are invariable or deterministic. There is variety, leeway, and change. But the absence of regulatory mechanisms is chaos. In the biological world, regulation evolves largely without conscious effort as we humans understand it. In contrast, we *Homo sapiens* intentionally create regulatory institutions to shape behavior and outcomes. It is among the most important actions we take to ensure our biological and social success as well as our emotional satisfaction.

In this study I hope I have provided a foundation for resurrecting an appropriately positive understanding of the necessary and beneficial role of public regulation. In contemporary political culture in North America, especially in the United States, it is popular to denigrate "Big Government" and the regulatory state. The attitude, I believe, is generated more by political rhetoric than by reason and experience. Regulations exist because we demand them. Every regulation evolves for a purpose, not by accident and usually not with insidious intent. Some regulations no longer serve their purpose, of course. Others may serve a purpose the majority no longer favors. But in all cases, there was intent and purpose behind every aspect of the regulatory framework we have created. Understanding the genesis of regulations is critical to intelligently reforming them.

Surprisingly, our complex electric power system and the policies that shape it actually change far more often and far more significantly than I expected when I started this research project. The power system is so large, so institutionalized, and so resilient in the face of perturbations that it appears from the outside as a monolithic, immovable object. But in fact it is a very dynamic, constantly evolving system. Certainly there is great inertia to it, and vested interests often successfully resist changes. Nevertheless, it is empowering to realize how commonly this ubiquitous socio-technological system changes in response to social desires and altered circumstances. As with biological evolution, change in the Northwest power system has often come in periods of rapid adaptation followed by periods of relative stability: punctuated equilibrium, as the ecologists call it. Not surprisingly, the periods of more rapid and substantive change are usually triggered by crises: economic, military, social, and environmental. Crises provide opportunities for public evaluation and structural reforms.

Governing institutions are our tool for reforming troubled institutions or solving persistent problems in the marketplace or seeking fairness and justice in economic transactions. We govern the electric system by establishing and periodically revising the rules, rights, privileges, and goals that structure the production and consumption of electric power. The process of governing this system, or any similar system, is fraught with conflict and mixed results. Governing institutions can appear inconsistent, unreliable, and ineffective because they represent the plural values and shifting balance of power of a democracy. Their mandates *are* inconsistent, more so than a business aimed unflinchingly at one goal: capital accumulation. The capabilities of any governing agency are limited by resources, by knowledge, by policy, and by forces that oppose the agency's mission. A failure to achieve objectives or solve problems in those instances is not a failure of regulation per se, but rather an expression of the limitations and pitfalls of democratic governance. But these messy, inconsistent governing institutions are the only effective tool we have for intentionally designing our social relations, for seeking balance between private wealth and public welfare, between individual liberty and social responsibility.

Ironically, the more inclusive our governing institutions, the more convoluted and potentially more conflicted the governing process becomes. Building and administering the electric power system were easier when only a few limited interests dominated decision-making, such as when producers simply sold power at a profit to willing buyers. But all kinds of dissatisfactions, inefficiencies, conflicts, and failures attended that era of electric power history. Consequently, through our governing institutions we redefined electric power as a public service and adopted a more complex set of authorities, rules, and objectives guiding electrical development. As a result, outcomes improved and more people were served and satisfied.

But the era of simple goals has passed. Laws and procedures providing much greater public participation in governance have advanced significantly since the 1970s. The public interests and values represented at the electric power decision-making table have greatly expanded: fisheries; tribal treaty rights; conservation; and energy efficiency (on the "demand" side rather than the supply side); promotion of clean, renewable energy; reduction in CO_2 emissions; access to the grid for independent power producers; competition among suppliers; and many other aims now influence the structure and function of the Northwest power system, making its governance that much more complicated. But we should celebrate rather than lament that complexity, unless we want to return to those simpler eras when only a few interests directed the electric power system toward a few narrow outcomes.

Regarding recent struggles over the Northwest power system and salmon recovery, geographer Eve Vogel notes that what appeared to be politicized, overly complex, inefficient, and expensive decision-making processes were in fact the everyday expression of democratic governance. Institutions for regional governance of the power system have evolved since the 1930s in the Columbia Basin, she says, to serve an increasingly complex array of goals and interests. Contemporary struggles, she argues, reflect an effort to unite the region in "participatory stewardship of a shared resource." While Vogel acknowledges that northwesterners have often failed to achieve their hopeful goals in river and power development, she nevertheless concludes that electric power and river basin decision-making in the Pacific Northwest in recent decades has become more "socially and ecologically inclusive . . . [resting] on wide participation and a real attempt to take responsibility for diverse needs within the region."[7] In this sense, complexity, stewardship, and shared governance represent progress. Let's keep it going.

Notes

INTRODUCTION

1. Thomas P. Hughes, *Networks of Power: Electrification in Western Society, 1880–1930* (Baltimore: Johns Hopkins University Press, 1983), ch. XIII; David E. Nye, *Electrifying America: Social Meanings of a New Technology* (Cambridge: MIT Press, 1990), x.

2. Vera Springer, *Power and the Pacific Northwest: A History of the Bonneville Power Administration* (Washington, DC: Bonneville Power Administration, 1976); Gus Norwood, *Columbia River Power for the People: A History of the Policies of the Bonneville Power Administration* (Portland: BPA, 1981); Gene Tollefson, *BPA and the Struggle for Power at Cost* (Portland: Bonneville Power Administration, 1987); Milton Conover, *The Federal Power Commission: Its History, Activities, and Organization* (Baltimore: Johns Hopkins Press, 1923).

3. Craig Wollner, *Electrifying Eden: Portland General Electric, 1889–1965* (Portland: Oregon Historical Society, 1990); Cecil Maiden, *Lighted Journey: The Story of BC Electric* (Vancouver: British Columbia Electric Company, 1948); Edward J. Crosby, *The Story of the Washington Water Power Company and Its Part in the History of Electric Service in the Inland Empire, 1889–1930* (Spokane: WWP, 1931); Patricia E. Roy, "Regulating the British Columbia Electric Railway: The First Public Utilities Commission in British Columbia," *BC Studies,* no. 11 (Fall 1971): 3–20; Steve Blewett, *A History of the Washington Water Power Company, 1889–1989* (Spokane: WWP, 1989); John Dierdorff, *How Edison's Lamp Helped Light the West: The Story of Pacific Power & Light Company and Its Pioneer Forebears* (Portland: Pacific Power & Light Co., 1971); Jeremy Mouat, *The Business of Power: Hydroelectricity in South Eastern British Columbia, 1897–1997* (Victoria, BC: Sono Nis Press, 1997); Susan Stacy, *Legacy of Light: A History of Idaho Power Company* (Boise: Idaho Power Company, 1991); Norman F. Stone, *Bountiful McKenzie: The Story of the Eugene Water and Electric Board* (Eugene, OR: Parkstone Co., 1986); Robert C. Wing and Robert C. Cumbow, *A Century of Service: The Puget Power Story* (Bellevue, WA: Puget Sound Power & Light Co., 1987).

4. Norwood, *Columbia River Power for the People*; Tollefson, *BPA and the Struggle for Power at Cost*; Ken Billington, *People, Politics, and Public Power* (Seattle: Washington Public Utility District Association, 1988); Karl Boyd Brooks, *Public Power, Private Dams: The Hell's Canyon High Dam Controversy* (Seattle: University of Washington Press, 2006); D. Clayton Brown, *Electricity for Rural America: The Fight for REA* (Westport, CT: Greenwood Press, 1980); Frank J. Busch, *Power for the People: Montana's Cooperative Utilities* (Missoula: University of Montana Press, 1976); Marquis Childs, *The Farmer Takes a Hand* (Garden City, NY: Doubleday and Co., 1952); David Cushman Coyle, ed., *Electric Power on the Farm. The Story of Electricity: Its Usefulness on Farms, and the Movement to Electrify Rural America* (Washington, DC: GPO, 1936); Wesley Arden Dick, "Visions of Abundance: The Public Power Crusade in the Pacific North-

west in the Era of J. D. Ross and the New Deal" (Ph.D. diss., University of Washington, 1973); Neil B. Freeman, *The Politics of Power: Ontario Hydro and Its Government, 1906–1995* (Toronto: University of Toronto Press, 1996); M. L. Ramsay, *Pyramids of Power: The Story of Roosevelt, Insull, and the Utility Wars* (New York: Da Capo Press, 1975); Richard Rudolph and Scott Ridley, *Power Struggle: The Hundred Year War over Electricity* (New York: Harper and Row, 1986); David Schap, *Municipal Ownership in the Electric Utility Industry—A Centennial View* (New York: Praeger, 1986).

5. Katrine Barber, *Death of Celilo Falls* (Seattle: University of Washington Press, 2005); Matthew Evenden, *Fish versus Power: An Environmental History of the Fraser River* (Cambridge: Cambridge University Press, 2004); Mark Fiege, *Irrigated Eden: The Making of an Agricultural Landscape in the American West* (Seattle: University of Washington Press, 1999); Philip Funigiello, *Toward a National Power Policy: The New Deal and the Electric Utility Industry, 1933–1941* (Pittsburgh: University of Pittsburgh Press, 1973); Paul Pitzer, *Grand Coulee: Harnessing a Dream* (Pullman: Washington State University Press, 1994); Roberta Ulrich, *Empty Nets: Indians, Dams, and the Columbia River* (Corvallis: Oregon State University Press, 1999); Eve Vogel, "The Columbia River's Region: Politics, Place, and Environment in the Pacific Northwest, 1933–Present" (PhD diss., University of Oregon, 2007); Richard White, *The Organic Machine: The Remaking of the Columbia River* (New York: Hill and Wang, 1995).

6. The best example of this genre is Sam H. Schurr et al., *Electricity in the American Economy: Agent of Technological Progress* (Westport, CT: Greenwood Press, 1990). Others that emphasize innovation and progress include the classic work by Hughes, *Networks of Power*; Thomas P. Hughes, *American Genesis: A Century of Invention and Technological Enthusiasm, 1870–1970* (Chicago: University of Chicago Press, 1989); Louis C. Hunter, *A History of Industrial Power in the United States, 1780–1930*, vols. 1–3 (Charlottesville: University Press of Virginia, 1986, and Cambridge: MIT Press, 1991); Forrest McDonald, *Insull* (Chicago: University of Chicago Press, 1962); John F. Wasik, *The Merchant of Power: Samuel Insull, Thomas Edison, and the Creation of the Modern Metropolis* (New York: Palgrave Macmillan, 2006); Sidney Alexander Mitchell, *S. Z. Mitchell and the Electrical Industry* (New York: Farrar, Straus, and Cudahy, 1960); Richard A. Pence, ed., *The Next Greatest Thing* (Washington, DC: National Rural Electric Cooperative Association, 1984); and many company-sponsored histories such as those cited above by Blewett, Crosby, Dierdorff, Maiden, and Wollner, as well as many government reports boosting the power industry, especially during the first half of the twentieth century, such as the Federal Power Commission's *National Power Surveys*, Leo Denis's *Electric Generation and Distribution in Canada* (Ottawa: Commission of Conservation in Canada, 1918); Arthur V. White and Charles Vick's *Water Powers of British Columbia* (Ottawa: Canada Commission of Conservation, 1919); and Canada's Dominion Bureau of Statistics, *Census of Industry, Central Electric Stations in Canada* (Ottawa, annual reports from 1917 to 1930).

7. See above citations for Billington, Brown, Busch, Childs, Dick, Norwood, and Pence; plus Aaron Jones, *Inland: The First 40 Years* (Forest Grove, OR: Inland Power & Light, 1977).

8. See especially Billington and Rudolph and Ridley cited above, plus Marc Mess-

ing, H. Paul Friesema, and David Morell, *Centralized Power: The Politics of Scale in Electricity Generation* (Cambridge, MA: Oelgeschlager, Gunn and Haip, 1979).

9. This perspective is most often expressed in commercial industry trade journals, such as *Electrical World* and *Electrical West,* and in ads and editorials in the popular press. But occasionally there are scholarly books defending this position, such as Ernest R. Abrams, *Power in Transition* (New York: Charles Scribner's Sons, 1940) and investor-owned company histories such as those written by Dierdorff and Crosby. Articulate expressions of both pro and con positions regarding for-profit systems versus public power are captured in the published debates sponsored by the National Civic Federation early in the twentieth century. See, for example, *Shall the Government Own and Operate the Railroads, the Telegraph and Telephone Systems? Shall the Municipalities Own Their Utilities?* (New York: National Civic Federation, 1915).

10. I believe the best historical analysis of the contemporary transition in the electric power system is Richard F. Hirsh, *Power Loss: The Origins of Deregulation and Restructuring in the American Electric Utility System* (Cambridge: MIT Press, 1999). I discuss his basic thesis in the epilogue to this book. See also Ed Smeloff and Peter Asmus, *Reinventing Electric Utilities: Competition, Citizen Action, and Clean Power* (Washington, DC: Island Press, 1997).

11. Policy history is one of the most common and traditional styles of historical analysis; there are many good examples. Classic older works that are still valuable today include Charles McKinley, *Uncle Sam in the Pacific Northwest: Federal Management of Natural Resources in the Columbia River Valley* (Berkeley: University of California Press, 1952); Funigiello, *Toward a National Power Policy*; Brown, *Electricity for Rural America*; John V. Krutilla, *The Columbia River Treaty: The Economics of an International River Basin Development* (Baltimore: Johns Hopkins University Press, 1967); Neil A. Swainson, *Conflict over the Columbia: The Canadian Background to an Historic Treaty* (Montreal: McGill-Queen's University Press, 1979). Several more contemporary examples of good policy history are Jay Brigham, *Empowering the West: Electrical Politics before FDR* (Lawrence: University Press of Kansas, 1998); Brooks, *Public Power, Private Dams*; Brian Q. Cannon, "Power Relations: Western Rural Electric Cooperatives and the New Deal," *Western Historical Quarterly*, vol. 31, no. 2 (Summer 2000): 133–160; Wesley Arden Dick, "When Dams Weren't Damned: The Public Power Crusade and Visions of the Good Life in the Pacific Northwest in the 1930s," *Environmental Review*, vol. 13, nos. 3/4 (1989): 113–153; Robin Fisher, "The Decline of Reform: British Columbia Politics in the 1930s," *Journal of Canadian Studies*, vol. 25, no. 3 (1990): 74–89; Karl Froschauer, *White Gold: Hydroelectric Power in Canada* (Vancouver: University of British Columbia Press, 1999); Harry H. Stein, "Fighting for Aluminum and for Itself: The Bonneville Power Administration, 1939–1949," *Pacific Northwest Quarterly*, vol. 99, no. 1 (Winter 2007/2008): 3–15; and Vogel, "The Columbia River's Region." Less scholarly but still helpful are Norwood, *Columbia River Power for the People* and Tollefson, *BPA and the Struggle for Power at Cost* (the latter a generally reliable treasure trove of details and images).

12. Nye, *Electrifying America*; Nye, *Consuming Power: A Social History of American Energies* (Cambridge: MIT Press, 1998). See also James Carey and John Quirk, "The

Mythos of the Electronic Revolution," *American Scholar*, vol. 39, no. 2 (1970): 219–241 and no. 3 (1970): 395–424; Ruth Schwartz Cowan, *More Work for Mother: The Ironies of Household Technology from the Open Hearth to the Microwave* (New York: Basic Books, 1983); Christina Hardyment, *From Mangle to Microwave: The Mechanization of Housework* (Cambridge, UK: Polity Press, 1988); Patricia Ann Kellner, "From the Pastoral to the Promethean: Electric Power Regimes in the Pacific Northwest and the Rhetoric of the Electrical Sublime" (Ph.D. diss., UCLA, 1994); Harold Platt, *The Electric City* (Chicago: University of Chicago Press, 1991); Ronald C. Tobey, *Technology as Freedom: The New Deal and the Electrical Modernization of the American Home* (Berkeley: University of California Press, 1996); and Blair Tothill, "Living Electrically: The British Columbia Electric Railway Company and the Development of the Domestic Electric Appliance Market in Victoria, 1919–1939" (master's thesis, University of Ottawa, 1997).

13. Hughes, *Networks of Power*, is perhaps the best example of this analytical approach. A scholar with wide-ranging talents, Hughes in this book focuses on organizational theory, the social construction of technological progress, and intentional system evolution. Canadian scholars Jeremy Mouat and Bruce Stadfeld adopt some of this institutional approach, too. See Mouat, *The Business of Power*, and Bruce Stadfeld, "Electric Space: Social and Natural Transformations in British Columbia's Hydroelectric Industry to World War II," Ph.D. thesis, University of Manitoba, 2002. Of course the most common examples of organizational history, though not the most sophisticated ones, are utility company histories. Two of the better ones are Wollner and Dierdorff, cited above.

14. The best work in this genre is Evenden, *Fish versus Power*. Mouat and Stadfeld (cited above) also pay extra attention to the environment in their works. Good studies by environmental historians from south of the international border include Richard White's *Organic Machine*, Mark Fiege's *Irrigated Eden*, Eve Vogel's dissertation "The Columbia River's Region," and numerous essays in the book I edited with Dale Goble, *Northwest Lands, Northwest Peoples: Readings in Environmental History* (Seattle: University of Washington Press, 1999), especially the chapters written by Carl Abbott, Michael Blumm, Dale Goble, and Eugene Hunn. More specifically, on the topic of the impact of the hydropower system on rivers and fisheries, see Katrine Barber, *Death of Celilo Falls*; Michael C. Blumm, "Hydropower vs. Salmon: The Struggle of the Pacific Northwest's Anadromous Fish Resources for a Peaceful Coexistence with the Federal Columbia River Power System," *Environmental Law*, vol. 11, no. 43 (1981): 211–261; Jeff Crane, *Finding the River: An Environmental History of the Elwha* (Corvallis: Oregon State University Press, 2011); Jim Lichatowich, *Salmon without Rivers: A History of the Pacific Salmon Crisis* (Washington, DC: Island Press, 1999); Anthony Netboy, *The Salmon: Their Fight for Survival* (Boston: Houghton Mifflin, 1973); Joseph E. Taylor III, *Making Salmon: An Environmental History of the Northwest Fisheries Crisis* (Seattle: University of Washington Press, 1999); and Roberta Ulrich, *Empty Nets: Indians, Dams, and the Columbia River* (Corvallis: Oregon State University Press, 1999). A very useful collection of primary source documents with some commentary is available in Joseph Cone and Sandy Ridlington, *The Northwest Salmon Crisis: A Documentary History* (Corvallis: Oregon State University Press, 1996 and 2000).

15. John M. Findlay and Ken S. Coates, eds., *Parallel Destinies: Canadian-American Relations West of the Rockies* (Seattle: University of Washington Press, 2002), xviii–xix.

16. Chapter 8 of this book covers the Depression era. Michael Fellman's chapter "Sleeping with the Elephant" in Findlay and Coates, *Parallel Destinies*, confirms that Canadian prime minister Mackenzie King took a much more timid approach to solving the problems of the Great Depression than the U.S. government, even though he was a prominent Liberal and admirer of President Franklin Roosevelt. See Fellman, 286–287.

17. Findlay and Coates, *Parallel Destinies*. The Coates quote is from p. 4, and the Fellman quote is from p. 278.

CHAPTER 1. THE BRILLIANT SPECTACLE

1. David E. Nye, *Electrifying America* (Cambridge: MIT Press, 1990), 1.

2. Cecil Maiden, *Lighted Journey: The Story of BC Electric* (Vancouver: British Columbia Electric Company, 1948), 11–12.

3. *Oregonian*, May 29, 1879, 3. A public scene resulted again when Portland's smaller, poorer sister to the north, Seattle, got its first glimpse of electric light in 1881 when the SS *Willamette* docked with an arc lighting system on board. Richard E. York, "Puget Sound Power and Light Company," *Electrical West 75th Anniversary Issue*, vol. 129, no. 2 (August 1962): 226–232.

4. "Docks Lighted," *Oregonian*, September 7, 1880, 3, quoted in Gene Tollefson, *BPA and the Struggle for Power at Cost* (Portland: BPA, 1987), 14, 20–21.

5. Thomas P. Hughes, *Networks of Power, 1880–1930* (Baltimore: Johns Hopkins University Press, 1983), 75–77.

6. Tollefson, *BPA and the Struggle for Power at Cost*, 14, 20–21, citing Villard's published memoirs. There are some conflicting reports about what happened to this lighting system on the *Columbia*. Tollefson says it remained on the ship, providing illumination for fifteen years before being upgraded. But an early history of the Washington Water Power Company, written by Edward J. Crosby and published in 1931, claims that Villard's dynamo and arc light system from the steamship *Columbia* ended up in Spokane Falls in 1885. The dynamo was reportedly installed in the basement of a flour mill, and the lights illuminated the Spokane Falls commercial district on a franchise granted by the city council in 1885. Edward J. Crosby, *The Story of the Washington Water Power Company and Its Part in the History of Electric Service in the Inland Empire, 1889–1930* (Spokane: WWP, 1931), 7. However, Crosby's story is doubtful. The light system on the *Columbia* that came to Portland in 1880 had incandescent lights, not arc lights. Crosby may have meant the arc light system from the SS *California* that arrived in Portland in 1879. A later historian of WWP also doubts that the lighting system from the *Columbia* ended up in Spokane: Steve Blewett, *A History of the Washington Water Power Company, 1889–1989* (Spokane: WWP, 1989), 3.

7. David E. Nye, *Electrifying America*, 4.

8. Crosby, *The Story of the Washington Water Power Company*, 7.

9. For a U.S. Government Accountability Office (GAO) report on the trespass and compensation claims of the Spokane Tribe regarding Grand Coulee Dam, see http://www.gao.gov/htext/do4125t.html, accessed December 20, 2011.

10. Maiden, *Lighted Journey,* 20.

11. Nye, *Electrifying America,* 33–47.

12. On lighting the Northwest Industrial Exposition, see Crosby, 15. On electrifying the 1905 World's Fair in Portland, see Craig Wollner, *Electrifying Eden: Portland General Electric, 1889–1965* (Portland: Oregon Historical Society, 1990), 50–63; and E. Kimbark MacColl with Harry H. Stein, *Merchants, Money, and Power* (Athens, GA: The Georgian Press, 1988), 357, 361–362, 383–386.

13. Nye, *Electrifying America,* 55–56.

14. Maiden, *Lighted Journey,* 2–9.

15. Maiden, 14–17.

16. *Oregonian,* May 29, 1879, 3.

17. Robert C. Wing and Robert C. Cumbow, eds., *A Century of Service: The Puget Power Story* (Bellevue, WA: Puget Sound Power and Light Co., 1987), 6–7. Tollefson, 14, recounts this Seattle event but mistakenly says the date was 1885.

18. Tollefson, 28–32; Maiden 14–18, 20–22; Crosby, 9; Dierdorff, *How Edison's Lamp Helped Light the West* (Portland: Pacific Power & Light Co., 1971), 19–26.

19. Tollefson, 14; Dierdorff, 5–8.

20. Maiden, 22; Crosby, 9; Wing and Cumbow, *A Century of Service,* 20.

21. Nye, *Electrifying America,* 57.

22. On the coevolution of electrification, industrialization, and corporatization, see David Nye, *Consuming Power: A Social History of American Energies* (Cambridge: MIT Press, 1998), ch. 4; Martin V. Melosi, *Coping with Abundance: Energy and Environment in Industrial America* (Philadelphia: Temple University Press, 1985), chs. 3 and 4; and Hughes, *Networks of Power,* ch. 3.

23. Hughes, 75–77; Nye, *Consuming Power,* 124–125; Tollefson, 36–38.

24. Crosby, 9.

25. This profile of S. Z. Mitchell is primarily drawn from Dierdorff, 1–11, who draws much of his information about Mitchell from the biography written by Mitchell's son, Sidney Alexander Mitchell, *S. Z. Mitchell and the Electrical Industry* (New York: Farrar, Straus, and Cudahy, 1960). All the authors who have written about the development of electric power in the Northwest also profile S. Z. Mitchell and his important role in the industry. See Tollefson, 14, 28–32, 60–64; Wing, 2–7, 10, 14–16, 76–80; and Susan Stacy, *Legacy of Light: A History of Idaho Power Company* (Boise: Idaho Power Company, 1991), 13–16, 31–33, 45–48.

26. Dierdorff, 1–11.

27. Ibid.

28. See Maiden, 12–27 on McMicking, Abbott, Keefer, and Higgins; and 41–49 on Barnard and events during the 1891–1896 depression.

29. Hughes, 266. See also pp. 405–406 for more on the role of geography in power systems.

30. Bruce Wolverton, "The Water Power of the Pacific Northwest," *Pacific Monthly,* vol. XI, no. 3 (March 1904): 156. (The whole article runs from pp. 155 to 169.)

31. U.S. Geological Survey website, http://vulcan.wr.usgs.gov/Volcanoes/Washington/ColumbiaRiver/description_columbia_river.html Accessed May 12, 2012.

32. Bonneville Power Administration, "NW Hydro: Multi-Purpose Dams of the Pacific Northwest," http://www.bpa.gov/Power/PL/Columbia/2-multi.htm. Accessed May 12, 2012.

33. Wolverton, "The Water Power of the Pacific Northwest," 158.

34. Hughes, 81–126; Tollefson, 36; Melosi, *Coping with Abundance,* 60–61.

35. Wollner, *Electrifying Eden,* 23–26; Tollefson, 42; Gus Norwood, *Columbia River Power for the People: A History of the Policies of the Bonneville Power Administration* (Portland: BPA, 1981), 51.

36. Nye, *Consuming Power,* 8.

37. A classic monograph arguing this thesis on the loss of worker control over production during industrialization is David Montgomery, *The Fall of the House of Labor: The Workplace, the State, and American Labor Activism, 1865–1925* (New York: Cambridge University Press, 1987).

38. For illuminating discussions of muscle power and the shift to steam power, see Melosi, *Coping with Abundance,* ch. 1, and Nye, *Consuming Power,* chs. 1 and 2. On "Cities of Steam," see Nye, ch. 3.

39. On early applications of the electric motor and the use of electricity in factories, see Sam H. Schurr et al., *Electricity in the American Economy: Agent of Technological Progress* (Westport, CT: Greenwood Press, 1990), ch. 1. See also Melosi, 62–67.

CHAPTER 2. POWER INCORPORATED

1. Christopher Armstrong and H. V. Nelles, *Monopoly's Moment: The Organization and Regulation of Canadian Utilities, 1830–1930* (Philadelphia: Temple University Press, 1986), 7.

2. Sam H. Schurr et al., *Electricity in the American Economy* (Westport, CT: Greenwood Press, 1990), 209–210; David E. Nye, *Electrifying America* (Cambridge: MIT Press, 1990), 85–97.

3. Cecil Maiden, *Lighted Journey: The Story of BC Electric* (Vancouver: British Columbia Electric Company, 1948), 27–29.

4. U.S. Bureau of the Census, *Census of Electrical Industries, Central Electric Light and Power Stations, 1912* (Washington, DC: GPO, 1915), 19. Hereafter I will refer to this report as the *U.S. Census of Electrical Industries, 1912.* Canada Dominion Bureau of Statistics, *Census of Industry, 1917, Central Electric Stations in Canada* (Ottawa, 1919), xxiii–xxiv and Table 32, p. 27.

5. On trolley car amusement parks, see Nye, *Electrifying America,* 122–132.

6. The perspective on Portland and the quote from Mumford are from E. Kimbark MacColl with Harry H. Stein, *Merchants, Money, and Power* (Athens, GA: The Georgian Press, 1988), 259.

7. MacColl and Stein, *Merchants, Money, and Power,* 264–268.

8. On Corbett's and Failing's early financial and political careers in Portland, see MacColl and Stein, *Merchants, Money, and Power,* 27–31, 94–97, 109–112, 143–154, 163–168, 182–186, 193–196, 238–240, and 246–248. On Swigert, see MacColl and Stein,

268–269. For another perspective on these businessmen and on the connection among bankers, real estate developers, and owners of streetcar and electric companies, see Craig Wollner, *Electrifying Eden: Portland General Electric, 1889–1965* (Portland: Oregon Historical Society, 1990), 29–32. Wollner glorifies the role of the entrepreneur and minimizes attention to conflicts between private interests and the public interest.

9. MacColl and Stein, 268–272.

10. Wollner quote is from p. 29. See also MacColl and Stein, 270–271.

11. MacColl and Stein, 415–416; Wollner, 63–80, for details on the background to the Portland Railway, Light and Power merger and a discussion of its early years under Josselyn. MacColl's narrative is sympathetic to the city and guarded in its assessment of businessmen, while Wollner's is the opposite. They make interesting reading in combination. MacColl's is the more perceptive and wide-ranging analysis.

12. The quote is from Kenneth T. Jackson, *Crabgrass Frontier: The Suburbanization of the United States* (New York: Oxford University Press, 1987), 120. The men who put together the streetcar and electric company mergers in Portland were not the only ones with real estate interests served by their streetcar lines. MacColl provides other examples, such as two bank owners, Frank Dekum and Richard Durham, who used their own institution's funds to form the Oregon Land and Investment Company in the 1880s, which bought undeveloped lands north of Portland. In 1888, they incorporated the Portland and Vancouver Railway and secured franchises from east Portland, Albina, and Multnomah County to run a rail line from the urban areas through their properties to the Columbia River opposite Vancouver, Washington, making their real estate vastly more valuable and saleable. MacColl and Stein, 266–269.

13. Robert C. Wing and Robert C. Cumbow, eds., *A Century of Service: The Puget Power Story* (Bellevue, WA: Puget Sound Power and Light Co., 1987), 22–24. See also Western Washington University's Center for Pacific Northwest Studies (CPNS) website, http://west.wwu.edu/cpnws/findingaids/cpnws/puget/pugethist.htm (accessed May 12, 2011), for an organizational history summary. Many Puget Sound Power & Light Company records are archived at the CPNS (Bellingham, WA).

14. Edward J. Crosby, *The Story of the Washington Water Power Company and Its Part in the History of Electric Service in the Inland Empire, 1889–1930* (Spokane: WWP, 1931), 16–17; Steve Blewett, *A History of the Washington Water Power Company, 1889–1989* (Spokane: WWP, 1989), 9.

15. Maiden, *Lighted Journey*, 25–30, on the first streetcar line in Victoria. Quotes are both from p. 28.

16. Maiden, 30.

17. Maiden, 31–39.

18. Maiden, 40–43.

19. Patricia E. Roy, "Direct Management from Abroad: The Formative Years of the British Columbia Electric Railway," *The Business History Review*, vol. 47, no. 2 [Canada] (Summer 1973): 239–241. See also Maiden, 43–47, but note that Maiden, who was not a trained historian, is not always a reliable source. For example, he credits Barnard rather than Horne-Payne as the primary organizer and financier of BCER.

20. Wing and Cumbow, *A Century of Service*, 21; Maiden, 42–43; Wollner, 38.

21. MacColl and Stein provide a good example of the more critical perspective, as does Jay Brigham, *Empowering the West: Electrical Politics before FDR* (Lawrence: University Press of Kansas, 1998).

22. Wallis Nash, *The Farm, Ranch and Range in Oregon* (Portland, OR: The Lewis and Clark Centennial Exposition Commission for the State of Oregon, 1905), 30.

23. Wollner, 98.

24. Gene Tollefson, *BPA and the Struggle for Power at Cost* (Portland: Bonneville Power Administration, 1987), 32.

25. Nash, *The Farm, Ranch and Range in Oregon*, 30.

26. Tollefson, *BPA and the Struggle for Power at Cost*, 32.

27. This difficulty in distinguishing electric companies from urban railroad companies is reflected in an ongoing confusion in the *U.S. Census of Electrical Industries*. The census issued separate reports for "central electric light and power stations" and "street and electric railways." Reports of the amount of power produced by each industry are not easily comparable over the years, however, because electric railways often acquired power companies and vice versa, forcing the census takers to juggle the records of electricity production back and forth between the two industries. See the explanations and caveats in *U.S. Census of Electrical Industries, 1912*, 11–15.

28. Washington Water Power provides another example of the merging and then later reseparation of electric power and electric railway companies. The 1889 merger that created WWP included both types of businesses, and for the next two decades the bulk of WWP's electrical generation served its streetcar businesses. WWP's rail business peaked in 1910 with nearly 25 million passengers. By 1912 ridership was in a noticeable decline due to competition from automobiles, trucks, and jitneys. By 1915 passenger traffic was down to 2 million. Subsequent mergers failed to keep the company profitable, and the last electric streetcar ended operation in 1931. See Blewett, *A History of the Washington Water Power Company, 1889–1989*, 12–13.

29. Nye, *Electrifying America*, 97–113.

30. MacColl and Wollner both detail the conflicted relationship between PGE and the city of Portland during the early 1900s, although with opposing sympathies. Compare MacColl and Stein, 405–417 with Wollner, 71–80.

31. On corruption and city franchises specifically, see David Schap, *Municipal Ownership in the Electric Utility Industry—A Centennial View* (New York: Praeger, 1986), 19–23. For more on this topic, see Richard Rudolph and Scott Ridley, *Power Struggle: The Hundred Year War over Electricity* (New York: Harper and Row, 1986), 38–41, 192; Ed Smeloff and Peter Asmus, *Reinventing Electric Utilities: Competition, Citizen Action, and Clean Power* (Washington, DC: Island Press, 1997), 8–10; Thomas P. Hughes, *Networks of Power: Electrification in Western Society, 1880–1930* (Baltimore: Johns Hopkins University Press, 1983), 201–206; Nye, *Electrifying America*, 168–181.

32. Clyde L. King, "Municipal Ownership versus Adequate Regulation," *The Regulation of Municipal Utilities* (New York: D. Appleton, 1912), 23–24; quoted in Schap, fn. 17, p. 40.

33. Like the "muckrakers" two generations before, MacColl and Stein, in *Merchants, Money, and Power*, expose the seamier side of wealth and power in Portland's

history. See ch. 18, "Private Enterprise and Public Interest," 357–373 and 416–417, for a discussion of franchise corruption and other forms of oligarchic abuse in Portland during this time.

34. David Nye offers insightful explanations for why private utilities dominated the American electric power industry throughout the late nineteenth and early twentieth centuries: *Electrifying America*, 139–140, 142, 168–176, and 179–181.

35. Consolidation of the electric power industry and the decline of municipal systems after the turn of the century are discussed in Smeloff and Asmus, 10–11; Hughes, 276–278; Rudolph and Ridley, 41–44; and Tollefson, 54. For a thorough analysis of the rise and decline of municipal systems, 1882–1932, see Schap, chs. 1–3. For an interesting overview of the political economy of municipal electric systems with profiles of Boston, San Francisco, and Seattle, see Charles David Jacobson, *Ties That Bind: Economic and Political Dilemmas of Urban Utility Networks, 1800–1990* (Pittsburgh: University of Pittsburgh Press, 2000), 74–136. Finally, see chs. 4 and 5 in this book for more on municipal electric utilities in the Northwest.

36. Yearly data on U.S. municipal electric systems are summarized in Schap, 9–11, 21–28. The data for Figure 2.2 are taken from this source.

37. Tollefson briefly profiles the formation and early history of these municipal systems in the U.S. Northwest, 46–54.

38. Jeremy Mouat, *The Business of Power: Hydroelectricity in South Eastern British Columbia, 1897–1997* (Victoria, BC: Sono Nis Press, 1997), 53–56.

39. For a brief history of the launch of Tacoma's municipal water and power company, see David Wilma, "City of Tacoma Buys Tacoma Light and Water Co. on July 1, 1893," *HistoryLink.org Online Encyclopedia of Washington State History*, http://www.historylink.org/essays/output.cfm?file_id=5022. Accessed December 21, 2011.

40. For a brief history of the creation of Seattle City Light, see Walt Crowley, "Seattle Voters Approve Electric Utility Bonds on March 4, 1902, Leading to Creation of Seattle City Light," *HistoryLink.org Online Encyclopedia of Washington State History*, http://www.historylink.org/essays/output.cfm?file_id=2317. Accessed December 21, 2011.

41. Tollefson, 49; and David Wilma, "City of Tacoma Buys Tacoma Light and Water," HistoryLink.org (fn. 38).

42. Tollefson, ch. 8; Rudolph and Ridley, 32–36.

43. *U.S. Census of Electrical Industries, 1912*, 17 and 24.

44. Ibid. Quote is from p. 21.

45. Canada Dominion Bureau of Statistics, *Census of Industry, 1917, Central Electric Stations in Canada*, Table 5, p. 6; Table 15, p. 15; and Table 28, p. 25.

46. Schap, 23–36.

47. Tollefson notes on pp. 36 and 38 that in the financial Panic of 1893, J. P. Morgan, who had put together General Electric in 1892, bought up many additional electric utilities. Many Americans complained that the Morgan Banking House, and others like it, benefited from the three-year depression while the rest of the country suffered.

48. John A. Fairlie, *Essays in Municipal Administration* (New York: Macmillan, 1908), 262, quoted in Schap, p. v.

49. On the rise of corporations and the development of energy industries in the late nineteenth and early twentieth centuries, see David E. Nye, *Consuming Power* (Cam-

bridge: MIT Press, 1998), ch. 4; John G. Clark, *Energy and the Federal Government: Fossil Fuel Policies, 1900–1946* (Urbana: University of Illinois Press, 1987), ch. 1; and Rudolph and Ridley, *Power Struggle*, 31–36.

50. Hughes compares the evolving electric supply systems in Berlin, Chicago, and London, discussing how each city dealt with the tension between private enterprise and government regulation for public welfare, *Networks of Power*, chs. VII–IX. Europeans were much more accepting of public ownership and strong government regulatory institutions than Americans were.

51. H. V. Nelles, *The Politics of Development: Forests, Mines, and Hydro-Electric Power in Ontario, 1849–1941*, 2nd edition (Montreal: McGill-Queen's University Press, 2005), 242–244. The quote is on p. 244 and is taken from the February 18, 1903, edition of the Toronto *World*.

52. Statistics cited in this paragraph are from http://epe.lac-bac.gc.ca/100/205/301/ic/cdc/heirloom_series/volume4/318-321.htm, accessed May 12, 2012. For a scholarly history of Ontario Hydro, see Neil B. Freeman, *The Politics of Power: Ontario Hydro and Its Government, 1906–1995* (Toronto: University of Toronto Press, 1996); and Nelles, *The Politics of Development*, ch. 6, "Hydro as Myth," 213–255.

53. Rudolph and Ridley, 32–33.

54. On early business-government cooperation to bring stability, reliable profits, and a degree of fairness to the Canadian electric utility industry, see Armstrong and Nelles, 74–100.

55. Martin V. Melosi, *Coping with Abundance: Energy and Environment in Industrial America* (Philadelphia: Temple University Press, 1985), 75, argues that most federal regulation stems from the period 1870 to 1930 and came as a result of market failures due to unregulated monopolies. Jay Brigham, *Empowering the West: Electrical Politics before FDR* (Lawrence: University Press of Kansas, 1998), 5, points out how reformers shifted much of their attention from railroads to the electric utilities after World War I.

56. *U.S. Census of Electrical Industries, 1912*, 34–39.

57. On the significance of the Niagara project, see Tollefson, 42; Brigham, 11; Hughes, 135–139. On the development of long-distance transmission technology and its impact on the electric power industry in general, see Tollefson, 36; Hughes, 81–105 and ch. 5; and Melosi, 60–61.

58. Wing and Cumbow, 12–19. The authors of this company history said the *original* Westinghouse generators were still producing at their rated capacity of 6,000 kilowatts in 1987. See also the company's original periodical, *Puget Sound Electric Journal*, vol. IV, no. 9 (September 1914): 4–5, for a profile of both Snoqualmie Falls powerhouses.

59. Wing and Cumbow, 15.

60. Ibid.

61. Mouat, *The Business of Power*, 28–35.

62. Maiden, 53–55.

63. Jerome G. Kerwin, *Federal Water Power Legislation* (New York: Columbia University Press, 1926), 105–107. See also Samuel P. Hays, *Conservation and the Gospel of Efficiency: The Progressive Conservation Movement, 1890–1920* (Cambridge: Harvard University Press, 1959), 114–115.

64. Kerwin, 31.

65. Western States Tourism Policy Council, Federal Land in the West, http://www.dced.state.ak.us/wstpc/Publications/FedLandWest.htm. Last updated September 16, 2009.

66. Gus Norwood, *Columbia River Power for the People: A History of the Policies of the Bonneville Power Administration* (Portland: BPA, 1981), 21; Kerwin, 108–109.

67. Act of February 15, 1901 (31 Stat. 790). See Kerwin, 108–109, and Frank E. Smith, ed., *Conservation in the United States: A Documentary History,* vol. 2, *Land and Water 1900–1970* (New York: Chelsea House Publishers, 1971), 162–163, for a discussion of these various acts.

68. Gifford Pinchot, *The Fight for Conservation* (New York: Doubleday, Page and Co., 1910), 84–88.

CHAPTER 3. BECOMING MODERN

1. From the *Bulletin* of the Portland General Electric Company, January 1906, quoted in Craig Wollner, *Electrifying Eden: Portland General Electric, 1889–1965* (Portland: Oregon Historical Society Press, 1990), 46.

2. Gene Tollefson, *BPA and the Struggle for Power at Cost* (Portland: Bonneville Power Administration, 1987), 62.

3. David E. Nye, *Electrifying America* (Cambridge: MIT Press, 1990), 147–149 and 247–248. On the 160 utopian novels published between 1888 and 1900, Nye cites Kenneth M. Roemer, *The Obsolete Necessity: America in Utopian Writings, 1888–1900* (Kent, OH: Kent State University Press, 1981).

4. Jeremy Rifkin, *The Hydrogen Economy: The Creation of the Worldwide Energy Web and the Redistribution of Power on Earth* (New York: Penguin Books, 2002).

5. Data drawn from the U.S. Bureau of the Census, compiled in Edison Electric Institute, *Historical Statistics of the Electric Utility Industry, to 1970* (New York: Edison Electric Institute, 1970), 21.

6. The power produced by electric railway companies and by isolated industrial plants increased too, but not as quickly or in as sustained a manner. By the 1917 electrical census, electric railways were in decline, industrial power users were shifting toward purchasing rather than producing power, and the proportion of U.S. electricity generated by central stations rose, reaching a majority (53 percent) in the 1927 census: U.S. Bureau of the Census, *Census of Electrical Industries, Central Electric Light and Power Stations, 1927* (Washington, DC: GPO, 1930), 3 and Table 1, p. 7.

7. I roughly calculated the per capita rates by dividing kwh generation in 1902 and 1922 by population numbers for each state from the 1900 and 1920 decadal censuses. The results are reasonably accurate but not precise because of the two-year difference in dates.

8. Wollner, *Electrifying Eden*, 115.

9. Edward J. Crosby, *The Story of the Washington Water Power Company and Its Part in the History of Electric Service in the Inland Empire, 1889–1930* (Spokane: WWP, 1931), 26–27.

10. Norman F. Stone, *Bountiful McKenzie: The Story of the Eugene Water and Electric Board* (Eugene, OR: Parkstone Co., 1986), 22.

11. For Seattle's graduated rate structure after the turn of the century, see Seattle City Light, *Annual Report of the Lighting Department for 1911*, 22, 25. Many additional examples for any time period can be found in the annual reports of the state public utility commissions. Interestingly, the annual reports and company histories of Pacific Power & Light, BC Electric, and Puget Sound Power & Light—all private utilities—do not discuss rate structures in chapters and reports covering the first two decades of the twentieth century. In contrast, the municipally owned electric utility annual reports and histories of that early period often proudly discuss rates.

12. Nye, *Electrifying America*, 187. His chapter titled "Flexible Factory" is one of the best studies of the early electrification of the American workplace. Also, Martin Melosi, *Coping with Abundance: Energy and Environment in Industrial America* (Philadelphia: Temple University Press, 1985), ch. 3, covers electrification in industrial production. The classic compendium for those who want to know everything about energy in American industry is the multivolume work by Louis C. Hunter, *A History of Industrial Power in the United States, 1780–1930* (Charlottesville: University Press of Virginia, 1986, and Cambridge: MIT Press, 1991).

13. Crosby, *Story of the Washington Water Power Company*, 19–20.

14. For a useful table showing the chronology of electrification of mechanical power in industry, see Warren D. Devine, Jr., "Electrified Mechanical Drive: The Historical Power Distribution Revolution," in Sam H. Schurr et al., *Electricity in the American Economy: Agent of Technological Progress* (Westport, CT: Greenwood Press, 1990), 24. See also Melosi, 112–116 and Nye, 187, for summary national data on the increase in electrification from 1900 to 1930. Technically speaking, industry did not entirely replace steam power at the rate suggested by these raw numbers since at that time most of the electricity in factories and mines was generated by steam-driven dynamos on-site. In fact, on-site steam engines spinning electrical generators produced 60–70 percent of electrical power used in manufacturing plants in the decade between 1899 and 1909. So, more precisely, electric motors commonly replaced steam-driven shafts and belts for moving machines in factories during these decades, while steam-powered turbines remained important for generating electricity (Devine, 28).

15. Devine, 22.

16. Devine, 25–29. The quote is from p. 28.

17. Quoted in Wollner, 40.

18. Public Service Commission of Oregon, *Ninth Annual Report* [1916] (Salem: State Printing Department, 1916), 108.

19. Puget Sound Traction, Light, and Power Co., *The Electric Journal*, vol. 2, no. 5 (June 1912): 64–65, has an early profile of electricity and the Renton coal mines. This company was originally called the Seattle Electric Company; then changed its name in 1912 to Puget Sound Traction, Light, and Power; then in 1920 changed to Puget Sound Power & Light, keeping that moniker until 1997 when it merged with a large regional gas company and became Puget Energy. The title of its journal started out as the *Seattle Electric Journal*, then in May 1912 became *The Electric Journal*, then changed to *Puget Sound Electric Journal* in September 1912, keeping that name until 1931, when the company stopped publishing the journal and switched to an annual report.

20. Puget Sound Traction, Light, and Power, *Puget Sound Electric Journal*, vol. 5,

no. 3 (March 1915): 4–14. For more on the Renton coal mine and Renton industries, see the same journal, February 1915 issue, 4–10, and July 1915 issue, 2–9. For a historical overview of the electrification of coal mining in the United States, see Warren D. Devine, Jr., "Coal Mining: Underground and Surface Mechanization," in Schurr et al., *Electricity in the American Economy*, 181–208.

21. For a lively social history of Everett and its lumber industry, see Norman H. Clark, *Mill Town: A Social History of Everett, Washington* (Seattle: University of Washington Press, 1970).

22. L. E. Childs, "Industries of Everett: The Everett Pulp and Paper Company," *Puget Sound Electric Journal*, vol. 6, no. 6 (June 1916): 3–6.

23. Rome C. Saunders, "Industries of Sumner: Fleischmann Yeast Company," *Puget Sound Electric Journal*, vol. 6. no. 5 (May 1916): 3–6.

24. Jeremy Mouat, *The Business of Power: Hydroelectricity in South Eastern British Columbia, 1897–1997* (Victoria, BC: Sono Nis Press, 1997), 28–39.

25. Ibid., 42–44.

26. Ibid., 73–91.

27. The history of the aluminum pyramid is discussed in an article by George J. Binczewski, "The Point of a Monument: A History of the Aluminum Cap of the Washington Monument," published by the Minerals, Metals, and Materials Society, http://www.tms.org/pubs/journals/JOM/9511/Binczewski-9511.html. Accessed December 21, 2011.

28. Warren D. Devine, Jr., "Early Developments in Electroprocessing: New Products, New Industries," in Schurr et al., *Electricity in the American Economy*, 77–80.

29. Nye, *Electrifying America*, ch. 5, "Flexible Factory," 186–235.

30. For a long-term view of Portland mass transit, see Martha J. Bianco, "Private Profit versus Public Service: Competing Demands in Urban Transportation History and Policy, Portland, Oregon, 1872–1970" (Ph.D. diss., Portland State University, 1994), especially chs. II, V, and VI. Wollner, 80–95 and 115–124, profiles the rise and decline of Portland's trolley company from the company perspective. On the relationship between Portland and the private trolley system in the 1910s–1920s and for a discussion of the decline and ultimate demise of Portland's trolleys, see E. Kimbark MacColl, *The Growth of a City: Power and Politics in Portland, Oregon, 1915–1950* (Athens, GA: The Georgian Press, 1979), 97–135 and 602–606.

31. Bianco argues most forcefully for the third set of problems as the main reason electric mass transit declined in Portland after 1916. While acknowledging that the automobile hastened its demise, she attributes the root cause to internal financial problems (holding company consolidation and overcapitalization), mismanagement, competing demands among streetcar constituents, unfavorable relationships with the city government and state utility commission, the rigidity of the "five-cent fare," and a general inability to adapt. See especially pp. 25–55.

32. Wollner, 115–124; on PRL&P and PEPCO's motorbus business, see pp. 117 and 120. Also MacColl, *The Growth of a* City, 118 and 425–426.

33. Wollner, 97–98; and MacColl, 119–132.

34. Wollner, 115–124 and 175–180; MacColl, 425–426, 567–569, and 602–606; Bianco, chs. VII and VIII.

35. Steve Blewett, *A History of the Washington Water Power Company, 1889–1989* (Spokane: WWP, 1989), 12–13.

36. MacColl, *The Growth of a City,* 132–135. Quote is on p. 134. On Seattle's purchase of the streetcar system from Puget Sound Traction, Light, and Power, see also *Puget Sound Electric Journal,* vol. 9, no. 1 (January 1919).

37. Nye, *Electrifying America,* 113–114.

38. *Puget Sound Electric Journal,* vol. 5, no. 11 (November 1915): 4.

39. Ibid., 4–7.

40. Susan Stacy, *Legacy of Light: A History of Idaho Power Company* (Boise: Idaho Power Company, 1991), 9.

41. Stacy, 7–8.

42. Christina Hardyment, *From Mangle to Microwave: The Mechanization of Housework* (Cambridge, UK: Polity Press, 1988), 62–63.

43. John Dierdorff summarizes news from the PP&L company bulletins of 1913–1915 in *How Edison's Lamp Helped Light the West: The Story of Pacific Power & Light Company and Its Pioneer Forebears* (Portland: Pacific Power & Light Co., 1971), ch. 7.

44. Ibid., 44.

45. Melosi, *Coping with Abundance,* 112–113. Melosi notes here that in 1929, 90 percent of sewing machines and washing machines and 80 percent of vacuum cleaners, radios, and refrigerators were purchased on credit with monthly payments.

46. Melosi, 112.

47. Wollner, 113, quoting from PEPCO, *Synchronizer,* 1929.

48. Ruth Schwartz Cowan, *More Work for Mother: The Ironies of Household Technology from the Open Hearth to the Microwave* (New York: Basic Books, 1983).

49. Ronald C. Tobey, *Technology as Freedom: The New Deal and the Electrical Modernization of the American Home* (Berkeley: University of California Press, 1996). Jay Brigham applied Tobey's concepts in his study of pre–New Deal Seattle and Los Angeles: *Empowering the West: Electrical Politics before FDR* (Lawrence: University Press of Kansas, 1998).

50. Melosi, 113, citing Robert Sobel, *The Age of Giant Corporations* (Westport, CT: Greenwood Press, 1972), 56.

51. On reclamation in the West in general, see Donald J. Pisani, *To Reclaim a Divided West: Water, Law, and Public Policy, 1848–1902* (Albuquerque: University of New Mexico Press, 1992). On irrigation in southern Idaho, see Mark Fiege, *Irrigated Eden: The Making of an Agricultural Landscape in the American West* (Seattle: University of Washington Press, 1999).

52. See Fiege, introduction and ch. 1. See also Karl Boyd Brooks, *Public Power, Private Dams: The Hells Canyon High Dam Controversy* (Seattle: University of Washington Press, 2006).

53. Tollefson, 54.

54. Tollefson, 54, citing Michael C. Robinson, *Water for the West: The Bureau of Reclamation, 1902–1977* (Chicago: Public Works Historical Society, 1979), 29.

55. *U.S. Census of Electrical Industries, Central Electric Light and Power Stations, 1912,* 154–157.

56. Stacy, 31.

57. Stacy, 23–24.

58. Ibid., 24–33.

59. Ibid., 57.

60. For a discussion of this interesting controversy, see Cases F-235 and F-377 in Idaho Public Utilities Commission, *Annual Report 1920–1921*, 86–88.

61. Stacy, 55–59.

62. Alan Trachtenberg, *The Incorporation of America: Culture and Society in the Gilded Age* (New York: Hill and Wang, 1982). See also Sobel, *The Age of Giant Corporations.*

63. Robert H. Wiebe, *The Search for Order, 1877–1920* (New York: Hill and Wang, 1967).

64. Samuel P. Hays, *Conservation and the Gospel of Efficiency: The Progressive Conservation Movement, 1890–1920* (Cambridge: Harvard University Press, 1959).

CHAPTER 4. PROFIT AND WELFARE

1. The best historical study of the rise of public regulation in the Canadian utility industry is Christopher Armstrong and H. V. Nelles, *Monopoly's Moment: The Organization and Regulation of Canadian Utilities, 1830–1930* (Philadelphia: Temple University Press, 1986). The best historical summary of the development of public utility regulation in the United States, especially of electric utilities, is Richard F. Hirsh, *Power Loss: The Origins of Deregulation and Restructuring in the American Electric Utility System* (Cambridge, MA: MIT Press, 1999), 15–18.

2. The creation and dissolution of British Columbia's ill-fated first PUC in 1919–1920 are detailed in Patricia E. Roy, "Regulating the British Columbia Electric Railway: The First Public Utilities Commission in British Columbia," *BC Studies,* no. 11 (Fall 1971): 3–20. See also Armstrong and Nelles, 253–261.

3. Hirsh, *Power Loss,* 34.

4. Marc Messing, H. Paul Friesema, and David Morell, *Centralized Power: The Politics of Scale in Electricity Generation* (Cambridge, MA: Oelgeschlager, Gunn and Haip, 1979), 69; and Jay Brigham, *Empowering the West: Electrical Politics before FDR* (Lawrence: University Press of Kansas, 1998), 9–11.

5. Gus Norwood, *Columbia River Power for the People: A History of the Policies of the Bonneville Power Administration* (Portland: BPA, 1981), 22–23.

6. Gene Tollefson, *BPA and the Struggle for Power at Cost* (Portland: BPA, 1987), 56–73.

7. For a detailed discussion of increasing scale in electrical generation, see Messing, Friesma, and Morrell, *Centralized Power,* 1–13. The data in Table 4.2 for maximum size of generators and maximum transmission voltages are found in U.S. Federal Power Commission, *National Power Survey, 1964* (Washington, DC: U.S. GPO, 1964), 14.

8. Regarding the drop in prices for electricity, John Dierdorff discusses Pacific Power & Light's 1911–1912 rate reduction programs in chapter 8 of his informative company history, *How Edison's Lamp Helped Light the West: The Story of Pacific Power & Light Company and Its Pioneer Forebears* (Portland: Pacific Power & Light Co., 1971).

This was not an isolated phenomenon. The average rate per kwh across the United States dropped from 9.9 cents in 1906 to 5.8 cents in 1920. See Edison Electric Institute, *Historical Statistics of the Electric Utility Industry, to 1970* (New York: Edison Electric Institute, 1970), p. 165.

9. On Pacific Power & Light consolidations after 1910, see Dierdorff, *How Edison's Lamp Helped Light the West,* chs. 3–5.

10. See Susan Stacy, *Legacy of Light: A History of Idaho Power Company* (Boise: Idaho Power Company, 1991), 26–32, on the Idaho Power Company. On the consolidations that led to the formation of Utah Power and Idaho Power, see also Tollefson, *BPA and the Struggle for Power at Cost,* 63–65, citing the U.S. Federal Trade Commission Report, *Utility Corporations,* gov. doc. 92, Parts 23–24 (Washington, DC: GPO, 1930); and Sidney Alexander Mitchell, *S. Z. Mitchell and the Electrical Industry* (New York: Farrar, Straus, and Cudahy, 1960).

11. See Robert C. Wing and Robert C. Cumbow, *A Century of Service: The Puget Power Story* (Bellevue, WA: Puget Sound Power & Light Co., 1987), 20–25 and 40–49 on the Stone & Webster consolidation that created Puget Sound Power & Light in 1912. See also Tollefson, 71–72.

12. On the rivalry between Seattle City Light and Puget Sound Power & Light, see Jay Brigham, *Empowering the West,* 104–120. On rate competition between these two companies, see chapter 7.

13. Tollefson, 71.

14. Tollefson, 65; Edward J. Crosby, *The Story of the Washington Water Power Company and Its Part in the History of Electric Service in the Inland Empire, 1889–1930* (Spokane: WWP, 1931), 21–22; Steve Blewett, *A History of the Washington Water Power Company, 1889–1989* (Spokane: WWP, 1989), 15–26.

15. This process of targeted technological innovation to solve practical problems in electric systems is dissected insightfully by Thomas P. Hughes, *American Genesis: A Century of Invention and Technological Enthusiasm, 1870–1970* (Chicago: University of Chicago Press, 1989), chs. IV and VI.

16. Dierdorff, 28.

17. Dierdorff, 28–29.

18. Dierdorff, 32–33.

19. Dierdorff, 30–31.

20. Dierdorff, 34 and 54.

21. Tollefson, 49–52.

22. Still a classic interpretation of America's stressful shift from an agrarian nation to an urban industrial nation is Robert H. Wiebe, *The Search for Order, 1877–1920* (New York: Hill and Wang, 1967).

23. Brigham, 11.

24. For an example of the Progressive perspective, see Morris Llewellyn Cooke, *Public Utility Regulation* (New York: Ronald Press, 1924).

25. Brigham profiles Ely's ideas about electric utilities on p. 5.

26. Richard Rudolph and Scott Ridley, *Power Struggle: The Hundred Year War over Electricity* (New York: Harper and Row, 1986), 38.

27. National Civic Federation, *Shall the Government Own and Operate the Railroads, the Telegraph and Telephone Systems? Shall the Municipalities Own Their Utilities?* (New York: National Civic Federation, 1915), 31–48.

28. Ibid., 60–75 for J. W. Sullivan's complete essay. The quotes above are on p. 75.

29. Seattle City Light, *Annual Report of the Lighting Department* (Seattle, Washington, 1911), 13.

30. *Annual Report of the Lighting Department* (Seattle, Washington, 1911), 22. The city could not acquire either site, as Puget Sound Power & Light gained control of them first.

31. *Annual Report of the Lighting Department* (Seattle, Washington, 1912 and 1913), 16.

32. *Annual Report of the Lighting Department* (Seattle, Washington, 1911), 22–25. See also Brigham, 104, in regard to a *San Francisco Examiner* investigative report of 1917 that documented these rate disparities.

33. Much of chapter 5 of Brigham's *Empowering the West* profiles the controversies between these two private and public utilities battling for turf in Seattle, as well as the larger public-versus-private debate in Washington state in the 1910s and 1920s. It is a very thorough analysis.

34. The comparative electric rate study is contained in the 1912 *Proceedings of the Northwest Electric Light and Power Association*, 62, and reprinted in Wesley Arden Dick, "Visions of Abundance: The Public Power Crusade in the Pacific Northwest in the Era of J. D. Ross and the New Deal" (Ph.D. diss., University of Washington, 1973), 249. Also Brigham, 103.

35. Oregon State, *Annual Report of the Public Service Commission*, 1915, pp. 100–113.

36. Charles David Jacobson, *Ties That Bind: Economic and Political Dilemmas of Urban Utility Networks, 1800–1990* (Pittsburgh: University of Pittsburgh Press, 2000), 122.

37. In 1918 specifically, 84 percent of British Columbia's power was hydroelectric. Leo Denis, *Electric Generation and Distribution in Canada* (Ottawa: Commission of Conservation of Canada, 1918), Table V, p. 281. Hydropower today still represents over 90 percent of BC Hydro's power supply. BC Hydro serves 94 percent of British Columbia's population.

38. Hydropower data for the United States derived from U.S. Bureau of the Census, *Census of Electrical Industries, Central Electric Light and Power Stations, for 1912*, Table 55, pp. 86–87 for the individual state data, and pp. 82–83 for the aggregate U.S. data. Aggregate data on Canada found in Denis, *Electric Generation and Distribution in Canada*, 1918, p. 2.

39. Information on British Columbia derives from Denis, *Electric Generation and Distribution in Canada*, 1918, pp. 212–230. Information on Northwest states derives from *U.S. Census of Electrical Industries*, 1912, pp. 89–90, 105–106, and 108–109.

40. Denis, *Electric Generation and Distribution in Canada*, 1918, pp. 212–230; and *U.S. Census of Electrical Industries*, 1912, pp. 89–90, 105–106, and 108–109.

41. Secondary sources cover this historical development quite well. See Hirsh, 11–31; Rudolph and Ridley, *Power Struggle*, 38–41; Hughes, *American Genesis*, 207; Brigham, 6–7. Primary sources are also abundantly available. Most revealing are the annual reports of the public utility commissions detailing their authorities, mission, and regulatory activities.

42. Hughes, 207; Rudolph and Ridley, 38–41.

43. The classic biography of Insull is Forrest McDonald, *Insull* (Chicago: University of Chicago Press, 1962). A more recent biography is John F. Wasik, *The Merchant of Power: Samuel Insull, Thomas Edison, and the Creation of the Modern Metropolis* (New York: Palgrave Macmillan, 2006). For those interested in primary sources, Insull's vision for the utility industry, and his guilt or innocence, see Samuel Insull and William Eugene Keily, *Central-Station Electric Service: Its Commercial Development and Economic Significance as Set Forth in the Public Addresses (1897–1914) of Samuel Insull* (Chicago: Private Printing, 1915). This collection of speeches is available online at the Hathi Trust Digital Library. See also Samuel Insull and Larry Plachno, *The Memoirs of Samuel Insull* (Polo, IL: Transportation Trails, 1992). A muckraking treatment is M. L. Ramsay, *Pyramids of Power: The Story of Roosevelt, Insull, and the Utility Wars* (New York: Da Capo Press, 1975).

44. Many public utility commissions have brief historical resumés available on their websites that track the dates in which regulatory authorities were established. For Oregon, see http://www.puc.state.or.us/PUC/consumer/history.shtml. For Washington, see http://www.utc.wa.gov/aboutUs/Pages/history.aspx. See also Washington State, Public Service Commission of Washington, *First Annual Report* (Olympia, WA: E. L. Boardman Public Printer, 1912), 5, and *Second Annual Report* (1913), 6. The annual reports of the state commissions are available in many university research libraries and state archives. For the national context leading to the establishment of these commissions, see Hirsh, 15–31.

45. Public Service Commission of Washington, *First Annual Report* (Olympia: EL Boardman Public Printer, 1912 [covering the year 1911]), 6–9. Note: the year covered by the report is often not the same as the official year of publication, so in future citations I always identify the year covered by the report.

46. Washington Public Service Commission, *Annual Report*, 1912, pp. 5–17.

47. Washington Public Service Commission, *Annual Report*, 1912, pp. 21–22, 29–30.

48. Washington Public Service Commission, *Annual Report*, 1914, pp. 6–7.

49. Washington Public Service Commission, *Annual Report*, 1916, p. 93.

50. Washington Public Service Commission, *Annual Report*, 1916, pp. 100–102.

51. Idaho State, Public Utilities Commission, *Annual Report*, FY 1913–1914, pp. 141–144.

52. Idaho Public Utilities Commission, *Annual Report*, FY 1913–1914, pp. 25–26, 76–77.

53. Idaho Public Utilities Commission, *Annual Report*, FY 1913–1914, pp. 79–84.

54. Idaho Public Utilities Commission, *Annual Report*, FY 1913–1914, pp. 72–84.

55. Ibid., 85–87.

56. Ibid., 89–91.

57. Ibid., 93–95.

58. Idaho Public Utilities Commission, *Annual Report*, FY 1913–1914, p. 98, and *Annual Report*, FY 1920–1921, p. 34. For an example from a WWP company history of how rate schedules changed over time in the early 1900s with no mention of the role of utility commissions, see Crosby, *Story of the Washington Water Power Company*, 26–27. The rate schedules cited by Crosby do not distinguish between Washington and Idaho

rates, do not include industrial rates for mines and mills, and do not reflect all the different classes of uses and rates found in the rate schedules filed with the PUC.

59. Oregon State, Public Service Commission, *Annual Report*, 1915, pp. 63–99. See especially p. 66.

60. The struggle between BC Electric and Vancouver and the interventions of the provincial legislature are summarized from Armstrong and Nelles, 254–268, and Roy's article "Regulating the British Columbia Electric Railway."

61. Armstrong and Nelles, 254–268, and Roy, "Regulating the British Columbia Electric Railway."

62. Ibid.

63. Quote from Armstrong and Nelles is on p. 262.

64. Brigham, 7, notes the diversity of opinion among scholars regarding whether PUCs have been effective in accomplishing their objectives. Scholars critical of PUCs tend to be public-power advocates. Rudolph and Ridley, for example, argue that state regulation did more harm than good by depressing the movement for municipally owned electric systems (Rudolph and Ridley, 40–41). Melosi also discusses the variable effectiveness of PUCs on pp. 76–77 and 125–126.

CHAPTER 5. "LET EVERY INCH OF WATER DO ITS DUTY"

1. Thomas P. Hughes, *Networks of Power: Electrification in Western Society, 1880–1930* (Baltimore: Johns Hopkins University Press, 1983), chs. X and XIII.

2. A valuable two-volume compendium of primary source documents on the history of U.S. conservation is available in many university libraries: Frank E. Smith, ed., *Conservation in the United States: A Documentary History*, vols. 1 and 2, *Land and Water* (New York: Chelsea House, 1971). Search engines occasionally list the book under the title *Land and Water*. Many of the primary source documents assessed and cited in this chapter are found in this compilation. Researchers also have easy electronic access to an extensive documentary history of American conservation published online by the Library of Congress as part of its American Memory project, http://memory.loc.gov/ammem/amrvhtml/conshome.html. Accessed December 21, 2011.

3. The quote and the story of the origins of conservation policy are from Gifford Pinchot's autobiography, *Breaking New Ground* (New York: Harcourt Brace & Co., 1947), 322–326. This passage on the origins of conservation is also reprinted in Smith, *Conservation in the United States*, vol. 2, pp. 19–23.

4. Library of Congress, American Memory site, http://memory.loc.gov/ammem/amrvhtml/cnchron4.htm. Accessed December 21, 2011.

5. The first report of the National Conservation Commission (1908) is reprinted in Smith, *Conservation in the United States*, vol. 2, pp. 95–110. The three-volume 1909 report of the National Conservation Commission is available online at the American Memory site, http://memory.loc.gov/cgi-bin/query/r?ammem/consrvbib:@FIELD%28NUMBER%28vg38%29%29. Accessed December 21, 2011.

6. A brief summary of the founding of the Conservation Commission of Canada and references to its first reports are found in Arthur V. White and Charles Vick, *Wa-*

ter Powers of British Columbia (Ottawa: Commission of Conservation Canada, 1919), 1. The quote from the 1987 Water Clauses Consolidation Act is on p. 71.

7. For a brief summary of Pinchot's life, see the U.S. Forest Service's Grey Towers Historic Site web page, http://www.fs.fed.us/gt/local-links/historical-info/gifford/gifford.shtml. Grey Towers is the Pinchot family mansion in Pennsylvania, donated to the government for use as a national historic site. For a full biography of Pinchot, see Char Miller, *Gifford Pinchot and the Making of Modern Environmentalism* (Washington, DC: Island Press, 2001).

8. Gifford Pinchot, *The Fight for Conservation* (New York: Doubleday, 1910, and Seattle: University of Washington Press Americana Library edition, 1967), 26–27.

9. Samuel Hays, *Conservation and the Gospel of Efficiency* (Cambridge: Harvard University Press, 1959, 1969), especially chs. VI–IX and XIII; and H. V. Nelles, *The Politics of Development: Forests, Mines, and Hydro-Electric Power in Ontario, 1849–1941*, 2nd edition (Montreal: McGill-Queen's University Press, 2005), 154–307. The quote from Nelles is on p. 248.

10. Books on the progressive conservation movement are legion, but most do not say much about electric power. Similarly, most histories of electric power do not say much, if anything, about how conservation intersects with it. This is curious since conservation and electrification were so deeply entwined in the early to mid-twentieth century. Besides the two monographs mentioned above, additional historical studies that address the role of conservationists vis-à-vis electric power include Martin Melosi, *Coping with Abundance*, chs. 3 and 4; Jay Brigham, *Empowering the West: Electrical Politics before FDR* (Lawrence: University Press of Kansas, 1998), ch. 1; John G. Clark, *Energy and the Federal Government* (Urbana: University of Illinois Press, 1987), ch. 2; and Gus Norwood, *Columbia River Power for the People*, chs. 2 and 3. The terms *progressivism* and *conservation* were more ubiquitous in the United States, but the principles and practices were widespread in both countries.

11. Smith, *Conservation in the United States*, vol. 2, "Report of the National Conservation Commission," December 7, 1908, pp. 95–110. Quotes are on p. 95.

12. Ibid., 96.

13. On the typology of waste, ibid., 97.

14. Quoted in White and Vick, *Water Powers of British Columbia*, 15.

15. Smith, *Conservation in the United States*, vol. 2, p. 97.

16. Leo Denis, *Water Powers of Canada* (Ottawa: Commission of Conservation of Canada, 1911), 1–22.

17. Nelles, 248–249.

18. John Muir, *The Yosemite* (Berkeley: University of California Press [reprint], 1988), 196–197.

19. U.S. Department of Commerce and Labor, Bureau of Corporations, *Report of the Commissioner of Corporations on Waterpower Development in the United States* (Washington, DC: Government Printing Office, 1912), 9–16.

20. The three companies were BC Electric Railway, Western Canada Power Company, and West Kootenay Power and Light. This percentage for concentrated ownership of water power in British Columbia was difficult to derive for 1912. My figure of

60 percent comes from extrapolating data from Denis, *Water Powers of Canada*, table following p. 22, tables on pp. 325–328, and text comments on pp. 317–319. The table following p. 22 shows total developed water power for each of Canada's provinces, while the tables on pp. 325–328 show data for each power plant on each river in British Columbia. Two large hydropower facilities in the province (North Arm and Stave Falls) were under construction in 1911; therefore, the total developed water-power tables did not list their horsepower. But the text on p. 317 indicates that the North Arm plant under construction near Vancouver had 33,000-horsepower capacity, while the subsequent 1919 Commission of Conservation report on *Water Powers of British Columbia* by White and Vick indicates that the Stave Falls plant had a capacity of 39,000 horsepower; see the table on "Power Plants of British Columbia," data for Western Canada Power Company, Stave Falls, 140. I added the horsepower for these two plants to the total listed in 1911 and then calculated the percentage of the new total controlled by the three companies. Note: Denis calls it the Western Power Company of Canada.

21. The information regarding Pinchot's testimony before Congress and the quote from the Commission of Conservation are both in White and Vick, *Water Powers of British Columbia* (1919), 38–40.

22. Martin V. Melosi, *Coping with Abundance: Energy and Environment in Industrial America* (Philadelphia: Temple University Press, 1985), 82.

23. Adamson is quoted by Jerome G. Kerwin, *Federal Water Power Legislation* (New York: Columbia University Press, 1926), 111. Also see Melosi, 79–83, on how utility companies and other investors sought to rapidly acquire the best water-power sites after the turn of the century and the federal response.

24. White and Vick, *Water Powers of British Columbia*, 70–73.

25. Act of June 17, 1902 (Public Law No. 161). The full text of the act can be read at the Library of Congress American Memory website, http://memory.loc.gov/cgi-bin/query. Accessed December 21, 2011.

26. Act of April 16, 1906 (34 Stat. 116). For the original text and legal commentary, see http://www.usbr.gov/power/Legislation/twnsites.pdf. Accessed December 21, 2011.

27. Marc Reisner, *Cadillac Desert: The American West and Its Disappearing Water*, revised edition (New York: Penguin, 1993), discusses hydropower as a cash machine for federal reclamation on pp. 134, 272–273. I discuss this issue in more detail in chapter 9, on the 1930s.

28. A collection of primary source documents on the early effort to establish a coordinated U.S. national water resources policy, including speeches by Francis Newlands (author of the Reclamation Act of 1902), President Roosevelt, and a report from the Inland Waterways Commission, is found in Smith, *Conservation in the United States*, vol. 2, pp. 69–94. Brigham, *Empowering the West*, discusses progressives and the water-power debate on pp. 13–16. Hays's classic *Conservation and the Gospel of Efficiency* traces evolving federal water policy in illuminating detail in chs. II, VI, and X.

29. Act of June 21, 1906 (34 Stat. 386).

30. For the original text, commentary, and context on the 1906 Dam Act and contemporaneous events, see Smith, *Conservation in the United States*, vol. 2, pp. 165–180; Kerwin, 111–112 and 128–129; Hays, 114–115; and Brigham, 13–14.

31. Denis, *Water Powers of Canada*, 303–308. Quote is on p. 308.

32. Ibid., 308–310.

33. Washington Department of Conservation and Development, Division of Water Resources, *First Biennial Report, April 1921–Sept. 1922* (Olympia: Frank Lamborn Public Printer, 1922), 3–22; *Third Biennial Report, 1925–1926*, 1–30. The 1917 Water Code is discussed in the *First Report* on pp. 3–5; the Federal Water Power Act is discussed on pp. 11–12.

34. Hays, 100–109. Roosevelt quote is on p. 106. The full text of the first and only report of Roosevelt's Inland Waterways Commission is in Smith, *Conservation in the United States*, vol. 2, pp. 83–94.

35. The Muscle Shoals controversy dragged on for over thirty years and is considered to be the fight that led to establishment of the Tennessee Valley Authority during the Depression. A detailed analysis of the first twenty years of the Muscle Shoals debate is in Kerwin, 266–283. For succinct overviews, see Hughes, 293–295, and Hays, 114–115. For the whole shebang, see Preston John Hubbard, *Origins of the TVA: The Muscle Shoals Controversy, 1920–1932* (New York: W. W. Norton & Co., 1968). A succinct 1966 article on the rise of federal regulation of water power covering this early period is Charles K. McFarland, "The Federal Government and Water Power, 1901–1913: A Legislative Study in the Nascence of Regulation," *Land Economics*, vol. 42, no. 4 (November 1966): 441–452. The Rainy River and James River vetoes are discussed on p. 445. Smith, *Conservation in the United States*, vol. 2, pp. 168–174, has the entire text of Roosevelt's veto of the James River project.

36. Brigham, 16.

37. Pinchot explains his reasoning for these water-power policies in *The Fight for Conservation*, 84–86.

38. Hays, 115–116.

39. Act of June 23, 1910 (36 Stat. 593). The text of the act is reprinted in Smith, *Conservation in the United States*, vol. 2, pp. 175–178.

40. Hays, 116–119 and 160–165; Brigham, 14–16; Melosi, 82.

41. The U.S. General Accounting Office (GAO) issued a report titled "The Evolution of Preference in Marketing Federal Power," GAO-01-373 (February 2001), which includes the legislative history and administrative evolution of the preference principle. Accessed at http://www.gao.gov/new.items/d01373.pdf.

42. Despite numerous attempts to gut the preference principle in the 1990s, it remains substantially in place at the time of this writing in 2012. In fact, the American Public Power Association celebrated the 100-year anniversary of the preference principle in the September/October 2006 issue of its magazine, *Public Power*. An editorialist revealed APPA's staunch commitment to and pride in sustaining this principle: "It is not an accident that preference has survived 100 years. It is a tribute to the determination, advocacy, foresight and legislative acumen of a century of public power leadership." See the APPA website for PDF access to this issue, http://www.appanet.org/newsletters/ppmagazinearchivelist.cfm?sn.ItemNumber=2108.

43. Hays, 161–162.

44. Seattle Electric Company [PSP&L], *Seattle Electric Journal*, vol. II, no. 1 (February 1912): 13.

45. The editor describing Ballinger's lecture fairly gushed: "Judge Ballinger discussed the situation in a very frank and convincing manner, leaving no uncertainty in

the minds of those who were lucky enough to hear him as to his broad-minded attitude toward water power development." Ibid.

46. The 1912–1920 statistics are drawn from Table 1 of the Edison Electric Institute's publication *Historical Statistics of the Electric Utility Industry—Through 1992* (Washington, DC: EEI, 1995), 3. The data is derived from the *U.S. Census of Electrical Industries, 1912, 1917,* and *1920.*

47. Bernard DeVoto quoted by Philip Fradkin in *A River No More,* 2nd revised edition (Berkeley: University of California Press, 1996), 83.

48. Once again, Hays had lasting insight when he described this centralizing tendency and the decentralizing counterforces that opposed it, in his 1969 preface to *Conservation and the Gospel of Efficiency,* vii–xiii. The best contemporary analysis of the Sagebrush Rebellion is R. McGreggor Cawley, *Federal Land, Western Anger: The Sagebrush Rebellion and Environmental Politics* (Lawrence: University Press of Kansas, 1993).

49. Hays, 1969 preface, xiii.

50. Leo Denis, *Electric Generation and Distribution in Canada* (Ottawa: Commission of Conservation in Canada, 1918), 6; the author states, "It is most gratifying to note the rapid increase in the use of stored water in connection with our hydro-electric developments." He goes on to enumerate a number of large hydropower reservoirs in Canada and the benefits of improved water use and power-generation efficiencies permitted by those reservoirs, including two in British Columbia: Goldstream and the Jordan River. No concerns for fish or navigation are expressed, and the whole enterprise is judged a "beneficial practice."

51. White and Vick, *Water Powers of British Columbia* (1919), 21.

52. Jim Lichatowich, *Salmon without Rivers: A History of the Pacific Salmon Crisis* (Washington, DC: Island Press, 1999), 180–182.

53. Lichatowich, 170–173.

54. The interesting story of the different fates and histories of the Fraser and Columbia rivers is detailed best in Lichatowich, ch. 8, "A Story of Two Rivers." Matthew Evenden has the best analysis of the historical contest between hydropower interests and fisheries interests on the Fraser River: *Fish versus Power: An Environmental History of the Fraser River* (Cambridge: Cambridge University Press, 2004). Hugh W. McKervill entertainingly tells the story of British Columbia's early salmon industry in *The Salmon People* (Vancouver, BC: Gray's Publishing, Ltd., 1967).

55. See Theodore Steinberg, *Nature Incorporated: Industrialization and the Waters of New England* (Cambridge: Cambridge University Press, 1991); Jeff Crane, "Finding the River: The Destruction and Restoration of the Kennebec and Elwha Rivers," Ph.D. diss. (Pullman: Washington State University, 2004).

56. Joseph E. Taylor III, *Making Salmon: An Environmental History of the Northwest Fisheries Crisis.* Seattle: University of Washington Press, 1999, 68–98. The quote from Taylor is the title of his chapter 3. See also Lichatowich, 111–123; Joseph Cone and Sandy Ridlington, *The Northwest Salmon Crisis: A Documentary History* (Corvallis: Oregon State University Press, 1996), 51–56.

57. The quotes from the Washington Fish Commissioner's report are from Lichatowich, 130–131.

58. Arthur V. White, *Fishways in the Inland Waters of British Columbia* (Ottawa: Commission of Conservation of Canada, 1918), 10, cites the U.S. Bureau of Fisheries conclusion about dam height and successful fishways. White footnotes examples of a few dams in Oregon that claimed successful fishways in a forty-foot and a fifty-foot dam but then cites examples of Washington having difficulties protecting salmon runs on rivers with taller dams.

59. Washington Department of Conservation and Development, Division of Water Resources, *First Biennial Report, April 1921–Sept. 1922*, 10–11.

60. Northwest Power Planning Council, *Staff Issue Paper: Hydropower Responsibility for Salmon and Steelhead Losses in the Columbia River Basin* (Portland, OR: Northwest Power Planning Council, 1986). See also Michael C. Blumm, *Sacrificing the Salmon: A Legal and Policy History of the Decline of Columbia Basin Salmon* (Den Bosch, the Netherlands: Book World Publications, 2002).

61. This story of the Elwha is taken primarily from Lichatowich, 131–135, and Jeff Crane, *Finding the River: An Environmental History of the Elwha* (Corvallis: Oregon State University Press, 2011), 41–86.

62. Lichatowich 131–135, and Crane, 41–86.

63. Ibid.

64. Leslie H. Darwin, *Thirtieth and Thirty-First Annual Reports of the State Fish Commissioner to the Governor of the State of Washington, April 1, 1919 to March 31, 1921* (Olympia, WA: Frank M. Lamborn Public Printers, 1921), 14.

65. Lichatowich, 134–135; Crane, 80–86.

66. Much has been written about the Elwha restoration effort, especially in the popular media. *High Country News* provided excellent in-depth coverage in Adam Burke's essay, "River of Dreams," September 24, 2001; and Michelle Blank's report "Elwha River Dams Move Closer to Destruction," July 16, 2008. Both are available electronically on the *High Country News* website, http://www.hcn.org.

67. A number of sources profile this critical case of the Quesnel Dam. The most detailed and valuable recent source is Evenden, *Fish versus Power*, 69–72. Two valuable primary sources are Arthur V. White, *Fishways in the Inland Waters of British Columbia*, 5–6; and *Annual Report of the Commissioner of Fisheries of British Columbia*, 1909, 1–13. See also Lichatowich, 175; and McKervill, 171–172. These sources sometimes differ in dates and salmon numbers, but the story is the same. Where there are factual differences between the sources, I have used the data in the more contemporary publication by Arthur White and the Commissioner of Fisheries report.

68. Ibid., especially White, *Fishways in the Inland Waters of British Columbia*, 6.

69. Evenden, 19–36; White, 6.

70. White and Vick, 21.

71. Ibid., 36–43.

72. Evenden, 72–83.

73. J. P. Babcock quoted in Bruce Stadfeld, "Electric Space: Social and Natural Transformations in British Columbia's Hydroelectric Industry to World War II," Ph.D. thesis (Winnipeg: University of Manitoba, 2002), 83.

74. Lichatowich, 147–148.

75. Lichatowich, 148, 201.
76. Brigham, 143–148.
77. Ibid., 144–145.

CHAPTER 6. WAR-ACCELERATED TRANSITIONS I

1. E. W. Rice, Jr., "Cooperation an Essential Element in the Winning of the War," in *Power*, vol. 47, no. 10 (March 5, 1918): 345–346.

2. Gerald D. Nash, *The American West Transformed: The Impact of the Second World War* (Lincoln: University of Nebraska Press, 1990).

3. Thomas P. Hughes, *Networks of Power: Electrification in Western Society, 1880–1930* (Baltimore: Johns Hopkins University Press, 1983, 1993), ch. XI, "War and Acquired Characteristics."

4. For a concise summary of the development of path-dependence theory, see Douglas Puffert's online article "Path Dependence" on the Economic History Association EH.net website, http://eh.net/encyclopedia/article/puffert.path.dependence. Accessed August 10, 2011.

5. Hughes, *Networks of Power*, 285–286.

6. Hughes, ch. XI. Regarding the "eventual region-wide coordination" in the United States, see Eve Vogel, "The Columbia River's Region: Politics, Place, and Environment in the Pacific Northwest, 1933–Present," Ph.D. diss. (Eugene: University of Oregon, 2007), especially chs. I–IV.

7. Numerous scholars who write about the history of electric power note that many progressive agendas moved forward in the 1920s, despite the pro-laissez-faire policies pursued by Presidents Harding, Coolidge, and Hoover. The conclusion to Jay Brigham's monograph, *Empowering the West: Electrical Politics before FDR* (Lawrence: University Press of Kansas, 1998), underscores how New Deal policies and programs had ample precedents in the Progressive Era *and* the 1920s, pp. 144–145.

8. Melosi, 89, 102.

9. U.S. data from EEI, *Historical Statistics of the Electric Utility Industry—Through 1992*, 3. Canada data from Canada Dominion Bureau of Statistics, *Census of Industry, Central Electric Stations in Canada, 1917*, 3–6; and *Central Electric Stations in Canada, 1920*, 14–15 and 24–27. Unfortunately, Canada began collecting and publishing detailed electrical industry statistics only in 1917, so comparative information before that is extremely difficult to derive. On the positive side, Canada published a *yearly* electrical census starting in 1917. In contrast, the United States published a census of electrical industries only every five years from 1902 through 1932.

10. H. V. Nelles, *The Politics of Development: Forests, Mines, & Hydro-Electric Power in Ontario, 1849–1941*, 2nd edition. (Montreal: McGill-Queen's University Press, 2005), 348.

11. Jeremy Mouat, *The Business of Power: Hydroelectricity in South Eastern British Columbia, 1897–1997* (Victoria, BC: Sono Nis Press, 1997), ch. 4.

12. Cecil Maiden, *Lighted Journey: The Story of BC Electric* (Vancouver: British Columbia Electric Company, 1948), 83–84.

13. Patricia E. Roy, "Regulating the British Columbia Electric Railway: The First Public Utilities Commission in British Columbia," *BC Studies,* no. 11 (Fall 1971): 5–6.

14. Mouat, 73–84.

15. Ibid., 85–91, 100.

16. Mouat, 110–112; Rebecca M. Bratspies and Russell A. Miller, eds., *Transboundary Harm in International Law: Lessons from the Trail Smelter Arbitration* (Cambridge: Cambridge University Press, 2006).

17. I measured the contraction of British Columbia's electric power industry by the decline in the power-generation capacity of the province's central electric stations reported in the *Census of Central Electric Stations in Canada.* From 1917 to 1920, horsepower capacity in British Columbia's electric power industry declined from 232,648 to 213,609. The trend in the U.S. Northwest was in the opposite direction and at a much faster pace.

18. Maiden, 87.

19. Maiden, 89.

20. Mouat, 99–100.

21. Donald Worster, "Two Faces West: The Development Myth in Canada and the United States," in Carol Higham and Robert Thacker, eds., *One West, Two Myths: A Comparative Reader* (Calgary: University of Calgary Press, 2004).

22. Melosi, 89.

23. Craig Wollner, *Electrifying Eden: Portland General Electric 1889–1965* (Portland: Oregon Historical Society Press, 1990), 103.

24. Public Service Commission of Oregon, *Eleventh Annual Report, 1916–1917* (Salem: State Printing Department, 1918), data tables on pp. 188–196.

25. City of Seattle, Department of Lighting, *Annual Report for Year Ending December 31, 1922,* 11–12.

26. This well-known episode in Northwest hydroelectric history is recounted in a number of sources. The best, though biased, primary source is City of Seattle, Department of Lighting, *Annual Report for Year Ending December 31, 1922,* 14–22. Two trustworthy secondary sources are Brigham, 105–106, and Jacobson, 121.

27. Seattle City Light, *Annual Report for Year Ending December 31, 1922,* 20, 30.

28. Dominion Bureau of Statistics, *Central Electric Stations in Canada, 1920,* 4. Manitoba and Nova Scotia formed provincial power commissions in 1919, and New Brunswick formed its commission in 1920.

29. Data derived from Dominion Bureau of Statistics, *Census of Industry, Central Electric Stations in Canada,* 1917, 3–6; 1918, 1–3, 6–7; 1919, 18–21, 30–31; and 1920, 14–15, 24–27.

30. Ibid.

31. Hughes, 293–295; Melosi, 122–124.

32. Milton Conover, *The Federal Power Commission: Its History, Activities, and Organization* (Baltimore: Johns Hopkins Press, 1923), 57–62.

33. An appendix to Conover's monograph reproduces the original act of 1920, as well as extensive quotes from the hearing records. The commission's authorities and powers laid out in the act are found on pp. 90–103.

34. Ibid., 94.

35. *Third Annual Report of the Federal Power Commission* (Washington, DC: GPO, 1923), 1–10.

36. Ibid., 4.

37. Conover, 66; *Third Annual Report of the Federal Power Commission*: 1, 12.

38. *Third Annual Report of the Federal Power Commission*: 8–9.

39. Hunter et al., *A History of Industrial Power*, 368; Hughes, 292.

40. Dierdorff, 60, 64–65.

41. Gene Tollefson, *BPA and the Struggle for Power at Cost* (Portland, OR: Bonneville Power Administration, 1987), 65; Edward J. Crosby, *The Story of the Washington Water Power Company and Its Part in the History of Electric Service in the Inland Empire, 1889–1930* (Spokane: WWP, 1931), 21–22; Steve Blewett, *A History of the Washington Water Power Company, 1889–1989* (Spokane: WWP, 1989), 15–26.

42. Dierdorff, 68–70.

43. Tollefson, 75.

44. Puget Sound Power & Light Company, *Hydro-Electric Development* (Seattle: PSP&L, 1920), last page (not numbered).

45. Hughes, 296.

46. Tollefson, 136; Norwood, 52–54.

47. *Third Annual Report of the Federal Power Commission* (Washington, DC: GPO, 1923), 6.

48. Hunter et al., 368.

49. Hunter et al., 368–369; William S. Murray et al., *A Superpower System for the Region between Boston and Washington*, U.S. Geological Survey Professional Paper 123 (Washington, DC: GPO, 1923); William S. Murray et al., *Superpower: Its Genesis and Future* (New York: McGraw Hill, 1925).

50. Janet B. Ettinger, Morris Llewellyn Cooke, Judson C. Dickerman, and Gifford Pinchot, *Report of the Giant Power Survey Board to the General Assembly of the Commonwealth of Pennsylvania* (Harrisburg, PA: Telegraph Printing Co., 1925).

51. Hunter et al., 369–370.

52. *Third Annual Report of the Federal Power Commission*, 6.

53. Hughes, 286.

CHAPTER 7. ROARING TWENTIES

1. *Electrical West* is a monthly trade journal for the commercial electric power industry, serving the North American West. In 1928 the journal launched a regular feature titled "Electricity Is Cheap—Use More of It," where it published monthly short speeches from the "Three Minute Speaking Contest" sponsored at the annual meeting of the Pacific Coast Electrical Association. See p. 103 of the August 1928 issue.

2. For United States data, Federal Power Commission, *National Power Survey: Statistics of Electrical Generating Plants in the United States, 1920–1935*, Power Series no. 6 (Washington, DC: GPO, 1936), 2–4. For Canada, Dominion Bureau of Statistics, *Census of Industry, 1927: Central Electric Stations in Canada* (Ottawa: F. A. Acland Printer, 1930): 7; and *Census of Industry, 1928*, 16–17.

3. U.S. regional and state-by-state statistics for kilowatt-hour generation, number

of customers, per capita consumption of electricity, horsepower per capita, population growth rates, and other interesting statistics are found in the tables of the electrical censuses compiled every five years. The *U.S. Census of Electrical Industries, 1927*, p. 18 and Tables 7 and 11 shows how electrical horsepower per capita in the United States increased at a far faster rate than did the population; p. 19, Table 12 shows hows electric sales to customers doubled from 1922 to 1927; pp. 41–42, Table 28 shows horsepower per capita by state, with the top four states being California, Idaho, Washington, and Montana, in that order.

4. *U.S. Census of Electrical Industries, 1927*, 2, 8.

5. *U.S. Census of Electrical Industries, 1927*, pp. 17–18, Tables 7 and 8, and pp. 82–84.

6. Jeremy Mouat, *The Business of Power: Hydro-Electricity in South Eastern British Columbia, 1897–1997* (Victoria, BC: Sono Nis Press, 1997), 101–105.

7. There is a rich literature on technological enthusiasm, which flourishes especially in times when rapid technological innovation integrates with successful marketing to create a transformational historical moment. Writing in the midst of such a transformational era, Lewis Mumford helped articulate an intellectual foundation for technological enthusiasm married to a cautious pessimism. Most relevant to the 1920s is his book *Technics and Civilization* (New York: Harcourt Brace & Co., 1934). Two more recent classics on the subject by historians are Thomas Hughes, *American Genesis: A Century of Invention and Technological Enthusiasm, 1870–1970* (Chicago: University of Chicago Press, 1989); and David E. Nye, *American Technological Sublime* (Cambridge: MIT Press, 1994). For a focused study of this topic applied to the Pacific Northwest, see Patricia Ann Kellner, "From the Pastoral to the Promethean: Electric Power Regimes in the Pacific Northwest and the Rhetoric of the Electrical Sublime," Ph.D. diss. (University of California, Los Angeles, 1994).

8. *U.S. Census of Electrical Industries, 1927*, p. 61, Table 43, for percentage of hydropower by state; p. 40 for the quote about water power as a base load; and p. 39 for the data on hydropower and population in the eleven western states. For the data on Canada, see *Census of Industry, 1927*, 38–39, Table 14.

9. State of Washington, *Third Biennial Report of the Department of Conservation and Development, April 1, 1925, to September 30, 1926* (Olympia, WA, 1927), 7.

10. British Columbia Department of Lands, *Water Powers, British Columbia* (1930), foreword, p. 5.

11. Regarding the FPC's inability to meet its obligations to audit applicants' finances to establish accurate valuations, fair rates, and reasonable capital costs, see *Eighth Annual Report of the Federal Power Commission* (Washington, DC: GPO, 1928), 1–7.

12. *Sixth Annual Report of the Federal Power Commission*, fiscal year ending June 30, 1926 (Washington, DC: GPO, 1926), 1. By the end of the decade, the Federal Power Commission had received over 1,000 applications (about 100 per year) for hydropower project licenses. Many applications were purely speculative, poorly planned, unable to get state-level approval, or unable to secure financing. Nevertheless, the FPC licensed 449 of those projects by 1930. These licensed projects totaled 2.6 million horsepower—a quarter of the entire hydropower capacity of the United States. Fully one third of all these projects were in the three Pacific coast states. One third of Idaho's hydropower

capacity in 1930 was produced at dams approved by the FPC in the 1920s. *Tenth Annual Report of the Federal Power Commission* (Washington, DC: GPO, 1930), 1–6.

13. FPC, *National Power Survey, 1920–1935*, 2–4; and Dominion Bureau of Statistics, *Census of Industry, 1928: Central Electric Stations in Canada*, 16–17. The 1928 Canadian electrical census attributes the jump in municipal power in 1923 to the Hydro Electric Power Commission of Ontario's acquiring the privately owned Toronto Power Company, p. 7. The growth of public and municipal power in Canada is also mentioned in the 1923 *Electrical Census of Canada*, 7.

14. Dominion Bureau of Statistics, *Census of Industry, 1928: Central Electric Stations in Canada*, 16–17 for the data on Canada. On the tenacity of Northwest municipal systems, see Tollefson, 52–54. For data on publicly owned versus privately owned systems in the United States on a state-by-state basis from 1920 to 1935, see FPC, *National Power Survey* (1936), 61–124. On the national trend of investor-owned utilities buying municipal systems, see *U.S. Census of Electrical Industries, 1927*, 63–64, Table 49. See also Rudolph and Ridley, *Power Struggle*, 38–41.

15. Some key sources on the proliferation of electrical appliances and sales campaigns in the 1920s David E. Nye, *Electrifying America: Social Meanings of a New Technology* (Cambridge: MIT Press, 1990), 261–286; Harold Platt, *The Electric City* (Chicago: University of Chicago Press, 1991), 237–267; Ruth Schwartz Cowan, *More Work for Mother: The Ironies of Household Technology from the Open Hearth to the Microwave* (New York: Basic Books, 1983), 172–190; and Ronald Tobey, *Technology as Freedom* (Berkeley: University of California Press, 1996), 1–35. Specifically on British Columbia, see Bruce Stadfeld, "Electric Space: Social and Natural Transformations in British Columbia's Hydroelectric Industry to World War II," Ph.D. thesis (University of Manitoba, 2002), 183–203.

16. *Census of Industry, 1926, Central Electric Stations in Canada*, 65. The chart on this page showing electric rates in B.C. communities actually displays average monthly bills for differing levels of consumption. I chose the middle-level rate of 40 kwh per month and divided the monthly bill by 40 to get the equivalent cost per kwh. Note that the 1925 and 1926 censuses include an appendix that discusses domestic electric rates in Canada and how they had declined since 1913. The 1926 census revised and updated the 1925 data. The full appendix takes up pp. 42–66 in the 1926 census. These special reports on domestic electric rates were not repeated again in the 1920s.

17. While electric rates are rarely mentioned in the electrical censuses, examples of rate structures abound in the annual reports of the public utility commissions. Utility company reports also often include rate information. In addition to these primary sources, many secondary sources provide anecdotal rate information, such as Gene Tollefson's *BPA and the Struggle for Power at Cost* (Portland: Bonneville Power Administration, 1987) and John Dierdorff's *How Edison's Lamp Helped Light the West* (Portland: Pacific Power & Light Co., 1971).

18. The rate information for WWP is from Tollefson, *BPA and the Struggle for Power at Cost*, 65. Rate information for PP&L comes from Pacific Power & Light Co., *Reclassification of Electric Plant: Statements A to I Inclusive* (Portland, OR: Geo. M. Allen & Son, Greenleaf Press, n.d. [1940?]), 45. This report, written in the late 1930s in response to a request from the Federal Power Commission, cites numerous examples

of rate reductions from the early 1900s to 1936 for the communities it served. Dierdorff (cited above) provides many similar examples. In ch. 8 of his book, Tollefson lists the rates charged by many of the municipally owned utilities, which were usually quite a bit lower than rates charged by privately owned utilities (47–54).

19. City of Seattle, Department of Lighting, *Annual Report for 1922*, 8–9. I refer in the text to this department as Seattle City Light, or City Light. Quote is on p. 9. Puget Sound Power & Light's monthly *Electric Journal*, published for its employees, sought to refute these claims that City Light had forced it to reduce its rates in Seattle, but its claims are weak and misleading. The main arguments are presented in a September 1923 column by Norwood Brockett, titled "Untangling the Wires," p. 2. In it he makes erroneous statements about the claims being made by City Light regarding PSP&L's electric rates, after which he argues that consolidations, plant efficiencies, and increased consumption were the reasons PSP&L reduced its rates. These factors, of course, did contribute to rate reductions, but they do not explain the exact timing of those reductions or the exact amount of the reductions in Seattle. Those can only be attributed to competition from City Light. Subsequent columns in October and November of 1923 explain why PSP&L had hired lobbyists in Olympia and why electric rates in small towns were higher than in cities. Several of Brockett's columns in 1924 also "explain" public power bills in the legislature.

20. Seattle City Light, *Annual Report for 1922*, 9, for the sliding scale rates from 1915 to 1922. For the *average* cost per kwh for residential service from 1918 to 1935, see Seattle City Light, *Annual Report for 1935*, table on p. 65. The comparisons of average costs per kwh for municipal versus commercial utilities in 1931 are on p. 64.

21. "Marketing Is Problem of Industry Today," *Electrical West*, vol. 61, no. 3 (September 1928): 126.

22. "That One More Kw.-hr. per Domestic Consumer," *Electrical West*, vol. 61, no. 4 (October 1928): 187.

23. Customer statistics derived from *U.S. Census of Electrical Industries, 1927*, 73 (Table 54) and 75 (Table 56).

24. *Electrical West*, vol. 61, no. 4 (October 1928): 196–197, 205. Steve Blewett, *A History of Washington Water Power Company, 1889–1989: Building on a Century of Service* (Spokane: WWP, 1989), 30.

25. "Electricity Is Cheap—Use More of It," *Electrical West*, vol. 61, no. 2 (August 1928): 103. W. M. Fisher is the author and prize-winner for his Three Minute Speech at the PCEA convention, reprinted in part here.

26. "Electricity Is Cheap—Use More of It," *Electrical West*, vol. 61, no. 2 (August 1928): 103.

27. *Electrical West*, vol. 62, no. 7 (June 1929): 580. (Note that for unexplained reasons, the June issue was published as no. 7 rather than no. 6. Also, *Electrical West* started the new volume in July.)

28. *Electrical West*, vol. 63, no. 1 (July 1929): 15.

29. Ibid.

30. Cecil Maiden, *Lighted Journey: The Story of the B.C. Electric* (Vancouver: British Columbia Electric Company, 1943), 90. On the topic of domestic use of electricity in Victoria, see the excellent M.A. thesis by Blair Tothill, "Living Electrically: The British

Columbia Electric Railway Company and the Development of the Domestic Electric Appliance Market in Victoria, 1919–1939" (University of Ottawa, 1997).

31. Maiden, 91–125.

32. Canadian *Census of Industry, 1927, Central Electric Stations in Canada*, 11–12, and 28–29 (Table 8). Because the census counted a household with electric service as one customer and cited data for "number of customers per 100 of population," the raw numbers in the table must be multiplied by the average household size to get an approximate figure for the percentage of households with electricity. Average household sizes are listed on p. 12 of the census.

33. *Electrical West*, vol. 62, no. 7 (June 1929): 579. (The June issue was published as no. 7 rather than no. 6.)

34. For examples of agricultural electrification involving grain elevators, electric threshing machines, and incubators, see *Electrical West*, vol. 62, no. 2 (February 1929): 88; vol. 62, no. 3 (March 1929): 142–143; and vol. 62, no. 7 (June 1929): 572.

35. Examples of articles on farm electrification in the *Puget Sound Electric Journal*, published monthly in Seattle by Puget Sound Power & Light Co., include an article on "Poultry Farming in Whatcom County" (May 1922): 4–8; an article on the use of electricity in orchards (June 1925): 3–8; an article on farm electrification demonstrations at the Western Washington Fair (November 1925): 9–11; an extensive article about developing rural businesses in the Puget Sound region (December 1926): 3–10; an article titled "Electricity on the Dairy Farm" (March 1927): 3–7; an article on the Puyallup agricultural research station (May 1927): 3–7; and the regular series of articles in the 1929–1931 issues on "Farm Electrification" and "Rural Electrification."

36. T. A. Wood, "Some Observations Regarding the Use of the 'Electric Plow,'" *Electrical West*, vol. 60, no. 12 (June 1928): 607. The author was the power sales engineer for Pacific Gas and Electric.

37. The 1925 *Annual Report of the Minister of Mines* in British Columbia reported that about 70 percent of West Kootenay Power and Light's electrical generation was consumed by COMINCO's zinc plant. Cited in Mouat, *The Business of Power*, 100, 116.

38. Maiden, *Lighted Journey*, 98.

39. Ellery R. Fosdick, "A New Electrolytic Zinc Reduction Plant for the West," *Electrical West*, vol. 61, no. 6 (December 1928): 315–316.

40. Maiden, *Lighted Journey*, 119.

41. C. W. Colvin, "Alouette Is Canada's Largest Automatic Hydro-Electric Generating Plant," *Electrical West*, vol. 61, no. 2 (August 1928): 73–75. Colvin was the assistant electrical engineer for BC Electric Railway Co.

42. *Electrical West*, vol. 62, no. 2 (February 1929): 86.

43. *U.S. Census of Electrical Industries, 1927*, 2, 4, 6, 11, and 63–64.

44. Judson King, *The Conservation Fight: From Theodore Roosevelt to the Tennessee Valley Authority* (Washington, DC: Public Affairs Press, 1959), 170. King cites a 1927 study by Senator Thomas Walsh (Democrat of Montana) published in the *Congressional Record*, in which he lists 1,500 mergers from the previous two years and charges the industry with violating the Sherman Anti-Trust Act.

45. Federal Trade Commission, *Electric-Power Industry—Control of Power Companies* (Washington, DC: GPO, 1927), xxviii.

46. Cecil Maiden, *Lighted Journey*, 89–90.

47. Federal Trade Commission, *Electric-Power Industry—Control of Power Companies* (Washington, DC: GPO, 1927), xxviii.

48. Robert C. Wing and Robert C. Cumbow, *A Century of Service: The Puget Power Story* (Bellevue: Puget Sound Power & Light Co., 1987), 46–47.

49. Wing and Cumbow, *A Century of Service*, 46–47. Also see Tollefson, 56–57 and 71–72. The data on the Stone & Webster group's holdings in 1927 come from the Federal Trade Commission report *Control of Power Companies* (1927), xxxiii.

50. Tollefson, 56–57 and 71–72, 82. See also the FTC report *Control of Power Companies*, xxix–xxxii, for a summary description of the major operating companies and subsidiary holding companies under the control of Electric Bond and Share. More detail on the EBASCO holdings is elaborated in chs. IV–VIII of the report.

51. Tollefson, 83.

52. An excellent review of the political background and battles over the FTC's investigation of the electric utility industry is found in Philip Funigiello, *Toward a National Power Policy: The New Deal and the Electric Utility Industry, 1933–1941* (Pittsburgh: University of Pittsburgh Press, 1973), ch. 1: "Politics, Power, and the Federal Trade Commission."

53. Gus Norwood, *Columbia River Power for the People*, 23; Funigiello, *Toward a National Power Policy*, 12; *Congressional Digest*, vol. 6 (November 1927): 299.

54. U.S. Federal Trade Commission, *Electric-Power Industry—Control of Power Companies* (1927), xvii. Hereafter cited as FTC, *Electric-Power Industry* (1927).

55. FTC, *Electric-Power Industry* (1927), xvi.

56. This quote appears widely on the Internet and is published in Will Rogers, *The Autobiography of Will Rogers* (Boston: Houghton Mifflin, 1949), 373.

57. FTC, *Electric-Power Industry* (1927), xiii.

58. U.S. Census Bureau, *Census of Electrical Industries, 1927, Central Electric Light and Power Stations* (Washington, DC: GPO, 1930), 7–10. Hereinafter referred to as *Census of Electrical Industries, 1927.*

59. Jay Brigham, *Empowering the West: Electrical Politics before FDR* (Lawrence: University Press of Kansas, 1998), Table 5.2, p. 111.

60. *U.S. Census of Electrical Industries, 1927*, 21–23.

61. *U.S. Census of Electrical Industries, 1927*, 21–22, 24.

62. Dominion Bureau of Statistics, *Census of Industry, 1920, Central Electric Stations in Canada*, 16–17; *Census of Industry, 1928, Central Electric Stations in Canada*, 22–23.

63. Quote is from *Census of Industry, 1928, Central Electric Stations in Canada*, 9; see also Table 1, pp. 16–17, for data on capital, revenue, expenses, transmission line mileage, customers, and more for the nation as a whole, 1920–1928.

64. Amazingly, 98 percent of the power for sale in Canada in 1928 was hydroelectric power, while the comparable figure for the United States was just under 40 percent. *Census of Industry, 1928, Central Electric Stations in Canada*, 5, 8; Federal Power Commission, *National Power Survey* (1936), 1–4.

65. *Census of Electrical Industries, 1927*, 18.

66. FTC, *Electric-Power Industry* (1927), xiii–xiv and xxiii.

67. *Eighth Annual Report of the Federal Power Commission* (1928), 1–4.

68. Quoted in E. Kimbark MacColl, *The Growth of a City: Power and Politics in Portland, Oregon 1915–1950* (Portland, OR: The Georgian Press, 1979), 366.

69. "Washington's Frequently Rejected Public Ownership Proposal Again Alive," *Electrical West*, vol. 62, no. 1 (January 1929): 1–2. This summary of the failed public-power efforts in *Electrical West* referred to the various bills and initiatives as having "obvious socialistic intent." A citizen public-power initiative was attempted in 1922 in Washington but failed. A bill introduced in the legislature by Homer T. Bone in 1923 also failed. A second revised bill in 1924, which was flamboyantly called the "Bone Free Power Bill," went before the voters as a referendum in November and lost by a significant margin. Two more bills to expand public power were introduced in the legislature in 1925; both failed. In 1929 a Rural Water Power Bill was introduced, and the commercial utility interests expected it to meet the same fate. But the economic crisis in 1929 changed the political environment, and public-power advocates were finally successful in 1930. Brigham reviews the bitter media battles between private and public power in the 1920s on pp. 107–123. The story about PSP&L funding the Voters' Information League is on p. 122.

70. "The Biennial Legislative Turmoil," *Electrical West*, vol. 62, no. 3 (March 1929): 1.

71. Janet B. Ettinger, Morris Llewellyn Cooke, Judson C. Dickerman, and Gifford Pinchot, *Report of the Giant Power Survey Board to the General Assembly of the Commonwealth of Pennsylvania* (Harrisburg, PA: Telegraph Printing Co., 1925), xi–xii.

72. A detailed but concise, articulate, and convincing analysis of the predatory practices of the holding companies on the operating companies is provided by E. Kimbark MacColl in *The Growth of a City*, ch. 14, "The Private Utilities and Public Power." He draws from the FTC reports and also Portland newspapers, private archives, and interviews to explain how Portland-area utilities weathered the boom and bust. On pp. 418–420 he summarizes how unnecessary debts were manufactured by the holding companies and financed by bankers with ties to the holding companies in order to artificially increase capitalization to increase revenue and profits for the out-of-state financiers. He also reviews both defenders and detractors of these practices.

73. Pacific Power & Light Co., *Reclassification of Electric Plant*, ch. v, "Pacific Power & Light Company Acquisitions, Growth, and Development 1910–1929," 62–90. This document has a complete history of the company, including acquisitions, maps with transmission lines, charts showing electric rates over time, tax information, and other financial records. The chapter cited here shows how kwh consumption for each community increased over time and how rates declined over time.

74. This Walla Walla controversy is expertly detailed by Brigham, 112–113. He draws on more than a year of newspaper reports from Walla Walla to tell the story. The detailed report and final findings and orders of the Washington regulatory agency on this case are found in *Sixth Annual Report of the Department of Public Works of Washington, Divisions of Transportation and Public Utilities, Covering the Period from December 1, 1925, to November 30, 1926* (Olympia, WA, 1927), 112–123.

75. Tollefson, 80.

76. Berkeley Snow, "Building a Load at a *Net Profit*," *Electrical West*, vol. 62, no. 1 (January 1929): 11–12. The author was the Northwest editor of *Electrical West*, stationed in Portland.

77. As mentioned in an earlier endnote, a good source on the FTC investigations in the late 1920s and early 1930s is ch. 1 of Funigiello, *Toward a National Power Policy*.

78. Guy W. Talbot, "We Have No Apologies to Make," *Electrical West*, vol. 61, no. 6 (December 1928): 328–329.

CHAPTER 8. THE BUBBLE BURSTS, 1929–1932

1. The Hoover quote is from a 1932 book by Edward Angly, who compiled an interesting collection of poor prognostications about the economic crisis by Herbert Hoover and his advisors. His book is titled *Oh Yeah?* (New York: Viking Press, 1931, 1932). The epigram quote above and many others are reproduced on the website *History Matters,* sponsored by George Mason University and the City University of New York, http://historymatters.gmu.edu/d/5063/. Accessed December 22, 2011.

2. Susan Strange, *Casino Capitalism* (Manchester, UK: Manchester University Press, 1997).

3. On the Greenspan quote, see economist Robert J. Shiller, *Irrational Exuberance*, 2nd edition (New York: Broadway Books, 2005). For a succinct definition of *irrational exuberance* and a review of the Greenspan quote, see Shiller's website, http://www.irrationalexuberance.com/definition.htm.

4. Select stock prices for G.E. and other companies along with a composite graph of New York stock values from 1924–1932 are compiled in Kimbark MacColl, *The Growth of a City: Power and Politics in Portland, Oregon 1915–1950* (Portland, OR: The Georgian Press, 1979), 365, 392–393.

5. MacColl, 393.

6. John Kenneth Galbraith, *The Great Crash, 1929* (New York: Mariner Books, 2009), 178. Galbraith attributes the quote from the Harvard Economic Society to the *Weekly Letter*, November 23, 1929.

7. The quote and national statistics on economic decline are from Wollner, *Electrifying Eden*, 138. The statistics on Oregon are from MacColl, 394.

8. On the Canadian depression, see Pierre Berton, *The Great Depression: 1929–1939* (New York: Anchor, 2001).

9. "Stock Market Will Not Affect Western Conditions," *Electrical West*, vol. 63, no. 6 (December 1929): 293.

10. "No Time for 'Banker's Complex,'" *Electrical West*, vol. 67, no. 2 (August 1931): 51; "Depression Has Not Hit Government," *Electrical West*, vol. 67, no. 4 (October 1931): 160; "Who Said Depression?" *Electrical West*, vol. 67, no. 4 (October 1931): 168–169.

11. *Electrical West*, vol. 66, no. 2 (February 1931): 55–61.

12. A number of articles and "progress reports" on Rock Island Dam appear in the pages of the *Puget Sound Electric Journal*. Published monthly by the Seattle-based Puget Sound Power & Light Company, the journal included articles on all aspects of the company's projects and prospects. It is available in the special collections of the University of Washington, Seattle. On Rock Island Dam, see the December 1930 issue, 979, and the June 1931 issue, 455–456.

13. *Electrical West*, vol. 68, no. 2 (February 1932): 55–57. (Note that volume 67 started July 1931 and ended December 1931.) *Electrical West*, vol. 70, no. 2 (February 1933): 41–48.

14. The electrical production statistics come from Canada, *Census of Industry, 1930, Central Electric Stations in Canada* (Ottawa, 1932), 6, 7, 42. Information on BC Electric is from Cecil Maiden, *Lighted Journey*, 124–133. On the general effects of the Depression on British Columbia and its politics, see Robin Fisher, *Duff Pattullo of British Columbia* (Toronto: University of Toronto Press, 1991), chs. 7 and 8; and Robin Fisher, "The Decline of Reform: British Columbia Politics in the 1930s," *Journal of Canadian Studies*, vol. 25, no. 3 (1990): 74–89.

15. Fisher, "The Decline of Reform," 74–77.

16. Fisher, *Duff Pattullo of British Columbia*, 243–244.

17. Ibid. On West Kootenay Power and Light, see Mouat, 113–114.

18. This resume of the Foshay debacle is summarized in part from Tollefson, 84–86. Some of the dates and citations were not accurate in Tollefson; I have corrected them here based on information in the Federal Trade Commission report that Tollefson cites: FTC, *Utility Corporations. Letters from the Chairman of the Federal Trade Commission in Response to Senate Resolution no. 83, 70th Congress, a Monthly Report on the Electric Power and Gas Utilities*, Senate Doc. 92, no. 25 (July 1930), Exhibit No. 23, pp. 220–226. These were monthly FTC reports that were published from 1928 through 1937. From here on, I will refer to them as FTC, *Utility Corporations*, no. __. Each number denotes a monthly report.

19. The story of PEPCO's troubles during this period is covered in great detail in two sources: MacColl, *The Growth of a City*, 425–436, and Wollner, *Electrifying Eden*, 139–154.

20. On the collapse of the Insull empire in general, see Richard Rudolph and Scott Ridley, *Power Struggle: The Hundred Year War over Electricity* (New York: Harper and Row, 1986), 57–62. On the effects of the Insull collapse on the Northwest, see MacColl, 411, 425–436; Wollner, 146; Tollefson, 89–93. References to J. P. Morgan's United Corporation and its bitter competition with Insull's Central Public Service are covered in Philip Funigiello, *Toward a National Power Policy: The New Deal and the Electric Utility Industry, 1933–1941* (Pittsburgh: University of Pittsburgh Press, 1973), 13. Detailed examination of the troubled assets of Insull's bankrupt Central Public Service Corporation is found in the April 1933 report of the FTC, *Utility Corporations*, Senate Doc. 92, no. 52, Exhibit 5332, pp. 1020–1075.

21. Rudolph and Ridley, *Power Struggle*, 57-62; Funigiello, 13; and FTC, *Utility Corporations*, Senate Doc. 92, no. 52, Exhibit 5332, pp. 1020–1075.

22. MacColl, 405–410; Tollefson, 102–104.

23. Both Post and Dill quoted in Tollefson, 103.

24. Editorials in *Electrical West* against public-power initiatives in 1930 include the August issue, p. 56, and October issue, p. 164. The quotes in the text from the December issue come from the editorial "District Power—What Significance?" on p. 280.

25. MacColl, 414. *Opinions and Decisions of the Public Utilities Commissioner of Oregon, March 1, 1931 to December 31, 1931, Incl.* (Salem, OR, 1935), 1–97; and *Opinions and Decisions of the Public Utilities Commissioner of Oregon*, vol. II, *January 1, 1932 to December 31, 1932* (Salem, OR, 1935), 3–106.

26. The quote about "pillagers" is found in Wollner, 154. The details on execu-

tive salaries and benefits are from MacColl's very lengthy resumé of Commissioner Thomas's Portland speech, 414–421.

27. *Electrical West*, vol. 68, no. 1 (January 1932): 1.

28. Ibid.

29. Franklin Roosevelt's "Portland Speech" is available in a number of sources. Quick access online is available at http://newdeal.feri.org/speeches/1932a.htm. Accessed December 22, 2011.

30. Ibid.

31. Ibid.

32. Ibid.

33. *Puget Sound Electric Journal* (March 1931): 195.

34. *Opinions and Decisions of the Public Utilities Commissioner of Oregon, March 1, 1931, to December 31, 1931, Incl.* (Salem, OR, 1935), 1, 9, 34, 43.

35. Jay Brigham, *Empowering the West: Electrical Politics before FDR* (Lawrence: University Press of Kansas, 1998), 145.

CHAPTER 9. A NEW DEAL FOR THE NORTHWEST, 1933–1938

1. William Orton, "Culture and Laissez-Faire," *Atlantic Monthly*, vol. 155 (June 1935): 46. My thanks to Philip Funigiello for uncovering this gem in his book *Toward a National Power Policy: The New Deal and the Electric Utility Industry, 1933–1941* (Pittsburgh: University of Pittsburgh Press, 1973), xi.

2. The Pattullo era is covered ably by Martin Robin, *Pillars of Profit: The Company Province, 1934–1972* (Toronto: McClelland and Stewart, 1973), chs. I and II; see pp. 9–11 for Robin's summary of Pattullo's political orientation. Also Robin Fisher, *Duff Pattullo of British Columbia* (Toronto: University of Toronto Press, 1991), chs. 7 and 8; and a summary article by Fisher, "The Decline of Reform: British Columbia Politics in the 1930s," *Journal of Canadian Studies*, vol. 25, no. 3 (1990): 74–89. See also William Rayner, *British Columbia's Premiers in Profile: The Good, the Bad, and the Transient* (Surrey, BC: Heritage House, 2000), 146–157. For a discussion of the stasis in British Columbia's electric power regime in the 1930s and the wistful attention British Columbia's media paid to New Deal projects south of the border, see Matthew Evenden, *Fish versus Power: An Environmental History of the Fraser River* (Cambridge: Cambridge University Press, 2004), 78–81.

3. Franklin Roosevelt, Address at Oglethorpe University, May 22, 1932. Reprinted in *The Public Papers and Addresses of Franklin D. Roosevelt*, vol. 1, 1928–1932, (New York: Random House, 1938), 639. Available on the New Deal Network website, http://www.newdeal.feri.org/texts/457.htm

4. Franklin Roosevelt's "Portland Speech," available online at http://newdeal.feri.org/speeches/1932a.htm. Accessed October 26, 2011.

5. Pitzer, 169; U.S. Bureau of Reclamation, "Grand Coulee Dam Statistics and Facts," May 2010.

6. H. V. Carpenter, "The Future Markets for Columbia River Power," *Northwest Science*, vol. 9. no. 4 (November 1935): 8–11. These figures for kilowatt capacity for

Bonneville and Grand Coulee dams are now vastly outmoded, as additional powerhouses and more efficient turbines have since been installed. Grand Coulee's current generating capacity is 6,809 *megawatts*—3.5 times the capacity Carpenter cites.

7. Ibid.

8. Paul C. Pitzer, *Grand Coulee: Harnessing a Dream* (Pullman: Washington State University Press, 1994), 68–80; Vera Springer, *Power and the Pacific Northwest: A History of the Bonneville Power Administration* (Washington, DC: BPS, 1976), 11–28; Gus Norwood, *Columbia River Power for the People* (Washington, DC: U.S. Department of Energy, BPA, 1981), 29–40; Gene Tollefson, *BPA and the Struggle for Power at Cost* (Portland: BPA, 1987), 110–112, 118.

9. Pitzer, 39–59; Norwood, 41–45; Tollefson, 110–119. U.S. Army Corps of Engineers, *Columbia River and Minor Tributaries*, vols. 1 and 2 (Washington, DC: GPO, 1933–1934), House Doc. 103, 73rd Cong., 1st sess.

10. Pitzer (ibid.) lists the best-known superlatives lavished on Grand Coulee Dam in the 1930s and 1940s on the first page of his introduction, xi. Woody Guthrie helped establish this reputation for Grand Coulee by singing about the dam in at least three different songs in his *Columbia River Collection,* written and composed for the Bonneville Power Administration in 1941. One of those songs, "The Biggest Thing That Man Has Ever Done," has the following lyrics: "The big Grand Coulee Dam in the state of Washington / Is just about the biggest thing that man has ever done."

11. The text and an audio recording of FDR's inaugural speech are available on the *History Matters* website of George Mason University and the City University of New York, http://www.historymatters.gmu.edu/d/5057/.

12. Tollefson, 110, 112, 118; Pitzer, 169.

13. Several sources are particularly useful in exploring technological utopianism and the "rhetoric of the electrical sublime." One of the best primary sources is Stuart Chase, "A Vision in Kilowatts," *Fortune*, vol. 7, no. 4 (1933): 35+. A classic scholarly monograph broadly treating the subject is Howard Segal's *Technological Utopianism in American Culture* (Chicago: University of Chicago Press, 1985). David E. Nye insightfully examines the technological utopian impulse and its influence on the rural electrification movement in *Electrifying America: Social Meanings of a New Technology* (Cambridge: MIT Press, 1990), 303–314. For scholarly studies focused on the Northwest specifically, see Wesley Arden Dick, "Visions of Abundance: The Public Power Crusade in the Pacific Northwest in the Era of J. D. Ross and the New Deal," Ph.D. diss. (University of Washington, 1973); especially ch. II, "Visions in Kilowatts," named after the similarly titled 1933 article mentioned above by Stuart Chase. Also, Patricia Ann Kellner, "From the Pastoral to the Promethean: Electric Power Regimes in the Pacific Northwest and the Rhetoric of the Electrical Sublime," Ph.D. diss. (UCLA, 1994); Eve Vogel, "The Columbia River's Region: Politics, Place, and Environment in the Pacific Northwest, 1933–Present," Ph.D. diss. (University of Oregon, 2007), ch. II; and Richard White, *The Organic Machine: The Remaking of the Columbia River* (New York: Hill and Wang, 1995), chs. 2 and 3.

14. The Chase quote is pieced together from White, *The Organic Machine*, 58, and James W. Carey, *Communication as Culture: Essays on Media and Society*, revised edition (New York: Routledge, 2009), 99–100.

15. Quoted in James Carey and John Quirk, "The Mythos of the Electronic Revolution," *American Scholar*, vol. 39, no. 2 (1970): 235–236.

16. Patricia Ann Kellner, "From the Pastoral to the Promethean," chapters II and III.

17. Wesley Arden Dick, "When Dams Weren't Damned: The Public Power Crusade and Visions of the Good Life in the Pacific Northwest in the 1930s," *Environmental Review*, vol. 13, nos. 3/4 (1989): 113–153. The quote from Congressman Knute Hill is on p. 121. The quotes from Senator Clarence Dill are on pp. 122–123. Note that the journal *Environmental Review* is an earlier incarnation of the current journal *Environmental History*.

18. Author unidentified, "WANTED: Buyers for One Billion Kw/-hr," *Electrical West*, vol. 74, no. 2 (February 1935): 24–32. Quotes are from pp. 24 and 32.

19. Robert Ormond Case, "The Eighth World Wonder," *Saturday Evening Post* (July 13, 1935): 23, 34, 36.

20. Gus Norwood reviews the arguments of the critics in *Columbia River Power for the People*, 41, 43. The article titled "Dam of Doubt" referred to above and in Norwood's book is from the June 19, 1937, issue of *Collier's*.

21. Frank J. Taylor, "The White Elephant Comes into Its Own," *Saturday Evening Post* (June 5, 1943): 26–27, 46.

22. Franklin T. Griffith, "Putting Bonneville to Work," *Electrical West*, vol. 80, no. 5 (May 1938): 28.

23. A detailed but lively blow-by-blow of the political wrangling and complex compromises negotiated between supporters and defenders of New Deal power policies is expertly related in Philip Funigiello, *Toward a National Power Policy: The New Deal and the Electric Utility Industry, 1933–1941* (Pittsburgh: University of Pittsburgh Press, 1973). This book is very helpful in interpreting the politics of this period.

24. Robin, *Pillars of Profit*, 11.

25. The quote from Robin Fisher is from his article "The Decline of Reform," 74–75. Fisher discusses Pattullo's long-standing commitment to social welfare policies on pp. 78–79. A more extended discussion of the early years of Pattullo's reform-minded administration is in Fisher, *Duff Pattullo of British Columbia*, ch. 7. In addition to Martin Robin and William Rayner, cited above, another good source on Premier Pattullo and the Liberal Party in British Columbia in the 1930s is Margaret Ormsby, "T. Dufferin Pattullo and the Little New Deal," *Canadian Historical Review*, vol. 43, no. 4 (December 1962): 277–297. Ormsby states, "Long before most Canadians, and certainly before [Prime Minister MacKenzie] King, Pattullo was a Keynesian in accepting the role of the government to control the fluctuations of the business cycle" (296). She characterizes the B.C. Liberal Party platform as "far to the left" of the positions held by national Liberal Party leader MacKenzie King, though distinctly to the right of the socialist Cooperative Commonwealth Federation, which became an influential third party in this decade (277). Pattullo's support for unemployment insurance, old-age pensions, health insurance for the poor, and minimum-wage laws is discussed on p. 282. The phrase "socialized capitalism" was a Pattullo slogan that eventually went out of use by the Liberals. It is mentioned by Ormsby on p. 284; by Fisher, "The Decline of Reform," on p. 79; and by Rayner on p. 149, and is the title to Martin Robin's first chapter on Pattullo's premiership in *Pillars of Profit*.

26. Ormsby, 282–283; Fisher, *Duff Pattullo of British Columbia*, 248–274.

27. The Pattullo quote is from Ormsby, p. 283.

28. Ormsby, 287–288, 293–294. For more on British Columbia's legislative progress in 1934–1935, see Martin Robin, 13–15; Fisher, "The Decline of Reform," 80–81; and Rayner, 149. For more on British Columbia's weak financial and political position in these years and Pattullo's frustration with the dominion government in Ottawa, see Fisher, *Duff Pattullo of British Columbia*, 253–271, and Robin, 15–17.

29. Ormsby, 291–292; Fisher, "The Decline of Reform," 85–86; Rayner, 154.

30. On the plan to reform provincial-dominion relations, see Ormsby, 284–285, and Robin, 15–17. On the provincial health care system and the public utilities commission, see Ormsby, 293–294, and Robin, 20–21, 29–30, 38, 39. On the Liberal Party as the moderate voice between Conservatives and Socialists and the CCF's electoral successes, see Ormsby, 278; Fisher, "The Decline of Reform," 77, 80, 86; and Robin, 22–27.

31. On the Liberals' increasing abandonment of reform after the 1937 election, see Ormsby, 294–295; Fisher, "The Decline of Reform," 81, 83–87; and Robin, ch. II.

32. Funigiello, chs. II–IV.

33. *Fifteenth Annual Report of the Federal Power Commission* (Washington, DC: GPO, 1935), 1–2. The first five pages of this 1935 report cogently summarize the structural and policy reforms contained in PUHCA and the Federal Power Act relevant to the Federal Power Commission.

34. Funigiello, 14–15, 25, 77–80; *Fifteenth Annual Report of the Federal Power Commission*, 2.

35. *Fifteenth Annual Report of the Federal Power Commission*, 3.

36. For the percentage of U.S. farms without electricity in 1930, see U.S. Bureau of the Census, *Fifteenth Census of the United States: 1930, Agriculture, General Report*, (Washington, DC: GPO, 1931), vol. IV, ch. VII, Table 4, p. 518. For the percentage of the U.S. population classified as rural in 1930, see the same census report, ch. I, p. 19.

37. On the percentage of rural electrification in New Zealand, Europe, and other parts of the world, see David Cushman Coyle, ed., *Electric Power on the Farm: The Story of Electricity: Its Usefulness on Farms, and the Movement to Electrify Rural America* (Washington, DC: GPO, 1936), ch. VI, "Rural Electrification Abroad," 71–83. See also Nye, *Electrifying America*, 287.

38. David Nye discusses the regional differences in rates of rural electrification in United States and some of the reasons for them in *Electrifying America*, 299–301.

39. The data I cite are for "farms reporting dwellings lighted by electricity" in U.S. Bureau of the Census, *Fifteenth Census of the United States: 1930, Agriculture, General Report* (Washington, DC: GPO, 1931), vol. IV, ch. VII, Table 4, p. 518.

40. Coyle, *Electric Power on the Farm*, 72–73, notes that in 1921 the province of Ontario launched a vigorous rural electrification program that subsidized up to 50 percent of the cost of building power lines to sparsely populated areas. By 1935, some 20 percent of Ontario farms had electricity, half again as many as the U.S. average.

41. Data cited in the text for the percentage of households with electric service are from Canada Dominion Bureau of Statistics, *Census of Industry, 1930, Central Electric Stations in Canada* (Ottawa, 1932), 28–29. Table 8 lists number of customers per 100 population and counts a household as one customer. The average family size in British

Columbia was about four persons per household, yielding a 73.4 percent electrification rate for residential service. To estimate the urban percentage of British Columbia's population, I used figures from the 1931 B.C. Municipal Census to tally the populations of Vancouver, Victoria, and their neighboring towns, and compared the sum with the province's total population. See B.C. Municipal Census, accessed February 1, 2012, at http://www.bcstats.gov.bc.ca/data/pop/pop/mun/mun1921_2006.asp.

42. Journalist Marquis Childs exaggerates the character and speed of the changes wrought by the REA, calling it "an astonishingly swift social revolution" in his popular book *The Farmer Takes a Hand* (Garden City, NY: Doubleday and Co., 1952), 13. Even respected historian William E. Leuchtenburg claims that "the REA revolutionized rural life" in *Franklin D. Roosevelt and the New Deal: 1932–1940* (New York: Harper and Row, 1963), 158.

43. A good national survey of rural electrification prior to the New Deal is chapter V of Coyle, *Electric Power on the Farm*, 59–70. Coyle's 1936 book doubles as a primary source since it was written just after the creation of the REA and produced under contract for that agency. It clearly advocates for the mission of and need for the REA. Regarding David Nye's observation that hydropower regimes were among the first to provide rural electrification, see *Electrifying America*, 292. On the limited interest of commercial utilities in rural electrification, see p. 294.

44. D. Clayton Brown, *Electricity for Rural America: The Fight for REA* (Westport, CT: Greenwood Press, 1980), 3–12. See also Funigiello, 123–130.

45. *Puget Sound Electric Journal* (October 1928): 3–6. Tollefson mentions PSP&L's pioneering efforts in farm service in the Puget Sound area on p. 84.

46. *Puget Sound Electric Journal* (June 1929), first page of the section on "Rural Electrification" (page number unreadable).

47. Susan Stacy, *Legacy of Light: A History of Idaho Power Company* (Boise: Idaho Power Co., 1991), 97. The electrification percentages I cite in the text for Washington and Oregon are for 1937 and come from Tollefson, 172–174.

48. Tollefson, p. 75; quoting Howton from H. Hudlow, *A History of Electric Power in Franklin County* (Pasco, WA: Franklin County PUD, n.d.), 1.

49. References to these extra fees and minimum monthly bills for extending rural service are ubiquitous in the literature about rural electrification. These special charges also appear in the Public Utilities Commission reports. One example is found in the *Eighteenth and Nineteenth Annual Reports of the Public Utilities Commission of Idaho* covering the period July 1, 1930–June 30, 1932, pp. 43–44, section titled "Extension Rule" for Case no. F-833, Order no. 1293, related to Public Utilities Consolidated Corporation. The Idaho PUC ruled that if the customers in a proposed extension area agreed to pay monthly minimum bills that over three years, would recoup the cost of construction, then the utility would pay for the power line itself. If the minimum monthly bills did not fully pay for the extension in three years, the company was authorized to require an advance payment of the difference from the property owners.

50. Funigiello mentions on p. 126 that farmers organized service-extension appeals to the state PUCs. The evidence from Washington state comes from the *Ninth Annual Report of the Department of Public Works of Washington*, covering December 1, 1928–November 30, 1929, pp. 21, 23.

51. I found no discussion of rural electrification before World War II in the company history of BC Electric by Cecil Maiden (*Lighted Journey*) or the history of West Kootenay Power and Light by Jeremy Mouat (*The Business of Power*). There is no mention of rural electrification in the Canada *Census of Electrical Industries* annual reports of the 1920s. Karl Froschauer's *White Gold: Hydroelectric Power in Canada* (Vancouver: UBC Press, 1999) does not even have "rural electrification" or "farms" in its index. British Columbia did not see any concerted effort at rural electrification until after 1947, when the legislature created the B.C. Power Commission for that specific purpose. Other accounts of rural electrification in Canada deal mostly with the 1950s and after. See, for example, Joan Champ, "Rural Electrification in Saskatchewan in the 1950s," Research paper prepared for the Western Development Museum, 2001. Champ notes in section 2.2 (p. 3) of the paper that only 1 percent of Saskatchewan farms were electrified in 1949.

52. *Second Annual Report of the Public Utilities Commission for the Year Ended 31st December 1940* (Victoria, BC, 1941), 29–32.

53. U.S. Bureau of the Census, *Fifteenth Census of the United States: 1930, Agriculture, General Report* (Washington, DC: GPO, 1931), vol. IV, ch. VII, Table 4, p. 518.

54. William E. Leuchtenburg, *Franklin D. Roosevelt and the New Deal*, 157–158.

55. Funigiello, 128–129.

56. The "scoundrels" quote is from Funigiello, 132, citing letters between Cooke and Ickes. More detail about Secretary Ickes's hostility toward the commercial power industry is provided in D. Clayton Brown, *Electricity for Rural America*, 40–41. Brown quotes from a memoir by Cooke, who claimed that at a meeting where he tried to convince Ickes that a cooperative plan with industry was the best option, Ickes replied, "I'll have nothing to do with the sons-of-bitches." Both Funigiello and Brown detail REA's inception from 1933 to 1935. See Funigiello, 129–142, and Brown, 40–45. A popular and lavishly illustrated history of rural electrification is Richard A. Pence, ed., *The Next Greatest Thing* (Washington, DC: National Rural Electric Cooperative Association, 1984).

57. Brown, 47–57; Funigiello, 38–48; Pence, 60–65; Coyle, 85–95.

58. In 1944, after eight years of program implementation, Congress changed the payback period to thirty-five years and lowered the interest rate to 2 percent. These changes helped continue electrification efforts for the most financially challenging areas still without service in 1944. See Brian Q. Cannon, "Power Relations: Western Rural Electric Cooperatives and the New Deal," *Western Historical Quarterly*, vol. 31, no. 2 (Summer 2000): 133–160. The information on the 1944 revision of REA loan terms is on p. 150.

59. Brown, 58–66; Funigiello, 149–153; Pence, 65–75.

60. Tollefson, 180, 182.

61. Tollefson, 178.

62. On the phenomenon of "spite lines," which were common throughout the U.S., see Funigiello, 143; Childs, 65–66; Brown, 71; and Brian Q. Cannon, "Power Relations: Western Rural Electric Cooperatives and the New Deal," 141–142.

63. Tollefson, 178; and Katie Archambault, "Rural Electrification in Oregon: 1930–

1955," Honors thesis (McMinnville, OR: Linfield College), 2010, 31–33. These two sources do not always agree on some of the details of the early history of the Blachly-Lane co-op. Founding dates in particular are inconsistent. The co-op has an abbreviated history on its website, http://www.blachlylane.coop/about/history.php. Accessed October 24, 2011.

64. Aaron Jones, *Inland: The First 40 Years* (Forest Grove, OR: Inland Power & Light, 1977), 2–23.

65. Jones, *Inland: The First 40 Years*, 22–33.

66. Jones, 34–38. For more stories about the beginnings of various individual rural electric co-ops in the Northwest, see Tollefson, 172–183; Childs, 195–214; and Archambault, "Rural Electrification in Oregon: 1930–1955."

67. Childs, 213.

68. Rural electrification data cited by Tollefson, 173–174, who cites an article in the *Oregon Journal* from 1945.

69. Nye, *Electrifying America*, 320–321.

70. Nye, *Electrifying America*, 334–335.

71. Cannon, "Power Relations." The quote is from his article abstract on p. 133.

72. Cannon, 135, for the quote. See also pp. 159–160.

CHAPTER 10. INTEGRATING THE U.S. NORTHWEST POWER SYSTEM, 1937–1941

1. Richard Neuberger, *Our Promised Land* (Moscow: University of Idaho Press, 1989), 88. This book was originally published by Macmillan in 1938.

2. *First Annual Report of the Bonneville Administrator* (Washington, DC: GPO, January 1939), 4.

3. Matthew Evenden, *Fish versus Power: An Environmental History of the Fraser River* (Cambridge: Cambridge University Press, 2004), 80.

4. David Nye observes how most REA co-ops across the United States got their power from commercial utilities in *Electrifying America* (Cambridge: MIT Press, 1990), 314–321.

5. Neuberger, *Our Promised Land*, 98–99.

6. Gunther V. Fritsch, *HYDRO: The Story of Columbia River Power*, film, 34 min. (Bonneville Power Administration, 1939). Woody Guthrie's 1941 songs about Columbia River development are compiled in his *Columbia River Collection*, still available in record stores and online.

7. My discussion of the debate over the disposition of power from Bonneville Dam is drawn mostly from the following primary and secondary sources: *First Annual Report of the Bonneville Administrator* (Washington, DC: GPO, January 1939), covering the latter part of 1937 and all of 1938; Gus Norwood, *Columbia River Power for the People: A History of Policies of the Bonneville Power Administration* (Portland: BPA, 1981), chs. 7–15; Gene Tollefson, *BPA and the Struggle for Power at Cost* (Portland: BPA, 1987), chs. 15–17; Charles McKinley, *Uncle Sam in the Pacific Northwest: Federal Management of Natural Resources in the Columbia River Valley* (Berkeley: University of California

Press, 1952), ch. 4; and Philip Funigiello, *Toward a National Power Policy: The New Deal and the Electric Utility Industry, 1933–1941* (Pittsburgh: University of Pittsburgh Press, 1973), ch. VII.

8. Ibid. See especially Norwood, chs. 7 and 8, and Funigiello, 174–184.

9. Ibid. See also Charles McKinley, 159.

10. On this tension between efficiency and equity, see the introduction to John M. Whiteley, Helen Ingram, and Richard Perry, eds., *Water, Place, and Equity* (Cambridge, MA: MIT Press, 2008), 1–32.

11. On the rates debate, see Norwood, 79–86; Funigiello, 184–193; and Tollefson, 132.

12. Norwood presents a concise history of precedents for preference policies in his ch. 10, "The Preference Clause," 69–79. He traces legislative precedents back to 1906. See also the U.S. General Accounting Office (now called the Government Accountability Office) report, "The Evolution of Preference in Marketing Federal Power," GAO-01-373, at http://www.gao.gov/new.items/d01373.pdf. Accessed February 2001.

13. Funigiello, 176–184; McKinley, 159. For a discussion of the creation, efforts, and accomplishments of the Pacific Northwest Regional Planning Commission, see Eve Vogel, "The Columbia River's Region: Politics, Place, and Environment in the Pacific Northwest, 1933–Present," Ph.D. diss. (Seattle: University of Washington, 2007), 37–44, 78–97.

14. Bonneville Project Act of August 20, 1937, ch. 720, 50 Stat. 731, sec. 2(a). For a careful and detailed analysis of the concerns, debates, and compromises leading up to the passage of the Bonneville Project Act, see Vogel, ch. III, 98–138.

15. Norwood, 68.

16. The quote is from the Bonneville Project Act, sec. 2(b). The powers and duties of the new administrator are also succinctly summarized in *First Annual Report of the Bonneville Administrator*, 25–27. For political background and commentary on this, see Norwood, 63–64; McKinley, 160–162; Funigiello, 193; and Vogel, 102–134.

17. Ibid.

18. Bonneville Project Act, sec. 6.

19. Bonneville Project Act, secs. 5 and 6. See also *First Annual Report of the Bonneville Administrator*, 32.

20. Bonneville Project Act, sec. 4(a). The preference policy is discussed in detail in McKinley, 160–162, and Norwood, 69–78.

21. Bonneville Project Act, secs. 4(b), 4(c), and 4(d).

22. BPA's *First Annual Report of the Bonneville Administrator*, 27–33, provides details on the calculations that went into allocating Bonneville's costs and determining the initial power rates. It even includes several verbatim paragraphs from the FPC report of February 1938 explaining its cost allocations. The text's parenthetical details about hydropower's allocation of the "ultimate" cost of the dam come from p. 34 of a long feature article on the Bonneville project in *Electrical West*, vol. 80, no. 3 (March 1938): 33–48. Regarding average rates paid by consumers at that time, see *Third Annual Report of the Administrator of the Bonneville Power Administration* (Portland, OR: BPA, 1940), 40 and Plates III, IV, and V. The lowest residential service rate in the Northwest at that time for the first 50 kwh was Eugene, Oregon's municipal system at 2 cents/kwh. Portland's rate was 3.5 cents, Spokane's was 4 cents, Hood River and Pendleton paid

5 cents, and poor Astoria residents paid 6 cents. These numbers come from Plate III, following p. 40 of the 1940 BPA annual report.

23. The quotes from Roosevelt's 1932 "Portland Speech" appear in chapter 8 and are available at http://newdeal.feri.org/speeches/1932a.htm. Charles McKinley also comments on who gained and who lost with the 1937 Bonneville Project Act on p. 162.

24. There is no end of commentary on J. D. Ross in the historical literature. His premature death in March 1939 solidified his quasi-legendary status in regional memory. Gus Norwood provides one of the more detailed biographical précis available in his *Columbia River Power for the People*, 103–106. Interestingly, Ross continued to serve as superintendent of City Light while he carried on his duties as SEC commissioner and as Bonneville administrator. For part of that time he earned no salary from City Light.

25. Ross's argument that Bonneville power would quickly be absorbed by the market is elaborated in the *First Annual Report of the Bonneville Administrator*, 6–25. Figure 8 on p. 15 shows Oregon Public Utilities Commission's data comparing electric rates, consumption rates, and revenue. Ross died in March 1939, before Bonneville power had made an impact on regional electric rates.

26. For Ross's viewpoint on rates, see *First Annual Report of the Bonneville Administrator*, 30–33. For secondary sources on this, see Norwood, 79–86; McKinley, 167–170; and Funigiello, 198–199.

27. For the transmission master plan, see *First Annual Report of the Bonneville Administrator*, 39–48. For background and commentary on transmission funding and construction, see Norwood, 108–109, 111–117; Tollefson, 135–138; and Funigiello, 206–207. The unemployment rate of 20 percent in the Northwest is from Tollefson, 133.

28. Funigiello covers in great detail the disarray and controversy attendant to Frank Banks's tenure as administrator from April through August 1939; see pp. 208–216.

29. Paul J. Raver, "The Public Power Program of the Bonneville Power Administration," Public address at Reed College, Portland, May 1, 1940, p. 2. Raver's public addresses are collected and available at the BPA archives in Portland. Norwood also refers to this speech on p. 68. Tollefson comments on the differences in style between Ross and Raver on p. 144. Funigiello discusses how Raver disappointed the PUD leaders by staying more or less aloof from political battles.

30. The annual reports of the Bonneville administrator detail yearly accomplishments. A good summary of the early years is contained in *Third Annual Report of the Administrator of the Bonneville Power Administration* (1940); see pp. 4–5 on the installation of generators, p. 10 on the Executive Order giving BPA authority to market Grand Coulee power, and pp. 69–71 on transmission line accomplishments. Tollefson has a photograph-rich chapter on this amazing construction program on pp. 142–159.

31. *Third Annual Report of the Administrator of the Bonneville Power Administration* (1940), 3–4 and 10; [Fourth] *Annual Report of the Administrator of the Bonneville Power Administration, Fiscal Year Ended June 1941*, 71.

32. *Third Annual Report of the Administrator of the Bonneville Power Administration* (1940), 9.

33. Ibid., 4.

34. Ibid., 40–45, 55–57.

35. The "high-water mark" quote is from the title of Tollefson's ch. 20. On the PUD elections of 1936 and 1938, see pp. 122–123 and pp. 133–135; also Funigiello, 200–201.

36. Tollefson, ch. 20. See also Norwood, 99–100.

37. Funigiello, 206–207 and 212–213.

38. Funigiello, 219–220.

39. Calculated using the Bureau of Labor Statistics consumer price index.

40. The story about ballot initiative 139, anti-public-power activities by the private utilities, and the quotes from Commissioner John Scott are from Norwood, 98–101. The quote from the FPC on anti-public-power expenditures by private utilities is from the *Twentieth Annual Report of the Federal Power Commission* (Washington, DC: GPO, 1941), 7.

41. Ernest R. Abrams, *Power in Transition* (New York: Charles Scribner's Sons, 1940), 14.

42. *Nineteenth Annual Report of the Federal Power Commission* (Washington, DC: GPO, 1940), 13–14.

43. Abram, 16.

44. Early in the twentieth century, Congress authorized the U.S. Forest Service to make regular PILT disbursements to counties based on gross revenues from land management activities in those counties, such as timber sales and grazing leases. These payments steadily expanded over the decades until they became institutionalized for all federal lands in Public Law 94-565 on October 20, 1976. The Department of Interior has a website on PILT, http://www.doi.gov/pilt/summary.html. Accessed November 16, 2011. Canada also has a PILT program; see http://www.tpsgc-pwgsc.gc.ca/biens-property/peri-pilt/index-eng.html. Accessed November 16, 2011.

45. *Nineteenth Annual Report of the Federal Power Commission* (Washington, DC: GPO, 1940), 13.

46. *Fourth Annual Report of the Administrator of the Bonneville Power Administration* for fiscal year ended June 1941, 66–67.

47. Ibid., 67–68.

48. Vogel, "The Columbia River's Region," 1–14, 26, and 135. The Roosevelt quote is from Vogel, 135. The BPA quote is from *Fourth Annual Report of the Administrator of the Bonneville Power Administration* for fiscal year ended June 1941, 51.

49. *Fourth Annual Report of the Administrator of the Bonneville Power Administration* for fiscal year ended June 1941, 73–74.

50. Ibid., 52, 66–71.

51. Ibid., 52–71. The discussion of defense is on pp. 52–66.

52. Norwood, 101.

53. The quote attributed to Premier McBride was introduced in chapter 5 and is from Arthur V. White and Charles Vick, *Water Powers of British Columbia* (Ottawa: Commission of Conservation Canada, 1919), 15. The phrase about a "vast hydroelectric servant" is from the epigram for this chapter. The phrase about BPA being a regional chamber of commerce is from Norwood, 125.

54. Funigiello, 258.

55. Clayton Koppes, "Environmental Policy and American Liberalism: The Department of Interior, 1933–1953," *Environmental Review* [*Environmental History* since

1996], vol. 7, no. 1 (Spring 1983): 17–53. His main argument about the two strands of liberalism in FDR's administration is summarized on pp. 17–20.

56. Richard White, *The Organic Machine: The Remaking of the Columbia River* (New York: Hill and Wang, 1995), 69.

CHAPTER 11. WAR-ACCELERATED TRANSITIONS II

1. Cecil Maiden, *Lighted Journey: The Story of the B.C. Electric* (Vancouver: British Columbia Electric Co., 1948), 147–148.

2. *Annual Report of the Administrator of the Bonneville Power Administration for Fiscal Year Ended June 30, 1944* (Washington, DC: GPO, 1944), 41. The BPA reports during the war years (1942–1944) were incorporated into the *Annual Report of the Secretary of the Interior*. I obtained my copies of the reports for these and other years from the BPA Library in Portland.

3. The information on COMINCO is from Jeremy Mouat, *The Business of Power: Hydro-Electricity in South Eastern British Columbia, 1897–1997* (Victoria, BC: Sono Nis Press, 1997), 119–121.

4. Mouat, 118.

5. On interconnections ordered by the Dominion Power Controller and other forms of control over power generation in Canada during World War II, see Dominion Bureau of Statistics, "Power Generation and Utilization in Canada," *The Canada Year Book 1945* (Ottawa: Edmond Cloutier, 1947), 331–332 and 336–337.

6. Martin Robin, *Pillars of Profit: The Company Province, 1934–1972* (Toronto: The Canadian Publishers, 1973), 45.

7. Matthew Evenden, *Fish versus Power: An Environmental History of the Fraser River* (Cambridge: Cambridge University Press, 2004), 131–134. See also Province of British Columbia, *Seventh Annual Report Public Utilities Commission for Year Ended December 31, 1945* (Victoria, BC: Charles F. Banfield, Printer, 1946), 10. The PUC indicates that BC Electric bought "dump" power from PSP&L but contracted for 22,000 kw of "firm" power from BPA.

8. On the stimulating effect of the war "lifting" British Columbia out of the Depression, see Jean Barman, *The West beyond the West: A History of British Columbia*, 3rd edition (Toronto: University of Toronto Press, 2007), 276–278. On British Columbia's budget surpluses starting in 1940, see William Rayner, *British Columbia's Premiers in Profile* (Surrey, BC: Heritage House, 2000), 162. On the transformative effect of the war on the U.S. Pacific Northwest, see William Robbins, *Landscapes of Conflict: The Oregon Story, 1940–2000* (Seattle: University of Washington Press, 2004), 7–15.

9. Evenden, 130–131; Maiden, *Lighted Journey*, 149.

10. Data for electrical generation by state for 1940 and 1945 are derived from U.S. Bureau of the Census, *Statistical Abstract of the United States 1941*, Sixty-Third Edition (Washington, DC: GPO, 1942), Table 474, p. 439; and U.S. Bureau of the Census, *Statistical Abstract of the United States 1947*, Sixty-Eighth Edition, Table 528, p. 484.

11. *Annual Report of the Administrator of the Bonneville Power Administration for the Fiscal Year Ended June 1942*, 27.

12. Gene Tollefson, *BPA and the Struggle for Power at Cost* (Portland: BPA, 1987), 214–218; Robbins, *Landscapes of Conflict*, 9–10.

13. Tollefson, 216.

14. *Annual Report of the Administrator of the Bonneville Power Administration for the Fiscal Year Ended June 1942*, 27–29, 36; Tollefson, 212–218.

15. Tollefson, 250.

16. Harry Stein, "Fighting for Aluminum and for Itself: The Bonneville Power Administration, 1939–1949," *Pacific Northwest Quarterly*, vol. 99, no. 1 (Winter 2007/2008): 3–15. The Stein quote is on p. 3. On pp. 161–168 and 214–216, Tollefson summarizes and provides a variety of images related to the aluminum industry and its many processing facilities built in the Northwest.

17. Ibid. The quotes from Raver and Moment are from Stein, "Fighting for Aluminum and for Itself," 7. The quote from the 1942 *Annual Report of the Administrator of the Bonneville Power Administration* is from p. 31. See also Tollefson, 161–168 and 214–216.

18. BPA, *Report on the Columbia River Power System, Fiscal Year 1945* (Portland: BPA, 1945), 7, 10, 12. This is the "annual report" BPA had been producing since 1938, but under a new title.

19. Tollefson, 234.

20. Steve Blewett, *A History of the Washington Water Power Company, 1889–1989: Building on a Century of Service* (Spokane: Washington Water Power Co., 1989), 41.

21. See, for example, Blewett, 40; Puget Sound Power & Light Company, *Report to Stockholders for the Year 1941* (Seattle, WA, 1942), 8, 12; *Report to Stockholders for the Year 1942*, 14, 17; *Report to Stockholders for the Year 1943*, 13. Page 14 of the 1942 report mentions the company's link-up with the lumber mills that had their own steam-generation plants.

22. Seattle City Light, *Annual Report for 1942*, 6–7; *Annual Report for 1943*, 9–10.

23. Dominion Bureau of Statistics, "Power Generation and Utilization," *The Canada Year Book 1947* (Ottawa: Edmond Cloutier, 1947), 489, Table 8. The Canada census at that time lumped British Columbia and the Yukon together.

24. On the diversion of power to war needs see Dominion Bureau of Statistics, *The Canada Year Book 1945*, 331–332. For more on the Dominion Power Controller's rationing and power directing, see pp. 336–337. On the brown-outs in British Columbia, see Maiden, *Lighted Journey*, 149.

25. Mouat, 119–125.

26. Evenden, 149–178. The quote is on p. 177.

27. Historian Martin Melosi examined the relationship between energy abundance and U.S. energy consumption and policy in his book *Coping with Abundance: Energy and Environment in Industrial America* (Philadelphia: Temple University Press, 1985). Even earlier, historian David Potter wrote a classic interpretation of the influence of abundance on American culture in his still-relevant book *People of Plenty: Economic Abundance and the American Character* (Chicago: University of Chicago Press, 1954). I examine this theme in regard to water resources in the Northwest in my article "Developing a Plentiful Resource: Transboundary Rivers in the Pacific Northwest," a chapter in Richard Perry, Helen Ingram, and John Whiteley, eds., *Water, Place, and Equity: Tempering Efficiency with Justice* (Cambridge: MIT Press, 2008), 147–188.

28. For a discussion of interconnections and lessons learned during World War I and applied at the start of World War II, see Philip Funigiello, *Toward a National Power Policy: The New Deal and the Electric Utility Industry, 1933–1941* (Pittsburgh: University of Pittsburgh Press, 1973), 29–35.

29. Blewett, 40.

30. Not affiliated with the EBASCO utilities, Puget Sound Power & Light credited the War Production Board as the impetus for the Northwest Power Pool in its annual *Report to Stockholders 1942* (Seattle: Puget Sound Power & Light, 1943), 15. The BPA's annual report for FY 1943, p. 14, similarly credited the War Production Board's Limitations Order L-94 of May 1, 1942, for mandating the larger regional power pool. The information on BC Electric's intertie with the BPA grid is from Evenden, 133–134 and 140. See also Tollefson, 228–231.

31. *Annual Report of the Administrator of the Bonneville Power Administration for Fiscal Year Ended June 30, 1943*, 3. *Annual Report of the Administrator of the Bonneville Power Administration for Fiscal Year Ended June 30, 1944*, 52–54.

32. Michael C. Blumm, "The Northwest's Hydroelectric Heritage," in Dale D. Goble and Paul W. Hirt eds., *Northwest Lands, Northwest Peoples: Reading in Environmental History* (Seattle: University of Washington Press, 1999), 264–294; quote is on p. 271.

33. Funigiello, 236–245, discusses these mutual fears and political positioning at the war's start.

34. On the PUDs organizing themselves and pursuing condemnation proceedings during the war, see Tollefson, 238–241. On investor-owned utility perspectives on this, see John Dierdorff, *How Edison's Lamp Helped Light the West: The Story of Pacific Power & Light Company and Its Pioneer Forebears* (Portland: Pacific Power & Light Co., 1971), 147–173; Blewett, 40; and Pacific Power & Light, *Report to Stockholders 1942*, 21–27. The private utilities were very concerned about public-power system acquisitions and extensively discussed them in their annual reports.

35. Quote by WWP president Kinsey Robinson is from Blewett, 39. Right after the quote, Blewett goes on to explain how by the end of the 1930s, northwesterners had "perhaps never been more divided," and how BPA and several PUDs in eastern Washington were trying to break up WWP during the war years. PSP&L's newspaper advertisements on national unity are found in Puget Sound Power & Light, *Report to the Stockholders for the Year 1941*, back cover, and *Report to Stockholders 1942*, front cover and inside front matter (no page number).

36. For general facts on these PUD acquisitions, see Tollefson, 238–240. For the WWP perspective on the acquisitions, see Blewett, 40. For BPA's perspective on the advance of public-power systems in these years, see *Annual Report of the Administrator of the Bonneville Power Administration for Fiscal Year Ended June 30, 1944*, 54–57.

37. For BPA's perspective on the $90 million offer to PSP&L, see *Annual Report of the Administrator of the Bonneville Power Administration for Fiscal Year Ended June 30, 1944*, 57. For PSP&L's perspective on this, see *Report to Stockholders 1942*, 21–25, and *Report to Stockholders 1943*, 21–25. On pp. 228–231, Tollefson provides a lengthy discussion of Senator Bone's and Representative Hill's bills, the Roosevelt administration's positions on them, quotes from Raver and Ickes, and the responses of the private utility sector in the Northwest.

38. See Tollefson, 231 on the death of the Bone and Hill bills and Raver's New York junket. For PSP&L's perspective on the Seattle City Council resolution, see *Report to Stockholders 1943*, 25. The Seattle City Light *Annual Report* for 1944, 6, includes verbatim sections of the city council resolution and brief commentary. City Light's *Annual Report* for 1950, 11–14, contains a discussion of Seattle's acquisition of PSP&L's properties, effected March 15, 1951.

39. Dierdorff, 146–149. Dierdorff also critically details the various efforts to pass a Columbia Valley Authority bill in 1942–1944, the efforts of the PUDs to acquire PP&L properties, assisted by BPA, and Referendum 25 in 1944, all on pp. 156–162.

40. The PSP&L *Report to Stockholders 1943*, 27–31, contains a lengthy critique of Referendum 25 and extensive quotes from the regional media opposing the measure. The quotes from the *Bellingham Labor News* and the *Washington Teamster* are on p. 31. The company's stockholder reports celebrate rather than disguise its attempts to influence editorial boards and public opinion. Dierdorff discusses Referendum 25 on p. 160, Tollefson on p. 246.

41. The quote from Tollefson is from p. 231. On additional attempts to pass a CVA bill during and after World War II, see Tollefson, 248; Gus Norwood, *Columbia River Power for the People* (Portland: BPA, 1981), 160; and Funigiello, 220–224.

42. BPA, *Report on the Columbia River Power System, Fiscal Year 1945*, 12 and 19. U.S. Bureau of the Census, *Statistical Abstract of the United States*, Sixty-Third Edition [1941 report] (Washington, DC: GPO, 1942), Table 474, p. 439; and *Statistical Abstract of the United States*, Sixty-Eighth Edition [1947 report] (Washington, DC: GPO, 1947), Table 528, p. 484. Note: the Statistical Abstract for years 1944 and 1945 were combined into one for the sixty-sixth edition.

43. Dominion Bureau of Statistics, *The Canada Year Book 1945*, 343 and 353.

44. On rural electrification in British Columbia during the war, see Evenden, 124–129. The Evenden quote is on 120. On the role of the CCF in promoting public power and rural electrification, see Robin, *Pillars of Profit*, 51, 73, 78.

45. Dominion Bureau of Statistics, *The Canada Year Book 1945*, 342–352. The comparative figures on consumption and power rates for British Columbia versus Ontario are from the 1944 *Progress Report* of the BC Rural Electrification Committee, cited in Evenden, 126–127.

46. Evenden, 135–138.

47. Evenden, 134–138; Mouat, 130–131; Robin, 80.

48. The $50 million five-year upgrade is mentioned in Evenden, 136, who cites newspaper articles on BC Electric's announcement. The $1 million rural electrification program is mentioned in *Seventh Annual Report of the Public Utilities Commission for the Year Ended December 31, 1945*, 19.

49. Maiden, 166.

50. The quote from BC Electric vice president E. H. Adams is from Maiden, 151. A description of the valuation process, public hearings, final conclusions, and rate rebates is available in *Fifth Annual Report of the Public Utilities Commission for the Year Ended December 31, 1943*, 19–20. The valuation case and more rate reductions are again discussed in *Sixth Annual Report of the Public Utilities Commission for the Year Ended December 31, 1944*, 11.

51. Maiden, 151–152; *Sixth Annual Report of the Public Utilities Commission for the Year Ended December 31, 1944*, 11; *Seventh Annual Report of the Public Utilities Commission for the Year Ended December 31, 1945*, 11.

52. See for example, *Third Annual Report of the Public Utilities Commission for the Year Ended December 31, 1941*, 10.

53. *Sixth Annual Report of the Public Utilities Commission for the Year Ended December 31, 1944*, 10.

54. *Seventh Annual Report of the Public Utilities Commission for the Year Ended December 31, 1945*, 9.

55. Evenden, 120.

56. Evenden, 120.

57. Tollefson, 270.

CHAPTER 12. "ENDLESS LIMITLESS DEVELOPMENT": A TELESCOPED VIEW OF THE POSTWAR ERA

1. Martin Robin, *Pillars of Profit: The Company Province, 1934–1972* (Toronto: McClelland and Stewart, 1973), 84.

2. Matthew Evenden, *Fish versus Power: An Environmental History of the Fraser River* (Cambridge: Cambridge University Press, 2004), 139.

3. Public Law 534, ch. 665, 78th Cong., 2nd sess. (H.R. 4485), "Flood Control Act of 1944; Public Law 14, ch. 19, 79th Cong., 1st sess. (S. 35), "River and Harbor Act of 1945." Note: the U.S. Fish and Wildlife Service has a very useful website providing links to digital copies of every water resource development act passed by Congress between 1938 and 2000: http://www.fws.gov/habitatconservation/wrda.html. These are scanned images of the original text of the laws. Some of these projects are also mentioned in BPA's *Report on the Columbia River Power System for Fiscal Year 1945*, 26–27.

4. The list of dams being built in the Northwest in the postwar era comes from Marc Reisner, *Cadillac Desert: The American West and Its Disappearing Water*, revised edition (New York: Penguin, 1993), 164–165, supplemented by Internet research. Reisner's chapter 5, titled "The Go-Go Years," covers western water development from the 1930s to the 1950s. His lively, provocative study of western water politics remains one of the best critical historical interpretations in print.

5. An excellent Web resource on the 1948 flood is the site sponsored by the Center for Columbia River History, http://www.ccrh.org/comm/slough/vpflood.php. Richard White, *The Organic Machine: The Remaking of the Columbia River* (New York: Hill and Wang, 1995), 74–75, discusses the 1948 flood and its boosting effect on federal enthusiasm for more dams.

6. Public Law 516, ch. 188, 81st Cong., 2nd sess. (H.R. 5472), "River and Harbor and Flood Control Act of 1950." The provisions for the Columbia River Basin and northwestern states are on pp. 17–19 of Title II on Flood Control.

7. Evenden discusses the 1948 Fraser River flood on pp. 142–146. The quote from *The Province* is on p. 145, and the Evenden quote is on p. 144. The plan produced by the joint province-dominion Fraser Basin Board is mentioned on p. 183. The decision to develop the Peace and Columbia rivers in the late 1950s is discussed on pp. 179–182.

Jeremy Mouat also discusses the 1948 flood in the Kootenay region and its political repercussions in his book *The Business of Power: Hydro-Electricity in South Eastern British Columbia, 1897–1997* (Victoria, BC: Sono Nis Press, 1997), 139–140.

8. Gene Tollefson, *BPA and the Struggle for Power at Cost* (Portland: BPA, 1987), 262–267.

9. Tollefson, 262, 300.

10. The story of the Pelton Dam controversy is covered thoroughly in Craig Wollner, *Electrifying Eden: Portland General Electric, 1889–1965* (Portland: Oregon Historical Society Press, 1990), 224–245. Additional useful information and primary sources are available at the Oregon Historical Society's *Oregon History Project* website on the Pelton Dam controversy by Joshua Binus, http://www.ohs.org/education/oregon history/historical_records/dspDocument.cfm?doc_ID=886E82C5-ECC0-98BC-F8569 B2233D64B07. Accessed December 2, 2011.

11. Wollner, *Electrifying Eden*, 224–245, and the Oregon Historical Society's website on the Pelton Dam controversy. The quote from Chinook chairman Grant Elliott is from Wollner, p. 240. On the fish ladder, see Tollefson, p. 300. On the failure of fish passage at Round Butte, see http://www.deschutespassage.com/deschutes-passage-background.html. Accessed December 2, 2011.

12. The best source on the Hells Canyon controversy is Karl Boyd Brooks, *Public Power, Private Dams: The Hell's Canyon High Dam Controversy* (Seattle: University of Washington Press, 2006). He mentions the Oxbow incident on p. 17. Three new members joined the five-person Federal Power Commission in 1953–1954, and BPA administrator Paul Raver was replaced by William A. Pearl in 1954.

13. Evenden, 179–195; Mouat, 138. New construction of hydroelectric and diesel-fired power plants is described in the annual reports of the Public Utilities Commission. See, for example, *Twelfth Annual Report of the Public Utilities Commission for Year Ended December 31, 1950*, 8, and *Thirteenth Annual Report of the Public Utilities Commission for Year Ended December 31, 1951*, 8, for discussions of the Buntzen Lake, Chilliwack, and other BC Electric facilities.

14. Mouat, 141–142; Evenden, 195–203. On the effort to jointly develop the Columbia River, which eventually led to the 1961 Columbia River Treaty, see Neil A. Swainson, *Conflict over the Columbia: The Canadian Background to an Historic Treaty* (Montreal: McGill-Queen's University Press, 1979); and John V. Krutilla, *The Columbia River Treaty: The Economics of an International River Basin Development* (Baltimore, MD: Johns Hopkins University Press, 1967).

15. Mouat, 142–143; Evenden, 182; Robin, 209–211.

16. Robin, 227–236; Evenden, 182; Mouat, 143–144. For more on the motivations and results of the decision to create BC Hydro to promote development, see ch. 7, "Peace, Pulp, and Power Hunger," in Karl Froschauer, *White Gold: Hydroelectric Power in Canada* (Vancouver: UBC Press, 1999), 174–210.

17. Robin, 228–231. "Big Dam Government" is the whimsical title Robin assigned to his chapter IX. Karl Froschauer (ibid.) also argues that the decision to expropriate the two companies and pursue development through a provincial public-power authority was "influenced by the economic limitations of a privatized electricity supply," p. 175.

18. Tollefson, 272–278, profiles the eleven PUDs that acquired private systems and launched operations after 1944.

19. Gus Norwood, *Columbia River Power for the People: A History of Policies of the Bonneville Power Administration* (Portland: BPA, 1981). Primary source information about the Northwest Public Power Association and its member organizations can be gleaned from its semiweekly newsletter, *Public Power Bulletin*, launched in 1946 and published for several years as a typed mimeographed manuscript. In April 1949 (vol. III, no. 2) it became a professional typeset journal under the title *The Pacific Northwest Public Power Bulletin*, available in many of the regional libraries of the Northwest. Tollefson, 278–279, mentions NWPPA's hiring of Norwood and his influence on the organization. Another source on Northwest public power is Ken Billington, *People, Politics, and Public Power* (Seattle: Washington Public Utility District Association, 1988). Billington's study is a personal recollection full of data but lacking citations. Like Norwood, he wears his heart on his sleeve. Billington mentions Norwood and the early years of NWPPA on pp. 64–66.

20. Tollefson, 278–279. For references to other regional PUD and rural co-op associations in the Northwest, see Billington, 65, and Northwest Public Power Association, *Public Power Bulletin*, vol. II, no. 2 (January 15, 1947): 4.

21. Tollefson, 281; Rebecca R. Wise, ed., *Rural Electric Fact Book* (Washington, DC: National Rural Electric Cooperative Association, 1963), 3; Frank J. Busch, *Power for the People: Montana's Cooperative Utilities* (Missoula: University of Montana Press, 1976), 92.

22. Tollefson, 284; Billington, chs. 3 and 6. For a brief commentary on PNUCC's contributions to regional cooperation after 1946, see Michael Blumm, "The Northwest's Hydroelectric Heritage," in Dale Goble and Paul Hirt, eds., *Northwest Lands, Northwest Peoples: Readings in Environmental History* (Seattle: University of Washington Press, 1999), 271–272.

23. Part 7 of Norwood's history of the BPA is titled "Power Partnership" and includes two chapters covering the Eisenhower era, which was marked by the president's "partnership policy."

24. BPA, *Report on the Columbia River Power System, 1950* (Portland: BPA, 1950), 12–13 and 29–30; BPA, *1960 Annual Report on the U.S. Columbia River Power System* (Portland: BPA, 1960), 1–4, 9.

25. *Report on the Columbia River Power System, 1950* (Portland: BPA, 1950), 7.

26. Most of the data in this paragraph come from BPA's *Report on the Columbia River Power System, 1950*, 18, and *1960 Annual Report on the U.S. Columbia River Power System*, 16. Population statistics are from the *1949 Report on the Columbia River Power System*, 41–42.

27. BPA, *Report on the Columbia River Power System, 1950*, 38; and BPA, *1960 Annual Report on the U.S. Columbia River Power System*, 4–7.

28. Northwest Power Planning Council, *Staff Issue Paper: Hydropower Responsibility for Salmon and Steelhead Losses in the Columbia River Basin* (Portland, OR: Northwest Power Planning Council, 1986). See also Michael C. Blumm, "Hydropower vs. Salmon: The Struggle of the Pacific Northwest's Anadromous Fish Resources for a

Peaceful Coexistence with the Federal Columbia River Power System," *Environmental Law*, vol. 11, no. 43 (1981): 211–261.

29. "Dams of the Columbia Basin and Their Effects on the Native Fishery," Center for Columbia River History website (a Washington State University and Portland State University consortium), at http://www.ccrh.org/comm/river/dams5.htm. See also Anthony Netboy, *The Salmon: Their Fight for Survival* (Boston: Houghton Mifflin, 1973), 285–289; and Roberta Ulrich, *Empty Nets: Indians, Dams, and the Columbia River* (Corvallis: Oregon State University Press, 1999).

30. Paul Pitzer, *Grand Coulee: Harnessing a Dream* (Pullman: Washington State University Press, 1994), 219–230.

31. Ibid. On The Dalles Dam and the Celilo Indians, see Katrine Barber, *Death of Celilo Falls* (Seattle: University of Washington Press, 2005).

32. Portions of this discussion are summarized from my previously published article "Developing a Plentiful Resource: Transboundary Rivers in the Pacific Northwest," in R. Perry, H. Ingram, and J. Whiteley, eds., *Water, Place, and Equity: Tempering Efficiency with Justice* (Cambridge: MIT Press, 2008), 147–188. For more on David Sohappy and his legal battle for Native fishing rights, see the award-winning documentary *River People: Behind the Case of David Sohappy*, produced in 1990 by Michal Conford and Michele Zaccheo. See also the Columbia River InterTribal Fish Commission's brief summary at http://www.critfc.org/text/usvor.html.

33. Evenden, 2–6 and 84–117.

34. Evenden, 157–170 and 179–229. See also Jim Lichatowich, *Salmon without Rivers: A History of the Pacific Salmon Crisis* (Washington, DC: Island Press, 1999), 170–201.

35. Hirt, "Developing a Plentiful Resource," 172–178. Michael C. Blumm, *Sacrificing the Salmon: A Legal and Policy History of the Decline of Columbia Basin Salmon* (Den Bosch, the Netherlands: Book World Publications, 2002). On the recent movement for dam removal and river restoration in the United States, see Elizabeth Grossman, *Watershed: The Undamming of America* (New York: Counterpoint, 2002). On the Elwha River story, see Jeff Crane, "Finding the River: The Destruction and Restoration of the Kennebec and Elwha Rivers," Ph.D. diss. (Pullman: Washington State University, 2004). For more on the decline of salmon from hydropower development and the marginalization of Northwest tribes, see White, *The Organic Machine*, ch. 4; Barber, *The Death of Celilo*, chs. 3 and 4; Brooks, *Public Power, Private Dams*, 81–92; and Joseph E. Taylor III, *Making Salmon: An Environmental History of the Northwest Fisheries Crisis* (Seattle: University of Washington Press, 1999), 235–236.

36. Only the Salmon River in Idaho was free of obstruction in the 1950s.

37. U.S. Department of Interior, Bureau of Reclamation, *The Columbia River: A Comprehensive Report on the Development of the Water Resources of the Columbia River Basin* (Washington, DC: Government Printing Office, February 1947), 43.

38. Evenden, 267–276.

39. The two best book-length studies of the Columbia River Treaty, one from an American perspective and one from a Canadian perspective, are John V. Krutilla, *The Columbia River Treaty: The Economics of an International River Basin Development* (Baltimore: Johns Hopkins University Press, 1967); and Neil A. Swainson, *Conflict over the Columbia: The Canadian Background to an Historic Treaty* (Montreal: McGill-Queen's

University Press, 1979). The best discussion of how B.C. hydropower and salmon politics influenced the treaty outcome is Evenden, *Fish versus Power*, 179–183 and 213–229. See also Paul Hirt and Adam Sowards, "The Past and Future of the Columbia River," in Barbara Cosens, ed., *Transboundary River Governance in the Face of Uncertainty: The Columbia River Treaty*, forthcoming from Oregon State University Press.

40. On environmental policy reform in the United States, see Richard N. L. Andrews, *Managing the Environment, Managing Ourselves: A History of American Environmental Policy* (New Haven: Yale University Press, 1999), ch. 14; and George Hoberg, *Pluralism by Design: Environmental Policy and the American Regulatory State* (New York: Praeger, 1992). On Canadian environmental policy reform, see Melody Hessing, Michael Howlett, and Tracy Summerville, *Canadian Natural Resource and Environmental Policy: Political Economy and Public Policy*, 2nd edition (Seattle: University of Washington Press, 2005); and David Richard Boyd, *Unnatural Law: Rethinking Canadian Environmental Law and Policy* (Seattle: University Washington Press, 2003), chs. 2 and 3.

41. Richard F. Hirsh, *Power Loss: The Origins of Deregulation and Restructuring in the American Electric Utility System* (Cambridge: MIT Press, 1999), 55–59. Hirsh provides two very useful line graphs on pp. 57 and 59 showing these trends in efficiency and unit size from the 1880s to the mid-1990s.

42. BPA, *1965 Report, U.S. Columbia River Power System* (Portland: BPA, 1965), 1.

43. Bonneville Power Administration, *2010 BPA Facts*, available online at http://www.bpa.gov/corporate/About_BPA/.

44. Hirsh, 58–63; see Figure 3.3 on p. 62 for a graph showing the trend in actual and inflation-adjusted electric rates from 1962 to 1996 for the United States as a whole.

45. The best source on the WPPSS debacle is Daniel Pope, *Nuclear Implosions: The Rise and Fall of the Washington Public Power Supply System* (Cambridge: Cambridge University Press, 2008). Also of value is Daniel Jack Chasan, *The Fall of the House of WPPSS* (Seattle: Sasquatch Publishing, 1985); and D. Victor Anderson, *Illusions of Power: A History of the Washington Public Power Supply System* (Westport, CT: Praeger Publishers, 1985). A brief summary is available online at the Oregon Historical Society's website, http://www.historylink.org/index.cfm?DisplayPage=output.cfm&File_Id=5482.

46. I did not find any books written on the Trojan nuclear power plant, but several articles provide details and insight on the post-decommission legacy of the plant. See especially Paul Koberstein, "Trojan: PGE's Nuclear Gamble," *Willamette Week*, March 9, 2005, A1; and the *Oregon Encyclopedia* online developed and maintained by Portland State University, http://www.oregonencyclopedia.org/entry/view/trojan_nuclear_power_plant/.

47. British Columbia Clean Energy Act of 2010, available at http://www.leg.bc.ca/39th2nd/1st_read/gov17-1.htm#part5.

48. Blumm, "The Northwest's Hydroelectric Heritage," 274–288. See also BPA, *2010 Resource Program* (Portland: BPA, September 2010), chs. 5 and 6.

CHAPTER 13. CONCLUSION

1. Amory Lovins and L. Hunter Lovins, *Brittle Power: Energy Strategy for National Security* (Andover, MA: Brick House Publishing, 1982).

2. A reliable and prolific source of energy data is the U.S. Energy Information Administration (EIA): http://www.eia.gov/. Historical data on just about everything related to electricity are available in various publications. Regarding the rising costs of producing electricity in the 1970s as well as rising electric rates, see EIA, *The Changing Structure of the Electric Power Industry, 1970–1991,* DOE/EIA-0562 (Washington, DC: U.S. Department of Energy, March 1993), 36–37.

3. See, for example, Figure 7.1 in this book, showing nominal electric rates steadily declining from 1885 to 1930. The EIA notes that from 1932 to the start of World War II, nominal residential electricity prices fell by a third; from 1942 to 1950 they continued to drop at an average rate of 3 percent per year; and from 1950 to 1960 they dropped an average of 1 percent per year. See EIA, *The Changing Structure of the Electric Power Industry: An Update* (Washington, DC: U.S. Department of Energy, December 1996), Appendix A: "History of the U.S. Electric Power Industry, 1882–1991," 105–112, available online at http://www.eia.gov/cneaf/electricity/page/electric_kid/append_a.html. Accessed January 15, 2012.

4. Legions of books and articles have been written on the restructuring of the electric power industry since the 1970s. I believe the best place to start and the best historical perspective is provided by Richard Hirsh, *Power Loss: The Origins of Deregulation and Restructuring in the American Electric Utility System* (Cambridge: MIT Press, 1999).

5. Contemporary renewable energy policy is outside the scope of this study, but for those interested in the topic, I recommend perusing the DSIRE online database to learn about the variety of local, state, and federal policies related to grid-tied renewable energy. DSIRE stands for Database of State Incentives for Renewables and Efficiency. It is a well-organized and informative, nonpartisan website sponsored by the U.S. Department of Energy and the Interstate Renewable Energy Council: http://www.dsireusa.org/.

6. Hirsh, chs. 13 and 14.

7. Eve Vogel, "The Columbia River's Region: Politics, Place, and Environment in the Pacific Northwest, 1933–Present," Ph.D. diss. (Seattle: University of Washington, 2007), 7, 13–14.

Bibliography

NEWSPAPERS, MAGAZINES, AND OTHER POPULAR AND TRADE PERIODICALS

Blank, Michelle. "Elwha River Dams Move Closer to Destruction." *High Country News* (July 16, 2008), available at http://www.hcn.org.

Burke, Adam. "River of Dreams." *High Country News* (September 24, 2001), available at http://www.hcn.org.

Case, Robert Ormond. "The Eighth World Wonder." *Saturday Evening Post*, July 13, 1935, 23, 34–36.

Chase, Stuart. "A Vision in Kilowatts." *Fortune*, vol. 7, no. 4 (1933): 35+.

Electrical West (trade journal for the western U.S. electric light and power industry).

Oregonian. (Portland, OR, daily newspaper).

Orton, William. "Culture and Laissez-Faire." *Atlantic Monthly*, vol. 155 (June 1935): 746–753.

Taylor, Frank J. "The White Elephant Comes into Its Own." *Saturday Evening Post*, June 5, 1943, 26–27, 46.

Wolverton, Bruce. "The Water Power of the Pacific Northwest." *Pacific Monthly*, vol. XI, no. 3 (March 1904): 155–169.

GOVERNMENT DOCUMENTS AND NGO, CORPORATE, AND UTILITY COMPANY NEWSLETTERS, JOURNALS, AND ANNUAL REPORTS

British Columbia. *Annual Reports of the Commissioner of Fisheries of British Columbia.* Victoria, BC: various years.

British Columbia. *Annual Reports of the Public Utility Commission.* Victoria, BC: 1939–1961.

British Columbia, Department of Lands. *Water Powers, British Columbia.* Victoria, 1930.

Canada Dominion Bureau of Statistics. "Power Generation and Utilization." *The Canada Year Book* (various years). Ottawa: Edmond Cloutier Printers.

Canada Dominion Bureau of Statistics. *Census of Industry, Central Electric Stations in Canada.* Ottawa, annual reports from 1917 to 1930.

Denis, Leo. *Water Powers of Canada.* Ottawa: Commission of Conservation of Canada, 1911.

Denis, Leo. *Electric Generation and Distribution in Canada.* Ottawa: Commission of Conservation of Canada, 1918.

Champ, Joan. "Rural Electrification in Saskatchewan in the 1950s." Research paper prepared for the Western Development Museum, 2001.

Coyle, David Cushman, ed. *Electric Power on the Farm. The Story of Electricity: Its Usefulness on Farms, and the Movement to Electrify Rural America.* Washington, DC, 1936.

Edison Electric Institute. *The Electric Light and Power Industry in the United States.* Statistical Bulletin no. 4. New York: EEI, January 1937.

Edison Electric Institute. *Historical Statistics of the Electric Utility Industry, to 1970.* New York: EEI, 1970.

Edison Electric Institute. *Historical Statistics of the Electric Utility Industry—Through 1992.* Washington, DC: EEI, 1995.

Ettinger, Janet B., Morris Llewellyn Cooke, Judson C. Dickerman, and Gifford Pinchot. *Report of the Giant Power Survey Board to the General Assembly of the Commonwealth of Pennsylvania.* Harrisburg, PA: Telegraph Printing Co., 1925.

Idaho State. Public Utilities Commission. *Annual Reports,* 1914–1930.

Murray, William S., et al. *A Superpower System for the Region between Boston and Washington.* U.S. Geological Survey Professional Paper 123. Washington, DC: GPO, 1923.

Northwest Power Planning Council. *Staff Issue Paper: Hydropower Responsibility for Salmon and Steelhead Losses in the Columbia River Basin.* Portland, OR: Northwest Power Planning Council, 1986.

Northwest Public Power Association. *Public Power Bulletin,* 1946–1948, continued as *The Pacific Northwest Public Power Bulletin* starting April 1949.

Oregon State. *Annual Reports of the Public Utilities Commission/Public Service Commission,* 1911–1960.

Pacific Power & Light Co. *Annual Report to Stockholders.* Various years.

Pacific Power & Light Co. *Reclassification of Electric Plant: Statements A to I Inclusive.* Report prepared for the Federal Power Commission. Portland, OR: Geo. M. Allen and Son, the Greenleaf Press, n.d. [1940?].

The Public Papers and Addresses of Franklin D. Roosevelt, Vol. 1, *1928–32.* New York City: Random House, 1938.

Public Service Commission of Oregon. *Ninth Annual Report, 1916.* Salem: State Printing Department, 1916.

Puget Sound Power & Light Co. *Puget Sound Electric Journal.* Seattle: Puget Sound Power & Light Co., 1921–1931.

Puget Sound Power & Light Co. *Annual Report to Stockholders.* Seattle: Puget Sound Power & Light Co., 1935–1960.

Puget Sound Traction, Light, and Power Co. *The Electric Journal.* Seattle: Puget Sound Traction, Light, and Power Co., 1912.

Puget Sound Traction, Light, and Power Co. *Puget Sound Electric Journal.* Seattle: Puget Sound Traction, Light, and Power Co., 1915, 1916, 1919.

Rice, E. W., Jr. "Cooperation an Essential Element in the Winning of the War." *Power,* vol. 47, no. 10 (March 5, 1918): 345–346.

Seattle City Light. *Annual Reports of the Lighting Department.* Seattle: City of Seattle, various years since 1911.

U.S. Army Corps of Engineers. *Columbia River and Minor Tributaries,* vols. 1 and 2. Washington, DC: GPO, 1933–1934. House Doc. 103, 73rd Cong., 1st sess.

U.S. Bonneville Power Administration. Various public addresses of Administrator Paul J. Raver. Available at BPA Library, Portland, Oregon.

U.S. Bonneville Power Administration. Annual reports. *Report on the Columbia River Power System* (title varies). Portland: BPA, 1938–1965.

U.S. Bonneville Power Administration. *2010 Resource Program.* Portland: BPA, September 2010.

U.S. Bureau of the Census. *Census of Electrical Industries, Central Electric Light and Power Stations*, 1902, 1907, 1912, 1917, 1922, 1927, 1932.

U.S. Bureau of the Census. *Fifteenth Census of the United States: 1930, Agriculture, General Report*, vol. IV. Washington, DC: GPO, 1931.

U.S. Bureau of the Census. *Statistical Abstract of the United States.* Washington, DC: GPO, various dates after 1932.

U.S. Congress. *Congressional Digest.* Washington, DC: GPO, various years.

U.S. Congress. Various original water resource development acts passed from 1938 to 2000, http://www.fws.gov/habitatconservation/wrda.html.

U.S. Department of Commerce and Labor, Bureau of Corporations. *Report of the Commissioner of Corporations on Waterpower Development in the United States.* Washington, DC: GPO, 1912.

U.S. Department of Interior, Bureau of Reclamation. *The Columbia River: A Comprehensive Report on the Development of the Water Resources of the Columbia River Basin.* Washington, DC: GPO, February 1947.

U.S. Energy Information Administration, Department of Energy. *The Changing Structure of the Electric Power Industry, 1970–1991.* DOE/EIA-0562. Washington, DC: U.S. Department of Energy, March 1993.

U.S. Energy Information Administration, Department of Energy. *The Changing Structure of the Electric Power Industry: An Update.* Washington, DC: U.S. Department of Energy, December 1996.

U.S. Energy Information Administration, Department of Energy. *Annual Energy Review 2010.* Washington, DC: U.S. Department of Energy, October 2011.

U.S. Federal Power Commission. *Annual Reports.* Washington, DC: GPO, various years since 1920.

U.S. Federal Power Commission. *National Power Survey: Statistics of Electric Generating Plants in the United States, 1920–1935.* Power Series No. 6. Washington, DC: GPO, 1936.

U.S. Federal Power Commission. *Utility Corporations. Letters from the Chairman of the Federal Trade Commission in Response to Senate Resolution no. 83, 70th Congress, a Monthly Report on the Electric Power and Gas Utilities.* (Monthly FTC reports that lasted from 1928 through 1937.)

U.S. Federal Power Commission. *National Power Survey, 1964.* Washington, DC: GPO, 1964.

U.S. Federal Trade Commission. *Electric-Power Industry—Control of Power Companies.* Washington, DC: GPO, 1927.

U.S. Government Accountability Office [formerly the General Accounting Office]. *The Evolution of Preference in Marketing Federal Power.* GAO-01-373, February 2001, available at http://www.gao.gov/new.items/d01373.pdf.

U.S. Government Accountability Office [formerly the General Accounting Office]. *Spokane Tribe's Additional Compensation Claim for the Grand Coulee Dam.* GAO-04-125T, October 2003. Accessed December 20, 2011, at http://www.gao.gov/htext/d04125t.html.

Washington State. *Annual Reports of the Public Service Commission.* Olympia, 1911–1920.

Washington State. *Annual Reports of the Department of Public Works of Washington, Division of Public Utilities.* Olympia, WA, 1920–1934.

Washington State. *Thirtieth and Thirty-First Annual Reports of the State Fish Commissioner to the Governor of the State of Washington, April 1, 1919 to March 31, 1921.* Olympia, WA: Frank M. Lamborn Public Printers, 1921.

Washington State. *Biennial Reports of the Department of Conservation and Development.* Olympia, WA, 1921–1948.

Washington State. *Reports of the Department of Public Service, Division of Public Utilities.* Olympia, WA, 1934–1944.

Washington State. *Reports of the Department of Public Utilities.* Olympia, WA, 1944–1948.

White, Arthur V. *Fishways in the Inland Waters of British Columbia.* Ottawa: Commission of Conservation of Canada, 1918.

White, Arthur V., and Charles Vick. *Water Powers of British Columbia.* Ottawa: Canada Commission of Conservation, 1919.

WEB RESOURCES

American Public Power Association. Newsletter archives. http://www.publicpower.org/Media/daily/ArchiveList.cfm.

BC Hydro. Energy in BC. http://www.bchydro.com/energy_in_bc.html.

Bonneville Power Administration. Publications. http://www.bpa.gov/corporate/pubs.

Bonneville Power Administration. NW Hydro: Multi-Purpose Dams of the Pacific Northwest. http://www.bpa.gov/Power/PL/Columbia/2-multi.htm.

Bonneville Power Administration. Commonly requested GIS maps. http://transmission.bpa.gov/LanCom/Geographic_Information_Services/maps.cfm.

British Columbia. BC Stats. http://www.bcstats.gov.bc.ca/index.asp.

Center for Columbia River History, a consortium of Washington State University–Vancouver and Portland State University. http://www.ccrh.org.

Columbia River InterTribal Fish Commission. http://www.critfc.org.

Hathi Trust Digital Library. http://www.hathitrust.org.

"History Matters." George Mason University and the City University of New York. http://historymatters.gmu.edu.

HistoryLink.org Online Encyclopedia of Washington State History. http://www.historylink.org

Library of Congress. American Memory Project. "The Evolution of the Conservation Movement." http://memory.loc.gov/ammem/amrvhtml/conshome.html.

New Deal Network. http://newdeal.feri.org.

Oregon Historical Society. "Oregon History Project." http://www.ohs.org/education/oregonhistory.

Portland State University. *Oregon Encyclopedia of History and Culture.* http://www.oregonencyclopedia.org.

U.S. Energy Information Administration. http://www.eia.gov.

U.S. Forest Service. Grey Towers Historic Site. http://www.fs.fed.us/gt/local-links/historical-info/gifford/gifford.shtml.

U.S. Geological Survey. Columbia River Basin, Washington. http://vulcan.wr.usgs.gov/Volcanoes/Washington/ColumbiaRiver/description_columbia_river.html.

Western States Tourism Policy Council. Federal Land in the West. http://www.dced.state.ak.us/wstpc/Publications/FedLandWest.htm.

PUBLISHED BOOKS AND SCHOLARLY ARTICLES, UNPUBLISHED THESES AND DISSERTATIONS

Abrams, Ernest R. *Power in Transition.* New York: Charles Scribner's Sons, 1940.

Anderson, D. Victor. *Illusions of Power: A History of the Washington Public Power Supply System.* Westport, CT: Praeger Publishers, 1985.

Andrews, Richard N. L. *Managing the Environment, Managing Ourselves: A History of American Environmental Policy.* New Haven: Yale University Press, 1999.

Archambault, Katie. "Rural Electrification in Oregon: 1930–1955." Honors thesis, Linfield College, 2010.

Armstrong, Christopher, and H. V. Nelles. *Monopoly's Moment: The Organization and Regulation of Canadian Utilities, 1830–1930.* Philadelphia: Temple University Press, 1986.

Barber, Katrine. *Death of Celilo Falls.* Seattle: University of Washington Press, 2005.

Barman, Jean. *The West beyond the West: A History of British Columbia,* 3rd edition. Toronto: University of Toronto Press, 2007.

Berton, Pierre. *The Great Depression: 1929–1939.* New York: Anchor, 2001.

Bianco, Martha J. "Private Profit versus Public Service: Competing Demands in Urban Transportation History and Policy, Portland, Oregon, 1872–1970." Ph.D. diss., Portland State University, 1994.

Billington, Ken. *People, Politics, and Public Power.* Seattle: Washington Public Utility District Association, 1988.

Binczewski, George J. "The Point of a Monument: A History of the Aluminum Cap of the Washington Monument." Published by the Minerals, Metals, and Materials Society, http://www.tms.org/pubs/journals/JOM/9511/Binczewski-9511.html. Accessed December 21, 2011

Blewett, Steve. *A History of the Washington Water Power Company, 1889–1989.* Spokane: WWP, 1989.

Blumm, Michael. "The Northwest's Hydroelectric Heritage." In Dale Goble and Paul Hirt, eds., *Northwest Lands, Northwest Peoples: Readings in Environmental History.* Seattle: University of Washington Press, 1999, 264–294.

Blumm, Michael C. "Hydropower vs. Salmon: The Struggle of the Pacific Northwest's Anadromous Fish Resources for a Peaceful Coexistence with the Federal Columbia River Power System." *Environmental Law,* vol. 11, no. 43 (1981): 211–261.

Blumm, Michael C. *Sacrificing the Salmon: A Legal and Policy History of the Decline of Columbia Basin Salmon.* Den Bosch, the Netherlands: Book World Publications, 2002.

Boyd, David Richard. *Unnatural Law: Rethinking Canadian Environmental Law and Policy.* Seattle: University of Washington Press, 2003.

Bratspies, Rebecca M., and Russell A Miller, eds. *Transboundary Harm in International*

Law: Lessons from the Trail Smelter Arbitration. Cambridge: Cambridge University Press, 2006.

Brigham, Jay. *Empowering the West: Electrical Politics before FDR.* Lawrence: University Press of Kansas, 1998.

Brooks, Karl Boyd. *Public Power, Private Dams: The Hells Canyon High Dam Controversy.* Seattle: University of Washington Press, 2006.

Brown, D. Clayton. *Electricity for Rural America: The Fight for REA.* Westport, CT: Greenwood Press, 1980.

Busch, Frank J. *Power for the People: Montana's Cooperative Utilities.* Missoula: University of Montana Press, 1976.

Cannon, Brian Q. "Power Relations: Western Rural Electric Cooperatives and the New Deal." *Western Historical Quarterly*, vol. 31, no. 2 (Summer 2000): 133–160.

Carey, James W. *Communication as Culture: Essays on Media and Society*, revised edition. New York: Routledge, 2009.

Carey, James, and John Quirk. "The Mythos of the Electronic Revolution." *American Scholar*, vol. 39, no. 2 (1970): 219–241 and no. 3 (1970): 395–424.

Carpenter, H. V. "The Future Markets for Columbia River Power." *Northwest Science*, vol. 9, no. 4 (November 1935): 8–11.

Cawley, R. McGreggor. *Federal Land, Western Anger: The Sagebrush Rebellion and Environmental Politics.* Lawrence: University Press of Kansas, 1993.

Chasan, Daniel Jack. *The Fall of the House of WPPSS.* Seattle: Sasquatch Publishing, 1985.

Childs, Marquis. *The Farmer Takes a Hand.* Garden City, NY: Doubleday and Co., 1952.

Clark, John G. *Energy and the Federal Government: Fossil Fuel Policies, 1900–1946.* Urbana: University of Illinois Press, 1987.

Clark, Norman H. *Mill Town: A Social History of Everett, Washington.* Seattle: University of Washington Press, 1970.

Cone, Joseph, and Sandy Ridlington. *The Northwest Salmon Crisis: A Documentary History.* Corvallis: Oregon State University Press, 1996 and 2000.

Conover, Milton. *The Federal Power Commission: Its History, Activities, and Organization.* Baltimore: Johns Hopkins Press, 1923.

Cooke, Morris Llewellyn. *Public Utility Regulation.* New York: Ronald Press, 1924.

Cowan, Ruth Schwartz. *More Work for Mother: The Ironies of Household Technology from the Open Hearth to the Microwave.* New York: Basic Books, 1983.

Crane, Jeff. "Finding the River: The Destruction and Restoration of the Kennebec and Elwha Rivers." Ph.D. diss., Washington State University, 2004.

Crane, Jeff. *Finding the River: An Environmental History of the Elwha.* Corvallis: Oregon State University Press, 2011.

Crosby, Edward J. *The Story of the Washington Water Power Company and Its Part in the History of Electric Service in the Inland Empire, 1889–1930.* Spokane: WWP, 1931.

Dick, Wesley Arden. "Visions of Abundance: The Public Power Crusade in the Pacific Northwest in the Era of J. D. Ross and the New Deal." Ph.D. diss., University of Washington, 1973.

Dick, Wesley Arden. "When Dams Weren't Damned: The Public Power Crusade and Visions of the Good Life in the Pacific Northwest in the 1930s." *Environmental Review*, vol. 13, nos. 3/4 (1989): 113–153.

Dierdorff, John. *How Edison's Lamp Helped Light the West: The Story of Pacific Power & Light Company and Its Pioneer Forebears.* Portland: Pacific Power & Light Co., 1971.

Evenden, Matthew. *Fish versus Power: An Environmental History of the Fraser River.* Cambridge: Cambridge University Press, 2004.

Fairlie, John A. *Essays in Municipal Administration.* New York: Macmillan, 1908.

Fiege, Mark. *Irrigated Eden: The Making of an Agricultural Landscape in the American West.* Seattle: University of Washington Press, 1999.

Findlay, John M., and Ken S. Coates, eds. *Parallel Destinies: Canadian-American Relations West of the Rockies.* Seattle: University of Washington Press, 2002.

Fisher, Robin. "The Decline of Reform: British Columbia Politics in the 1930s." *Journal of Canadian Studies,* vol. 25, no. 3 (1990): 74–89.

Fisher, Robin. *Duff Pattullo of British Columbia.* Toronto: University of Toronto Press, 1991.

Fradkin, Philip. *A River No More,* 2nd revised edition. Berkeley: University of California Press, 1996.

Freeman, Neil B. *The Politics of Power: Ontario Hydro and Its Government, 1906–1995.* Toronto: University of Toronto Press, 1996.

Froschauer, Karl. *White Gold: Hydroelectric Power in Canada.* Vancouver: University of British Columbia Press, 1999.

Funigiello, Philip. *Toward a National Power Policy: The New Deal and the Electric Utility Industry, 1933–1941.* Pittsburgh: University of Pittsburgh Press, 1973.

Galbraith, John Kenneth. *The Great Crash, 1929.* New York: Mariner Books, 2009.

Goble, Dale D., and Paul Hirt, eds. *Northwest Lands, Northwest Peoples: Readings in Environmental History.* Seattle: University of Washington Press, 1999.

Grossman, Elizabeth. *Watershed: The Undamming of America.* New York: Counterpoint, 2002.

Hardyment, Christina. *From Mangle to Microwave: The Mechanization of Housework.* Cambridge, UK: Polity Press, 1988.

Hays, Samuel P. *Conservation and the Gospel of Efficiency: The Progressive Conservation Movement, 1890–1920.* Cambridge, MA: Harvard University Press, 1959.

Hessing, Melody, Michael Howlett, and Tracy Summerville. *Canadian Natural Resource and Environmental Policy: Political Economy and Public Policy,* 2nd edition. Seattle: University of Washington Press, 2005.

Higham, Carol, and Robert Thacker, eds. *One West, Two Myths: A Comparative Reader.* Calgary: University of Calgary Press, 2004.

Hirsh, Richard F. *Power Loss: The Origins of Deregulation and Restructuring in the American Electric Utility System.* Cambridge, MA: MIT Press, 1999.

Hirt, Paul W. "Developing a Plentiful Resource: Transboundary Rivers in the Pacific Northwest." In R. Perry, H. Ingram, and J. Whiteley, eds., *Water, Place, and Equity: Tempering Efficiency with Justice.* Cambridge, MA: MIT Press, 2008, 147–188.

Hirt, Paul, and Adam Sowards. "The Past and Future of the Columbia River." In Barbara Cosens, ed., *Transboundary River Governance in the Face of Uncertainty: The Columbia River Treaty.* (Forthcoming from Oregon State University Press, 2012.)

Hoberg, George. *Pluralism by Design: Environmental Policy and the American Regulatory State.* New York: Praeger, 1992.

Hubbard, Preston John. *Origins of the TVA: The Muscle Shoals Controversy, 1920–1932.* New York: W.W. Norton and Co., 1968.

Hughes, Thomas P. *Networks of Power: Electrification in Western Society, 1880–1930.* Baltimore: Johns Hopkins University Press, 1983.

Hughes, Thomas P. *American Genesis: A Century of Invention and Technological Enthusiasm, 1870–1970.* Chicago: University of Chicago Press, 1989.

Hunter, Louis C. *A History of Industrial Power in the United States, 1780–1930,* vols. 1–3. Charlottesville: University Press of Virginia, 1986, and Cambridge, MA: MIT Press, 1991.

Insull, Samuel, and William Eugene Keily. *Central-Station Electric Service: Its Commercial Development and Economic Significance as Set Forth in the Public Addresses (1897–1914) of Samuel Insull.* Chicago: Private Printing, 1915.

Insull, Samuel, and Larry Plachno. *The Memoirs of Samuel Insull.* Polo, IL: Transportation Trails, 1992.

Jackson, Kenneth T. *Crabgrass Frontier: The Suburbanization of the United States.* New York: Oxford University Press, 1987.

Jacobson, Charles David. *Ties That Bind: Economic and Political Dilemmas of Urban Utility Networks, 1800–1990.* Pittsburgh: University of Pittsburgh Press, 2000.

Jones, Aaron. *Inland: The First 40 Years.* Forest Grove, OR: Inland Power & Light, 1977.

Kellner, Patricia Ann. "From the Pastoral to the Promethean: Electric Power Regimes in the Pacific Northwest and the Rhetoric of the Electrical Sublime." Ph.D. diss., UCLA, 1994.

Kerwin, Jerome G. *Federal Water Power Legislation.* New York: Columbia University Press, 1926.

King, Judson. *The Conservation Fight: From Theodore Roosevelt to the Tennessee Valley Authority.* Washington, DC: Public Affairs Press, 1959.

Knittel, Christopher R. "The Adoption of State Electricity Regulation: The Role of Interest Groups." *The Journal of Industrial Economics,* vol. LIV, no. 2 (June 2006): 201–222.

Koppes, Clayton. "Environmental Policy and American Liberalism: The Department of Interior, 1933–1953." *Environmental Review,* vol. 7, no. 1 (Spring 1983): 17–53.

Krutilla, John V. *The Columbia River Treaty: The Economics of an International River Basin Development.* Baltimore: Johns Hopkins University Press, 1967.

Leuchtenburg, William E. *Franklin D. Roosevelt and the New Deal: 1932–1940.* New York: Harper and Row, 1963.

Lichatowich, Jim. *Salmon without Rivers: A History of the Pacific Salmon Crisis.* Washington, DC: Island Press, 1999.

Lovins, Amory, and L. Hunter Lovins. *Brittle Power: Energy Strategy for National Security.* Andover, MA: Brick House Publishing, 1982.

MacColl, E. Kimbark. *The Growth of a City: Power and Politics in Portland, Oregon, 1915–1950.* Athens, GA: The Georgian Press, 1979.

MacColl, E. Kimbark, with Harry H. Stein. *Merchants, Money, and Power: The Portland Establishment, 1843–1913.* Athens, GA: The Georgian Press, 1988.

Maiden, Cecil. *Lighted Journey: The Story of BC Electric.* Vancouver: British Columbia Electric Company, 1948.

McDonald, Forrest. *Insull.* Chicago: University of Chicago Press, 1962.

McFarland, Charles K. "The Federal Government and Water Power, 1901–1913: A Legislative Study in the Nascence of Regulation." *Land Economics*, vol. 42, no. 4 (November 1966): 441–452.

McKervill, Hugh W. *The Salmon People.* Vancouver, BC: Gray's Publishing, Ltd., 1967.

McKinley, Charles. *Uncle Sam in the Pacific Northwest: Federal Management of Natural Resources in the Columbia River Valley.* Berkeley: University of California Press, 1952.

Melosi, Martin V. *Coping with Abundance: Energy and Environment in Industrial America.* Philadelphia: Temple University Press, 1985.

Messing, Marc, H. Paul Friesema, and David Morell. *Centralized Power: The Politics of Scale in Electricity Generation.* Cambridge, MA: Oelgeschlager, Gunn and Haip, 1979.

Miller, Char. *Gifford Pinchot and the Making of Modern Environmentalism.* Washington, DC: Island Press, 2001.

Mitchell, Sidney Alexander. *S. Z. Mitchell and the Electrical Industry.* New York: Farrar, Straus, and Cudahy, 1960.

Montgomery, David. *The Fall of the House of Labor: The Workplace, the State, and American Labor Activism, 1865–1925.* New York: Cambridge University Press, 1987.

Mouat, Jeremy. *The Business of Power: Hydroelectricity in South Eastern British Columbia, 1897–1997.* Victoria, BC: Sono Nis Press, 1997.

Muir, John. *The Yosemite.* Berkeley: University of California Press, 1988.

Mumford, Lewis. *Technics and Civilization.* New York: Harcourt Brace and Co., 1934.

Nash, Gerald D. *The American West Transformed: The Impact of the Second World War.* Lincoln: University of Nebraska Press, 1990.

Nash, Wallis. *The Farm, Ranch and Range in Oregon.* Portland, OR: The Lewis and Clark Centennial Exposition Commission for the State of Oregon, 1905.

National Civic Federation. *Shall the Government Own and Operate the Railroads, the Telegraph and Telephone Systems? Shall the Municipalities Own Their Utilities?* New York: National Civic Federation, 1915.

Nelles, H. V. *The Politics of Development: Forests, Mines, and Hydro-Electric Power in Ontario, 1849–1941,* 2nd edition. Montreal: McGill-Queen's University Press, 2005.

Netboy, Anthony. *The Salmon: Their Fight for Survival.* Boston: Houghton Mifflin, 1973.

Neuberger, Richard. *Our Promised Land,* reprint edition. Moscow: University of Idaho Press, 1989. Originally published by Macmillan in 1938.

Norwood, Gus. *Columbia River Power for the People: A History of the Policies of the Bonneville Power Administration.* Portland: BPA, 1981.

Nye, David E. *Electrifying America: Social Meanings of a New Technology.* Cambridge, MA: MIT Press, 1990.

Nye, David E. *American Technological Sublime.* Cambridge, MA: MIT Press, 1994.

Nye, David E. *Consuming Power: A Social History of American Energies.* Cambridge, MA: MIT Press, 1998.

Ormsby, Margaret. "T. Dufferin Pattullo and the Little New Deal." *Canadian Historical Review,* vol. 43, no. 4 (December 1962): 277–297.

Pence, Richard A., ed. *The Next Greatest Thing.* Washington, DC: National Rural Electric Cooperative Association, 1984.

Perry, Richard, Helen Ingram, and John Whiteley, eds. *Water, Place, and Equity: Tempering Efficiency with Justice.* Cambridge, MA: MIT Press, 2008.

Pinchot, Gifford. *The Fight for Conservation.* New York: Doubleday, Page and Co., 1910.

Pinchot, Gifford. *Breaking New Ground.* New York: Harcourt Brace and Co., 1947.

Pisani, Donald J. *To Reclaim a Divided West: Water, Law, and Public Policy, 1848–1902.* Albuquerque: University of New Mexico Press, 1992.

Pitzer, Paul. *Grand Coulee: Harnessing a Dream.* Pullman: Washington State University Press, 1994.

Platt, Harold. *The Electric City.* Chicago: University of Chicago Press, 1991.

Pope, Daniel. *Nuclear Implosions: The Rise and Fall of the Washington Public Power Supply System.* Cambridge: Cambridge University Press, 2008.

Potter, David. *People of Plenty: Economic Abundance and the American Character.* Chicago: University of Chicago Press, 1954.

Ramsay, M. L. *Pyramids of Power: The Story of Roosevelt, Insull, and the Utility Wars.* New York: Da Capo Press, 1975.

Rayner, William. *British Columbia's Premiers in Profile: The Good, the Bad, and the Transient.* Surrey, BC: Heritage House, 2000.

Reisner, Marc. *Cadillac Desert: The American West and Its Disappearing Water,* revised edition. New York: Penguin, 1993.

Rifkin, Jeremy. *The Hydrogen Economy: The Creation of the Worldwide Energy Web and the Redistribution of Power on Earth.* New York: Penguin Books, 2002.

Robbins, William. *Landscapes of Conflict: The Oregon Story, 1940–2000.* Seattle: University of Washington Press, 2004.

Robin, Martin. *Pillars of Profit: The Company Province, 1934–1972.* Toronto: McClelland and Stewart, 1973.

Rogers, Will. *The Autobiography of Will Rogers.* Boston: Houghton Mifflin, 1949.

Roy, Patricia E. "Regulating the British Columbia Electric Railway: The First Public Utilities Commission in British Columbia." *BC Studies,* no. 11 (Fall 1971): 3–20.

Roy, Patricia E. "Direct Management from Abroad: The Formative Years of the British Columbia Electric Railway." *The Business History Review,* vol. 47, no. 2 [Canada] (Summer 1973): 239–259.

Rudolph, Richard, and Scott Ridley. *Power Struggle: The Hundred Year War over Electricity.* New York: Harper and Row, 1986.

Schap, David. *Municipal Ownership in the Electric Utility Industry—A Centennial View.* New York: Praeger, 1986.

Schurr, Sam H., Calvin C. Burwell, Warren D. Devine, Jr., and Sidney Sonenblum. *Electricity in the American Economy: Agent of Technological Progress.* Westport, CT: Greenwood Press, 1990.

Segal, Howard. *Technological Utopianism in American Culture.* Chicago: University of Chicago Press, 1985.

Shiller, Robert J. *Irrational Exuberance,* 2nd edition. New York: Broadway Books, 2005.

Smeloff, Ed, and Peter Asmus. *Reinventing Electric Utilities: Competition, Citizen Action, and Clean Power.* Washington, DC: Island Press, 1997.

Smith, Frank E., ed. *Conservation in the United States: A Documentary History,* vol. 1, *Land and Water, 1492–1900,* and vol. 2, *Land and Water, 1900–1970.* New York: Chelsea House Publishers, 1971.

Sobel, Robert. *The Age of Giant Corporations.* Westport, CT: Greenwood Press, 1972.

Springer, Vera. *Power and the Pacific Northwest: A History of the Bonneville Power Administration.* Washington, DC: BPA, 1976.

Stacy, Susan. *Legacy of Light: A History of Idaho Power Company.* Boise: Idaho Power Company, 1991.

Stadfeld, Bruce. "Electric Space: Social and Natural Transformations in British Columbia's Hydroelectric Industry to World War II." Ph.D. thesis, University of Manitoba, 2002.

Stein, Harry H. "Fighting for Aluminum and for Itself: The Bonneville Power Administration, 1939–1949." *Pacific Northwest Quarterly,* vol. 99, no. 1 (Winter 2007/2008): 3–15.

Steinberg, Theodore. *Nature Incorporated: Industrialization and the Waters of New England.* Cambridge: Cambridge University Press, 1991.

Stone, Norman F. *Bountiful McKenzie: The Story of the Eugene Water and Electric Board.* Eugene, OR: Parkstone Co., 1986.

Strange, Susan. *Casino Capitalism.* Manchester, UK: Manchester University Press, 1997.

Swainson, Neil A. *Conflict over the Columbia: The Canadian Background to an Historic Treaty.* Montreal: McGill-Queen's University Press, 1979.

Taylor, Joseph E., III. *Making Salmon: An Environmental History of the Northwest Fisheries Crisis.* Seattle: University of Washington Press, 1999.

Tobey, Ronald C. *Technology as Freedom: The New Deal and the Electrical Modernization of the American Home.* Berkeley: University of California Press, 1996.

Tollefson, Gene. *BPA and the Struggle for Power at Cost.* Portland: Bonneville Power Administration, 1987.

Tothill, Blair. "Living Electrically: The British Columbia Electric Railway Company and the Development of the Domestic Electric Appliance Market in Victoria, 1919–1939." Master's thesis, University of Ottawa, 1997.

Trachtenberg, Alan. *The Incorporation of America: Culture and Society in the Gilded Age.* New York: Hill and Wang, 1982.

Ulrich, Roberta. *Empty Nets: Indians, Dams, and the Columbia River.* Corvallis: Oregon State University Press, 1999.

Vogel, Eve. "The Columbia River's Region: Politics, Place, and Environment in the Pacific Northwest, 1933–Present." Ph.D. diss., University of Oregon, 2007.

Wasik, John F. *The Merchant of Power: Samuel Insull, Thomas Edison, and the Creation of the Modern Metropolis.* New York: Palgrave Macmillan, 2006.

White, Richard. *The Organic Machine: The Remaking of the Columbia River.* New York: Hill and Wang, 1995.

Wiebe, Robert H. *The Search for Order, 1877–1920.* New York: Hill and Wang, 1967.

Wing, Robert C., and Robert C. Cumbow, eds. *A Century of Service: The Puget Power Story.* Bellevue, WA: Puget Sound Power & Light Co., 1987.

Wise, Rebecca R., ed. *Rural Electric Fact Book.* Washington, DC: National Rural Electric Cooperative Association, 1963.

Wollner, Craig. *Electrifying Eden: Portland General Electric, 1889–1965.* Portland: Oregon Historical Society, 1990.

Worster, Donald. "Two Faces West: The Development Myth in Canada and the United States." In Carol Higham and Robert Thacker, eds., *One West, Two Myths: A Comparative Reader.* Calgary: University of Calgary Press, 2004.

Index